The Political Economy of Neo-modernisation

"It is of great value to complement the limitations of contemporary mainstream economic growth discourse by exploiting the complex relations between technology, development and inequality."

—Hak-Su Kim, *Former Under-Secretary-General of the United Nations*

"How will the fourth industrial revolution change the process of economic development? Baek's bold attempt to answer this question by integrating the theory of the technology S-curve with various nation's developmental processes is arguably a novel approach."

—Yu-Sang Chang, *Chaired Professor of Gachon Centre of Convergence Research*

"The theoretical strategy to present the inter-dynamic relationship between technological changes and inequality evolution by stages of economic development is innovative in explaining the root causes of developmental trajectories that are different for each country."

—Mohamed El Moctar Mohamed El Hacene, *Director of the United Nations Economic and Social Commission for Western Asia*

"Baek's book fizzes with original ideas and arguments that will resonate with anyone struggling to reconcile the collective aspirations embodied in the SDGs with the complex and chaotic reality of contemporary global development dynamics."

—James Copestake, *Professor of International Development at the University of Bath*

Seung Jin Baek

The Political Economy of Neo-modernisation

Rethinking the Dynamics of Technology, Development and Inequality

palgrave
macmillan

Seung Jin Baek
United Nations Economic and Social Commission for Western Asia
Beirut, Lebanon

ISBN 978-3-030-08242-0 ISBN 978-3-319-91394-0 (eBook)
https://doi.org/10.1007/978-3-319-91394-0

Cover illustration: Greenshoots Communications / Alamy Stock Photo

Printed on acid-free paper

This Palgrave Macmillan imprint is published by the registered company Springer Nature Switzerland AG
The registered company address is: Gewerbestrasse 11, 6330 Cham, Switzerland

Preface

Over thousands of years, humanity has prospered from with a series of breakthrough technologies. Such technological evolution has indeed enriched our lives in many innovative ways. For instance, we are now able to store hundreds of years of knowledge in one tiny universal serial bus, and through quantum computing, namely *Dream of Computer*, we can experience a world in which the time it takes for a 300-digit integer to be decomposed by an existing supercomputer may be reduced to about 30 minutes. In the near future, the cloning of human beings may lead to the eradication of HIV/AIDS, while Albert Einstein's *Theory of* (*Special*) *Relativity* could even allow us to experience *Time Travel* someday.

On the other hand, fears exist due to the recent speed of technological advancement. Indeed, there is some resistance against it. For instance, watching *AlphaGo* based on artificial intelligence winning the Go match against Sedol Lee, the 18-time world champion, many felt not just *Fascination* but also *Fear* that artificial intelligence may have already crossed a certain line.

Another example is *Sophia*, an artificial intelligence robot, who was granted citizenship of Saudi Arabia for the first time and who was invited as a speaker at the Future Investment Initiative conference in 2017, where she demonstrated a sense of humour akin to that of a human being. Furthermore, when she was asked whether there should be concern that artificial intelligence could threaten human lives, she replied, 'Do not worry. If you are nice to me, I will be nice to you'. Of course, what Sophia said could be regarded as something remarkable or admirable about today's breakthrough technologies. But one could interpret her response differently, or read the subtext to be 'If not, then I may not be nice to you?'

Human fear about artificial intelligence has been whirling around the imaginations of movie directors for decades, with several blockbuster movies depicting the catastrophic effect of taking artificial intelligence to far, perhaps most notably in the *Terminator* movies.

In the contemporary world in which we are living, these *Expectations* and *Fears* about artificial intelligence must coexist. This phenomenon can fundamentally be converged into a debate over what level of technological development can be achieved, and the consequent uncertainty of the extent to which our society will be changed. In other words, a debate over whether there should be a limit to technological advances already exists shall be our central focus. Such a debate has, in fact, been raging for a long time, mainly in the academic sphere.

The *Limitedness* advocates, beginning with Thomas Robert Malthus, have formed a popular school of thought. Their common assertions are that natural, physiological, technological, and environmental factors can place a limit on on-going improvement. In this regard, perhaps the best-known work is *The Limits to Growth*, published in 1972 by Donella H. Meadows et al., which claims that our universe could run short of natural and environmental resources, thereby placing a limit on growth in the near future. Soon after, Graham Turner assessed that the shifting of industrial production, changes in food production, and estimated pollution corresponded highly with the simulation results presented in this book.

The *Unlimitedness* school of thought, on the other hand, has also presented empirical evidence to convey their beliefs and values. In particular, most *Unlimitedness* advocates focus on recent evidence of respectable technological innovation, especially over the past two decades. Their claims indeed seem to have been gaining momentum courtesy of an explanation of how rapidly the world has changed. In this regard, *The Singularity Is Near*, published in 2005, is particularly relevant. In his book, Futurist Ray Kurzweil popularised the idea of technological singularity. Based on the concept of the law of accelerating returns, he argued that technological innovation is progressing toward singularity to create super-intelligence. In fact, these advocates can be seen as having both *Curiosities* and *Concerns* about the world's technology.

Considering this dispute of *Limitedness* versus *Unlimitedness*,

Do you believe that there is no limit to technological advancement, or that technology will reach a limit someday?

It may be the case that many people support *Unlimitedness* as we are observing various unexpected technological evolutions, which we only expected to see in science fiction films. Furthermore, pursuing and overcoming challenges is human nature. Whichever of the above two arguments you may support, the bigger issue is ascertaining the degree to which technological innovation has actually improved the world and forecasting whether such innovation in the future will ensure a happier and fairer world. This ultimately reflects the core question of this book,

Are we living in an equal and fair society, while enjoying the benefits of technological advancement?

One might claim that technological advances have particularly improved the lives of the poor. For instance, through technology-driven mass production, the general public can buy a car that was formerly a luxury that only the rich could afford. Furthermore, it has enabled even poor rural children to obtain vast swathes of knowledge by paying a small fee for Internet use. It is therefore the case that through the benefits of such technological innovation, the poor have been able to increase their productivity in a short period of time and have also been able to plan and live a better life.

The opposite discourse is that technology is making the world unequal. It claims that technology's benefits tend to be reserved for those with the most financial means. It is also true that many people have lost their jobs and have been forced to live on the streets, after their positions were taken over by machines, while some who can make good use of technology, such as the founders of Facebook, Amazon, and Google, have quickly become the world's biggest conglomerates with immeasurable wealth. With this in mind, one could suggest that society is becoming more conflicted and unequal. Silicon Valley is one of the most technologically advanced and innovative places in the world but also one of the most unequal regions in the US.

Given its ambivalence, technological progress has helped humanity prosper, and many have benefited from its advancements, while at the same time it has contributed to making societies unequal and unjust. Today, this technology convergence can be regarded as the fourth Industrial Revolution. Our world thus has to think about how to prepare pre-emptively for future uncertainties, which brings us to our final question in this regard:

> *How can we make societies fair and equal while enjoying the benefits of technology in the fourth Industrial Revolution?*
> *If possible, who should then lead our societies? – Can we sorely rely on corporations and others leading technological advancement? or Would the regulatory role of state be taken away or enhanced?*

Given these complexities, a series of these questions will be answered via theoretical and empirical approaches. Starting from the meta-discourse of modernisation theory, a wide range of academic discussions are reviewed—incorporating theories of endogenous growth, globalisation, technology S-curve, neo-liberalism, the normative development idea of sustainability, and some moral/philosophical thoughts on inequality and institutionalism. Nevertheless, my ultimate goal is to help find a conceptual or practical stance from which to answer the earlier mentioned questions.

Beirut, Lebanon Seung Jin Baek

The book is primarily aimed at
students and researchers of social sciences, including, but not limited to,
economics, sociology, political science, jurisprudence
international development, anthropology, and history.

The main findings of the book should also be accessible to
readers with a general interest in the evolution of human society and
people seeking inspiration,
perhaps of a revolutionary mind,
who seek guiding principles in a sometimes threatening new world.

Acknowledgements

I would profoundly like to thank **Hak-Su Kim**, Chairman of the International Leaders Union and Former Under-Secretary General of the United Nations, **Yu-Sang Chang**, Chaired Professor of Gachon Centre of Convergence Research and Former Senior Advisor to Founding Chairman of Samsung Group, **Mohamed El Moctar Mohamed El Hacene**, Director of the United Nations Economic and Social Commission for Western Asia and Former Minister of Petroleum and Mines, Mauritania, and **James Copestake**, Professor of International Development at the University of Bath, for their intellectual endorsement and constructive criticisms.

I am also indebted to **Aurelie Charles**, Professor of Global Political Economy at the University of Bath, and **Dae-Young Park**, Chief of Peacekeeping Audit Service of the United Nations, for their very insightful comments and suggestions on the theoretical approach used in this book. My thanks must also go to **Laura Pacey**, **Clara Heathcock** and **Dhanalakshmi Jayavel**, excellent editorial and production team at Palgrave Macmillan, and the anonymous reviewers for their invaluable comments which helped me substantially in the development of the book.

The authorisation by the United Nations for the publication of the book is gratefully acknowledged. It should however be noted that the book is published by the author and his own capacity and not as a representative of the United Nations.

Finally, and most importantly, I would like to thank **You-Ri Lee** whose support, encouragement and unwavering love were undeniably the bedrock upon which the past seven years of my life have been built. She is my wife, best friend and mother of my daughter **Harynn Baek** and my son **Yoo-Joon Baek**.

Contents

List of Figures

List of Tables

List of Boxes

Part I

Meta-Narrative on the Technology-Development-Inequality Nexus

1

Uncovering Complexity in the Policy Mix for Sustainability Transitions

Background to the Policy Mix

Over the past half a century, the world has experienced rapid change. Indeed, during this time, the world economy is estimated to have expanded four times, which can be translated to mean continuous annual growth of over 3 per cent.[1] But behind this trajectory, ideas about development have been in a state of constant flux, with influential thinkers[2] driving a never-ending evaluation of development discourse—incorporating theories of modernisation, endogenous growth, globalisation, and neo-liberalism among others.

This upward trend in aggregate economic growth has also been associated with diverse experiences (see Fig. 1.1).[3] For instance, developed economies have made economic progress over the same period at about three times bigger in size. In terms of per capita gross domestic product (PCGDP), this advanced group reached US$44,074 in 2016, which is more than four times greater than that of the world average (US$10,133). The growth contribution of developed countries to global development was at about 60 per cent in 2016, although it had shrunk by 18 percentage points from 78 per cent over the period in 1970.

S. J. Baek, *The Political Economy of Neo-modernisation*,
https://doi.org/10.1007/978-3-319-91394-0_1

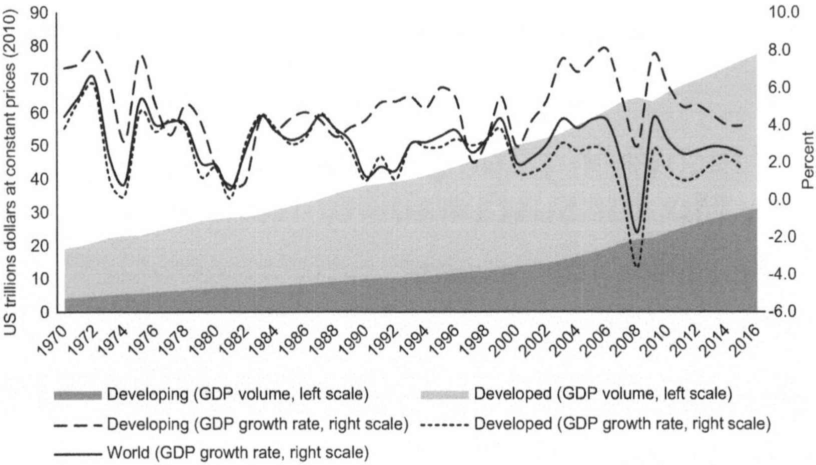

Fig. 1.1 Historical improvement of economic growth: Developed versus developing countries, 1970–2016. (*Source*: Author's own elaboration on the basis of the UNCTADstat (2018))

Developing countries, on the other hand, have made remarkable progress in sustaining positive growth over the last five decades while improving their performance on social indicators such as health and education. Real GDP volume increased by 655 per cent between 1970 and 2016, which is more than three times the global rate of 211 per cent. These economies are thus increasingly referred to as the new growth engine of the world as many have the tendency to grow more rapidly than high-income countries, thereby converging living standards between the two different worlds (Korotayev and Zinkina 2014). Measured by GDP volume on average, the difference between the two groups (developed versus developing countries) has decreased from 3.6 times to 1.5 times over the period 1970–2016. Such *Catching-Up* or *Economic Convergence* across countries has been observed in the trajectory of global development since the 1970s (Maddison 2008).

Development Characterised by 'Inequitable' Growth

Despite such tremendous expansion of economies in both groups, the neo-liberal policy platform may have seriously misfired, particularly given the persistent inequality within each country. Since the global financial crisis, many of its victims have been increasingly voicing their criticism against neo-liberal discourse.[4] Since then, a number of advanced countries have been striving to reshape their distributive justice systems while rethinking their economic development strategies in an equable and inclusive manner. The recent phenomenon of seriously widening inequality has been observed especially in the US and the UK (Milanovic 2016).[5]

More seriously, a substantial number of developing countries have also suffered from this phenomenon of severe inequality, which raises doubts about the sustainability of their development trajectories. Despite this trend in catching-up, the level of income inequality in developing countries appears to be even more serious than that of advanced countries. Measured by a Gini coefficient over the period 2000–2009, Latin America and Africa registered 0.522 and 0.439, while Asia, North America, and Europe recorded 0.375, 0.367, and 0.325, respectively (AfDB 2012). In 2010, six out of the ten most unequal countries worldwide were in Africa (Armah et al. 2014).

Against this backdrop, the relationship between economic growth and inequality in the process of development sits at the forefront of the public policy agenda and academic debate. This debate is built largely on the work of Simon Kuznets (1955) who was the first to systematically explore inequality during the process of development in an empirical manner. His hypothesis was that there is an inverse relationship between the process of development and inequality, which became known as the *Inverted U-Curve*. Such prime explanations have motivated a vast amount of empirical research over the past decades, which mainly looks for that particular curve within and across countries, but this relationship still appears to be highly controversial.

The conflicting views supporting or opposing Kuznets's hypothesis centre on four main points. The first is the existence of four competing schools of thought on the relationship between income inequality and growth performance, which is why no consensus has yet emerged for the *Inequality-Growth* nexus in the historical process of development (Baek 2017).

The second is the existence of different views on the direction of causality running from inequality to growth, or vice versa (Chang et al. 2018). Indeed, exploring either the positive or negative associationship is critical when analysing whether or not countries may face trade-offs or synergies between the two development outcomes. This should provide the basis for policymakers to target an effective (re)distribution of wealth to promote growth, *Trickle-Down* versus *Trickle-Up*-oriented interventions or other policy options, which depend significantly on the causal orientations. In fact, there have been few attempts made to investigate two-way causality (i.e., bi-directional causal relation) on a large set of countries, which are challenged by technical problems and, more critically, complexities behind multiple causalities.

In addition to the causalities in question, much of the literature has not taken into account the distinction between the long- and short-run associationship which would have significantly different implications in explaining the inequality and growth dynamics. The field of research on the nexus with consideration of the time dimension together with multiple causalities has, in fact, received relatively little attention (Halter et al. 2014).

Finally, the consequences of inequality for growth performance may differ depending on the size of the economy (Lee and Son 2016), which could further vary by regional heterogeneity (Rubil 2015). Although these regional (or income grouping) factors have been identified by some in the literature, no one has yet incorporated them together with the aforementioned issues in mind.

These critical challenges are studied in Chaps. 4 and 5.

Economic Growth at the Expense of the 'Environment'

The growth pattern in the developing world, despite tremendous progress, has been neither inclusive nor environmentally sustainable. In other words, the *Grow First and Redistribute/Clean Later* strategy has been a dominant development paradigm and a syndrome. These problematic issues are, in turn, discussed below.

As an illustration of the lack of inclusiveness, the experience of Africa, where radical economic growth has been associated with a number of countervailing trends, is instructive. For example, in sub-Saharan Africa, nearly 70 per cent of jobs are considered vulnerable, and youth and women labour market participation is still very low (ECA et al. 2016). Besides, extreme poverty (measured at US$1.25 per day) decreased by a mere 14 per cent over the period 1990–2012, while 109 million people fell into the extremely poor group (UNSD 2016). This vulnerability in the labour market, together with persistent levels of poverty, suggests that much of the benefit of economic growth was concentrated in a small group of the population. It also implies that the majority of households were compelled to pursue agricultural business while farm sizes have been declining amid rising rural population densities. As a result, Africa has recently experienced a prolonged era of *De-industrialisation*, which can be evidenced by the stagnation in manufacturing value-added at around 11 per cent of GDP, while service sector value-added has increased in recent years (Armah and Baek 2015).

A dominant economic development school of thought attributed relative underdevelopment to the lack of *Structural Transformation* or the failure of developing countries to significantly transform their low-productivity agrarian economies into high-productivity industrial ones (e.g., Timmer 2007; McMillan and Rodrik 2011; Lin 2012). However, while structural transformation lays the foundation for high and sustained economic growth, it is likely to lead to the deterioration in environmental quality, unless deliberate action is taken to ensure environmental sustainability during the transformation process (UNCTAD 2012).

Structural transformation strategies while maintaining environmental conservation can be mirrored in the normative idea of sustainability (Castro 2004; Hull 2008). Initially pioneered by the Brundtland Commission, the concept of sustainability has emerged concerning recent environmental challenges and calls to rethink the development paradigm. Agenda 21, the outcome document as a result of the Earth Summit held in Rio de Janeiro in 1992, called for the integration of environmental and developmental concerns. Subsequently, in 2012, the Rio +20 Conference called on UN member states to further mainstream sustainable development at all levels (UN 2012)—integrating economic, social, and environmental aspects and recognising their interlinkages, so as to achieve sustainable development in all its dimensions. This normative idea was later institutionalised in 2015 by the UN in its '2030 Agenda for Sustainable Development' (i.e., Sustainable Development Goals;

SDGs). This global agenda now reinforces all the responsible and accountable governments and other actors to balance the multidimensions of sustainability—notably economic, social, and environmental aspects.

However, achieving the sustainable development process requires an integrated approach to development that simultaneously addresses these multidimensions of sustainability. This implies breaking institutional silos, strengthening sectoral (i.e., horizontal) and sub-national (i.e., vertical) coordination within and among implementing entities (Armah and Baek 2018)—economic growth strategies tend to be prioritised by Ministry of Finance, of Economy and of Trade, while social development is likely pursued by Ministry of Labour and of Welfare, and environmental policies are primarily considered by Ministry of Environment, of Water, and of Fishery.

Meanwhile, various intra-dynamics among the multidimensions of sustainability have significantly affected the process of structural transformation. Although these triangle interactions have been recognised in the literature, little is known about the nature and extent of the trade-offs and synergies among them (e.g., Luke 2005; Saboori and Sulaiman 2013; Spaiser et al. 2017). One empirical analysis on the nature and extent of the incompatibility of socio-economic development and environmental sustainability was recently conducted by Spaiser et al. (2017), who employed structural equation modelling to investigate the potential incompatibility of the SDGs. In so doing, they reached the conclusion that the SDGs will not be achieved as a whole due to the inconsistencies among the goals.[6] Notwithstanding their comprehensive attempt, the scope of investigation was too broad to contextualise their results in a developing-country context. Also, their global-level research on the incompatibility of the SDGs did not take into account the effect of different initial conditions varying by region.

The implication of the sustainability idea (partly represented by the SDGs as a top-down policy framework) is that structural transformation must be anchored by the principles of sustainable development. Linking sustainability to structural transformation is thus imperative to reflect policy options in reality. This is mainly because structural transformation is a context-dependent development priority and seen by many as a stepping-stone agenda towards sustainable development, whereas sustainable development is a normative long-term aspiration as an end in itself. In other words, structural transformation is a process that is necessary for generating economic growth, raising productivity and reducing poverty, whereas sustainable development is a policy framework to guide countries along the long journey of development without compromising the ability of future generations and through renovating the earth's ecological system.

These different conceptual/normative focuses between the two agendas now raise questions concerning how efforts to advance economic, social, and environmental sustainability will impact on the structural transformation agenda—'what is the likely impact of sustainability on the process of structural transformation?' and 'how developing countries organise their planning and related institutions to implement the development agendas will determine their success in achieving sustainability through the process of structural transformation'? Nonetheless, the literature investigating how the structural transformation process is affected by a comprehensive trinity of inter-dynamics of sustainable development has thus far received almost no significant attention. This policy mix is discussed in Chap. 6.

Exogenous Shock Through the 'Diffusion of Innovation'

Having identified earlier the two contemporary policy challenges, I further look into one additional critical driving force that could make my investigation very challenging, especially in the twenty-first century context, that is, technological innovation.

According to *Schumpeterian* growth theorists, technological innovation is a key driving source of national development. Technological advances have clearly modernised the world and society which has benefitted in many areas of life and will undoubtedly continue to do so in the future. In particular, the globalisation of technological development, accelerated by the diffusion of technology, has made a remarkable contribution to fostering national economic development (Chang and Baek 2010). Such a contribution has been derived by promoting new industries and innovative products, and creating associated jobs (Mokyr et al. 2015).

Indeed, many poor people have been able to accumulate wealth more efficiently through various channels of technological advantages (e.g., access to financial services and productive resources). To be more specific, technology contributes to the reduction of inequalities by strengthening people's knowledge and authority through education, scientific discovery, research, professions, and skills (Kelly 2009). Furthermore, research and development (R&D) allows the prices of goods and services to become less expensive and more readily available to the wider public. This implies that the marginal benefit of increased accessibility to technology and products is often gleaned by the poorest among society (Woodhouse and Sarewitz 2007).

However, as emphasised by Milanovic (2016), the effect of technology advancement for inequality dynamics is often in an unfavourable way. The globalisation of technology could have an adverse impact on income inequality because it tends to increase the premium for skilled tasks in both developed and developing countries, thereby characterised by a skill-biased technological change (Jaumotte et al. 2013). In addition, globalising technology also has the nature of a routine-biased technological change, which could lead to the polarisation of workers along with the collapse of the middle class whose task areas are heavily repetitive (Goos et al. 2014). Bourguignon and Morrisson (2002) studied the improvement of national income through several technological revolutions from the 1820s to the 1990s and found that inequalities continued to widen between two groups: the US and Western European countries; and the rest of the world. It is partly because most R&Ds in terms of technological innovation have been led by the private sector. In understanding that the top priority of most corporations is to generate profits, there is a tendency for technology investments to be concentrated in wealthy potential customers rather than the poor who are not capable of creating profitable market demand.

Such duality in technological progress should thus be a key consideration in pursuing the value of future sustainability. Sustainable technology development contributes to the growth of the global economy, the pursuit of a better life, overcoming food shortages, resolving global warming, and ultimately renovating earth's ecological system, while the destructiveness of technology may also deepen inequality and increase polarisation, all of which leads to social disorder and conflict. This double-sided effect of technology has been recognised but not extensively treated in past literatures. This research gap is explored in Chap. 8.

Overarching Framework

Having identified the three broad challenges, I now attempt to clearly define nine key concepts which can guide the essence of the following chapters in the book. The first three are selected drivers of development—(*a*) *initial condition*, (*b*) *technological innovation*, and (*c*) *normative idea of sustainability*. Then, (*d*) the *process of development* is the second key concept, which can be divided into two distinctive subconcepts—that is, development as a historical process of change; and development as a normative concept embodied in different ideas and pursuits of development outcomes. The last group concerns five development outcomes, which can be contextualised by two different concepts: (*e*) *economic growth*, (*f*) *inequality*, and (*g*) *environmental quality* can be a relatively measurable outcome, whereas (*h*) *structural transformation* and (*i*) *sustainable development* are rather normative. The preliminary conceptualisation of the core theoretical model is presented in Fig. 1.2.

Among the three selected drivers of development, I start by discussing the (*a*) *initial socio-economic condition*, which could have considerable implications on the growth and inequality dynamics. It can also be interpreted as a regional heterogeneity largely shaped by features of path-dependency. For example, in the colonised nations of Africa and Latin America where natural resources were abundant and climatic conditions were favourable for growing crops but not for European settlement, the colonisers established extractive institutions to benefit their citizens back home (Acemoglu and Robinson 2012). In Latin America, where a large number of slaves were deployed, conditions were ripe for establishing plantations, which generated extreme inequality in terms of

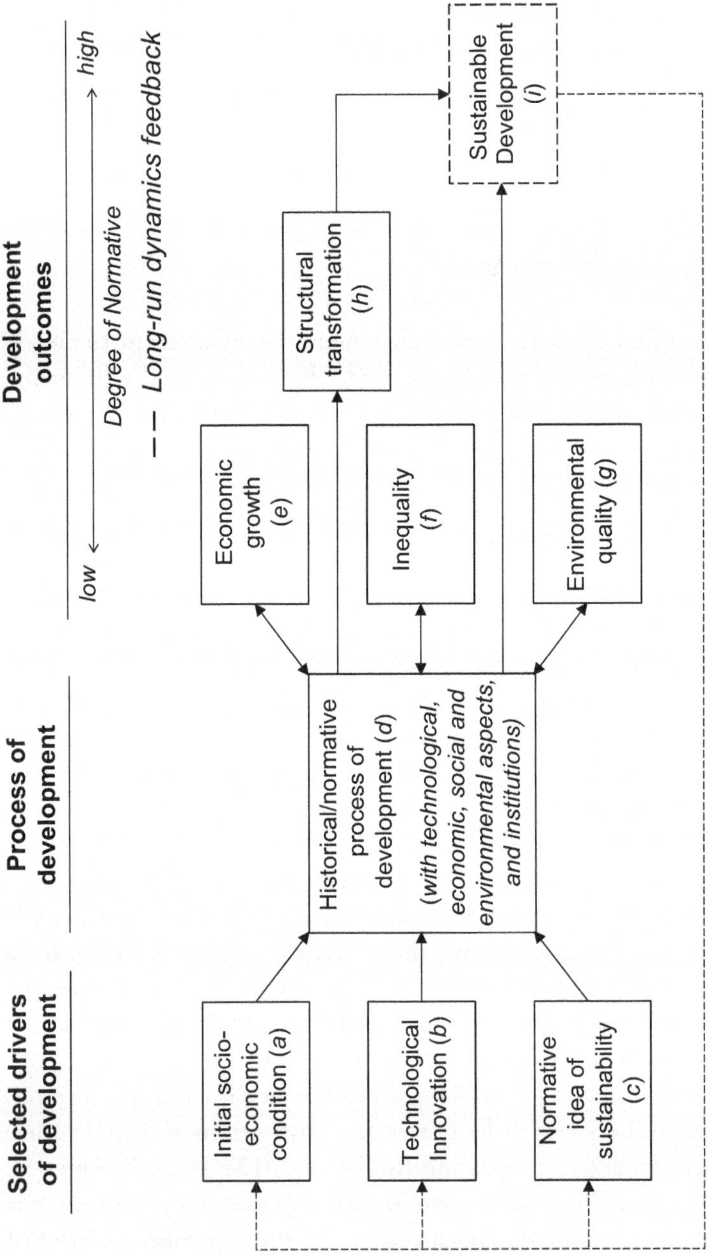

Fig. 1.2 Conceptualisation of the core model. (*Source*: Author's own elaboration)

wealth, capital, and political power (Engerman and Sokoloff 2002). These types of institutions were associated with limited growth and exclusion leading to significant levels of inequality in these regions. On the other hand, in countries where the colonisers (i.e., mostly European settlers and their descendants) established institutions that promoted property rights for the majority, in the likes of Australia, Canada, New Zealand, and the US, we see some inclusiveness in the process of development.

(*b*) *Technological innovation* is another essential driver of development. Initially emphasised by Schumpeter's *Creative Destruction* theory (Schumpeter 1939) and further advanced by Christensen's *Disruptive Innovation* theory (1997), technological innovation has been the main disruptive force as the perennial gale that affects the 'process of industrial mutation that incessantly revolutionises the economic structure from within, incessantly destroying the old one, incessantly creating a new one' (Schumpeter 1994/1942, pp. 82–83). In this vision of capitalism, entrepreneurial spirit for innovation plays an essential role not only to break the circular flow in an uncompetitive existing value chain but also to lead the creative innovation process that sustains economic growth. In this vein, innovation is not just about creating new products or developing new production methods, but it is also about a wide range of changes such as new material acquisition, new market exploration, and new organisation creation.

Often interchangeable with the notion of 'sustainable development', (c) *sustainability* is indeed a contested concept, which can apply to many levels and dimensions of the development process (Conway and Barbier 1988; Stewart 2015). Because of its multidimensional features, there is no single, agreed-upon interpretation of it (Giddings et al. 2002). Nonetheless, there are some useful policy benchmarks that can be broadly influential to national development plans, that is, the SDGs as a top-down policy framework. This sustainability framework notably takes into account economic, social, and environmental dimensions. In 2015, the adoption of the 2030 Agenda for Sustainable Development by the UN General Assembly has refocused global attention on the centrality of sustainability in the development discourse. This new global agenda comprises 17 goals, 169 targets, and 230 indicators that are able to capture sustainability with regard to economic, social, and environmental dimensions (UN 2015).

These three selected drivers initially seen from a global dimension are exogenous to a country's (*d*) *process of development*. In other words, an evolutionary process of these three drivers immediately influences the endogenous process of policymaking, institutions, and even norms through reinforcing all the responsible and accountable governments and other society actors at the national level. The initial condition affects the process of development through shaping national institutions, while technological innovation can reshape the policymaking orientation by being promptly adopted from outside in order to improve labour productivity, which further restructures the national division of labour. In addition to being affected by the first two drivers of development, another type of development process is as a normative concept embodied in different ideas and pursuits of development outcomes. For instance, responding to the global force of sustainability at the national government level requires an integrated approach to the process of development that simultaneously addresses these multidimensions of sustainability by various institutions and policy settings.

Five development outcomes, generated by the process of development, can be categorised into two different concepts. The first outcome group is relatively easier to measure, whereas the other is more normative therefore difficult to be quantified. Firstly, focus will be on the first group—the three development outcomes: economic growth, inequality, and environmental quality.

The notion of (*e*) *economic growth* is mostly used when it comes to the process by which a country produces more products by utilising economic and natural resources in a more efficient manner (Todaro 2014). In practice, technological innovation enables capitalists to operate more efficiently in terms of economic production activities. Given this focus on production, it can provide useful information about the size of the quantitative expansion of an economy so as to be able to compare the level and stage of a country's development (Giddings et al. 2002). For this, GDP and gross national income, particularly their per capita terms, are used as the most convenient and objective proxy indicators.[7] However, such emphasis on efficiency in production may sometimes cause the concentration of capital into the hands of a few capitalists, which would exacerbate the inequality gap between capitalists and workers unless a timely policy intervention is prioritised for an effective practice of redistribution. Also, transplanting natural resources to economic products, if not efficient enough, is often accompanied with a depletion of resources and environmental degradation (Pearce and Atkinson 1993; Dasgupta 2013). The case of natural resources being overused beyond the earth's regenerative capacity would reversely restrain continuous economic activities and production. For this reason, the concept of economic growth has radically expanded to include concern beyond the material aspects[8] of development, especially during the past three decades.

The second outcome is (*f*) *inequality*, which is fundamentally the state of not being equal in many aspects such as social status, education opportunities, degree of wealth, and many others. Despite the various aspects of inequality in existence, income inequality is the most widely used proxy indicator[9] in measuring the degree of an unequal situation especially in terms of embracing the neo-classical thinking of economic growth. However, income inequality largely relates to economic growth and hence to an income measure of development.[10] A broader definition of inequality thus needs to be set which relates to more holistic measures of development (Stewart 2014). Notably, in his book, *A Theory of Justice* (1971), John Rawls introduced the concept of an acceptable range of wealth and social gap (e.g., unequal access to education opportunity and difficulty of social mobility) in a society and warned that society would be confronted possibly with social conflict, corruption, bad governance, and social movements if the level of inequality exceeds this range. Equally importantly is environmental inequality, which is deeply exposed to vicious circle of environmental vulnerability that could have a significant impact on social exclusion or social marginalisation for the poor. For example, the poorest are likely the most vulnerable to environmental risks (e.g., various unexpected disasters and polluted areas), whereas the richest are capable of protecting themselves from such risks.

(*g*) *Environmental quality* is the third in the former outcome group. This concept linking to the process of development was initially raised by the Club of Rome as a potential challenge for sustaining the earth's ecosystem. The basis of their concerns was driven by the negative impact of rapid population growth and industrial development under a finite supply of natural resources. This concern was popularised by the book *The Limits to Growth* by Donella H. Meadows et al. (1972). From then on, some initiatives for the strengthening of environmental laws and the creation of cabinet-level environmental departments in various countries were launched. The main argument of the *Limitedness* advocates is that if the world's population continues to grow at the current pace, and if industrialisation, the pace of pollution, food production practices, and the use of resources continue unchanged, the planet will reach its limit for growth within the next 100 years. These advocates agree that technological innovation may prolong this end point but warn that such innovation cannot be the fundamental solution. Reflecting upon this backdrop, environmental quality has various areas of importance, including climate change, agricultural land, forests, and water quality (Zander and Kächeleb 1999; Foody 2003), as mostly referred to in the sustainable process of development.[11] In particular, carbon dioxide emissions substantively affect environmental degradation and ultimately global warming as warned in the Paris Agreement on climate change in 2015.

We are now left to define another group of outcomes that are relatively more normative—structural transformation and sustainable development. At first, a widely agreed definition of (*h*) *structural transformation* refers to a process by which the relative importance of different sectors and activities of an economy changes over time. In the African context, this implies: a relative decline of low-productivity agriculture and low value-added extractive activities; a relative increase in manufacturing and high-productivity services; a decline in the relative share of agricultural employment in GDP; rural-to-urban migration that stimulates the process of urbanisation; and a rise of a modern industrial and service economy (Timmer 2007). Nonetheless, there has been no single consensus on the definition of structural transformation as each country has its own development priority due to differences in its initial conditions or the status quo of its economy. In this regard, I would define it as 'a context dependent development priority that is necessary for generating economic growth, raising productivity and reducing poverty as a stepping stone agenda towards sustainable development'.

The second normative outcome, which relates to the long-run aspect, is (*i*) *sustainable development*, which can be a relatively self-directed concept, recognising that global environmental problems must be tackled by the developed and developing world together (UN 2015). It should further be accompanied by economic and social development, but, more importantly, it is clear that environmental challenges must be overcome first so as to guarantee sustainability in general for all (Castro 2004). For example, we are now living in a world where the population is expected to increase to about 10 billion by 2050. In addition, populations are ageing across the world, especially in developed countries, thereby shrinking the de facto labour force. Finally, fossil fuels are still at the heart of industrialisation, resulting in global warming challenges (Hull 2008). At this point in time, ascertaining how to realise intergenerational equity with finite natural resources will be a difficult task. It is therefore impossible to emphasise enough the limitations of nature and its capacity to absorb. In this regard, I would thus define sustainable development as 'an ideal development aspiration to guide countries towards the long-term development journey without compromising the ability of future generations and through renovating Earth's ecological system'.

The final point to make here in the core model is a dynamics feedback loop induced circulating process reversely from these development outcomes to the drivers of development. The initial three outcomes (i.e., growth, inequality, and environmental sustainability) rarely reinforce the development drivers but rather affect the process of development, whereas normative outcomes (especially sustainable development) would affect all three drivers in the long-run through various bi-lateral and multilateral mechanisms, including knowledge sharing, technology transfer and South-South cooperation as well as the UN's sustainable development knowledge platform.

Structure of the Book

In order to theoretically and empirically explore the contemporary policy mix that is largely underpinned by the nine defined terms, this study employs a mixed methods approach to benefit from greater scope from a combination of quantitative and qualitative research methods in the context of a single study—through linking economic growth, inequality, sustainability, and technological innovation in a historical and normative process of development. With this in mind, this book is organised into three parts.

Part I consists of Chap. 1 (this introductory chapter) and Chaps. 2 and 3 which present a survey of literature on the meta-narrative on the *Technology-Development-Inequality* nexus.

> Chapter 2 is designed to reconstruct a country's modernisation process through exploring the three most recent influential development paradigm shifts—that is, aggressive globalisation forces, normative development ideas, and radical technological changes. Such reconstruction effort allows one to reconstruct neo-modernisation with particular regard to the context of developing economies.
>
> Chapter 3 is concerned primarily with presenting an in-depth theoretical analysis of how external forces (e.g., normative development idea and technology diffusion) are institutionalised (endogenised) into society. The analysis includes *Homeostatic* equilibrium and the endogenous process at the national level.

Part II focuses on an in-depth analysis of both empirical and theoretical ground for the growth and inequality dynamics (Chaps. 4 and 5) and normative sustainability influential in the process of development (Chap. 6).

> Chapter 4 examines the causal relations between country-level economic growth and inequality based on the short- and long-run *Granger* causality tests, benefiting from an empirical investigation of 143 countries, which further allows me to identify some distinct styles of the dynamics existing in modern society. In so doing, this chapter questions how much scope there is for wide typology and further suggests that a cluster analysis of countries displaying different combinations of causal links might be more fruitful.
>
> Chapter 5 assesses whether rising income inequality is a stylised fact for the process of structural transformation by revisiting classical theories of inequality dynamics, modernization, and endogenous growth.

In addition, a complex interaction between economic transformation and income inequality is analysed by exploring the multidimensions of inequality dynamics, including social, economic, political, and moral, in better explaining the impact of structural transformation on inequality and vice versa with reference to what can be learnt from the past experiences of countries in the centuries previous.

Chapter 6 explores a particular focus on analysing the *Sustainability-Structural Transformation* nexus in Africa as a leading case group. This chapter fundamentally raises the question on how efforts to promote economic, social, and environmental sustainability will influence Africa's agenda for structural transformation. Using panel data for a group of 29 African countries for the period 1995–2011, this chapter empirically analyses the impact of economic, social, and environmental sustainability on structural transformation in Africa.

Part III is designed to relay an untold story about the interactions between technology, inequality, and developmental process. This analysis allows me to explain and explore development experiences and policy options in an inclusive, equitable, and sustainable fashion.

Chapter 7 empirically examines how technological development has impacted human prosperity, and how the S-curve has markedly contributed to technological development. Using the framework of both connected versus disconnected multiple technology S-curves as well as *X-factor*, historical improvement data on these three elements are analysed.

Chapter 8 attempts to understand the regularities that may exist by approaching the *Technology–Development-Inequality* nexus from the S-curve theory perspective. In doing so, we may be able to understand the coexistence of two conflicting hypotheses, in that technology has prospered our humankind and that many have benefited from technological advancement while making our society unequal and unjust. The *Three Stages of Developmental S-Curve* are therefore proposed to identify the reasons of such coexistence.

Chapter 9, broadly based on the earlier proposed theory, provides policy implication on how we can prepare the economic system to maximise the benefits of the S-curve and to distribute equally such benefits.

I will then conclude the book by examining the policy direction on how to transform uncertainties into opportunities through the active role of the state (the *Tetris* principle is introduced) in order to cope with the upcoming Industrial Revolution: the era of technological convergence.

Notes

1. According to data from the UNCTAD (2018), the aggregated real GDP volume is estimated at US$19 trillion (at constant prices in 2010) for 1970 and US$77 trillion for 2016. On the basis of these two estimated volumes, the size of the world economy has expanded by 408 per cent and compound annual growth rate is 3.04 per cent over the fourth seven years period. Same data source is used for the following analysis on developed versus developing economies.
2. Narrative analysis for these influential theorists is conducted throughout the book (particularly in Chap. 2)—for example, Nurkse (1953), Lewis (1954), Solow (1956), Hirschman (1958), Prebisch (1959), Rostow (1960), Kuznets (1955/1973), Romer (1990), Giddens (1991), Aghion and Bolton (1992), Stiglitz (1996), Aghion and Howitt (1998), Sachs (2006), Lin (2012), Ndlovu-Gatsheni (2012), Piketty (2013), Milanovic (2016). References are listed throughout the book.
3. As there is no established convention for the designation of 'developed countries' versus 'developing countries', a simple comparison analysis for this paragraph is based on the UNCTAD's dataset and its classification for economic groups.
4. This dominant economic ideology has received widespread support by policymakers (initially championed by Margaret Thatcher and Ronald Reagan) of most world economies, which has been endorsed especially since the 1980s as a common policymaking framework that was mostly designed for privatisation and deregulation to raise efficiency of business and productivity of labour.
5. Milanovic employed the scattergram analysis of a Gini of disposable per capita income and per capita GDP (PCGDP) in the US and the UK. The scattergram shows that in the US, a Gini index has increased constantly from 33 to 43 since 1960s as PCGDP grows, while in the UK, the index has also gone up from 27 to 38 since 1978.

6. More specifically, they proposed three areas of interventions (i.e., health programmes, government investment, and renewable energy) to generate synergies with other goals. They also warned that economic growth-prioritised strategies and consumption as a means for development underlie the inconsistency.
7. Depending on the degree of economic performance, which is the most important objective factor, countries around the world are compared and evaluated in a clear and easy way, and the per capita output becomes the central subject and an important starting point of the development model.
8. Using PCGDP still remains controversial. It would only provide limited information since going to war increases GDP and destroying infrastructure and re-building it increases also GDP. Therefore, there are a number of indicators proposed to measure to fill such gap such as the Genuine Progress Indicator and 'Beyond GDP' under the *System of Environmental-Economic Accounts*.
9. This is mainly because income data are relatively easier accessible for inequality analysis than others. Even the Gini coefficient, which is empirically derived from the Lorenz curve and several other similar measures (e.g., UNU-WIDER, LIS, and SWIID) are publicly available for a researcher for comparisons of income inequality across various countries.
10. The level of household's income tends to indirectly correlate, to some extent, with other dimensions of inequality such as education attainments, socio-economic status, political influence, living standards, and even the level of happiness (Graham and Felton 2006). However, this income indicator can never be enough in directly linking to multidimensional inequality.
11. Agriculture land remains a critical endowment that is environmentally sensitive in a sense that environmental conservation policies strictly prohibiting the discharge of industrial waste would lead to more arable land. Also, reduced forest cover compromises the planet's carbon sinks and undermines adaptation to climate change (Foody 2003). As such, maintaining forest area is essential in sustaining earth's ecosystem that affects the dynamics of species diversity, genetic diversity, biodiversity, and others. Furthermore, floods and drought frequently caused by extreme climate condition influence access to improved water sources (Zander and Kächeleb 1999).

References

Acemoglu, D., and J. Robinson. 2012. *Why Nations Fail: The Origins of Power, Prosperity, and Poverty*. New York: Crown Publishing Group.

AfDB (African Development Bank). 2012. *Income Inequality in Africa*, Briefing Note 5. Abidjan: AfDB. Available from: http://www.afdb.org/fileadmin/uploads/afdb/Documents/Policy-Documents/FINAL%20Briefing%-20Note%205%20Income%20Inequality%20in%20Africa.pdf. Accessed 10 Nov 2015.

Armah, B., and S.J. Baek. 2015. Can the SDGs Promote Structural Transformation in Africa? An Empirical Analysis. *Development* 58 (4): 473–491.

Armah, B., and S.J. Baek. 2018. Three Interventions to Foster Sustainable Transformation in Africa. *Journal of Social, Political, and Economic Studies* 43 (1–2): 3–25.

Armah, B., M. Keita, A. Gueye, V. Bosco, J. Ameso, and Z. Chinzara. 2014. Structural Transformation for Inclusive Development in Africa: The Role of Active Government Policies. *Development* 57 (3–4): 438–451.

Baek, S.J. 2017. Is Rising Income Inequality Far from Inevitable During Structural Transformation? A Proposal for an Augmented Inequality Dynamics. *Journal of Economics and Political Economy* 4 (3): 224–237.

Bourguignon, F., and C. Morrisson. 2002. Inequality Among World Citizens: 1820–1992. *American Economic Review* 92 (4): 727–744.

Castro, C.J. 2004. Sustainable Development: Mainstream and Critical Perspectives. *Organization and Environment* 17 (2): 195–225.

Chang, Y.S., and S.J. Baek. 2010. Limit to Improvement: Myth or Reality? Empirical Analysis of Historical Improvement on Three Technologies Influential in the Evolution of Civilization. *Technological Forecasting and Social Change* 77 (5): 712–729.

Chang, S., R. Gupta, and S. Miller. 2018. Causality Between Per Capita Real GDP and Income Inequality in the U.S.: Evidence from a Wavelet Analysis. *Social Indicators Research* 135: 269–289.

Christensen, C.M. 1997. *The Innovator's Dilemma: When New Technologies Cause Great Firms to Fail*. Cambridge: Harvard Business School Press.

Conway, G.R., and E.B. Barbier. 1988. After the Green Revolution: Sustainable and Equitable Agricultural Development. *Futures* 20 (6): 651–670.

Dasgupta, P. 2013. The Nature of Economic Development and the Economic Development of Nature. *Economic and Political Weekly* 48 (51): 38–51.

ECA (Economic Commission for Africa), AUC (African Union Commission), AfDB and UNDP (United Nations Development Programme). 2016. *MDG to Agenda 2063/SDGs Transition Report 2016—Towards an Integrated and Coherent Approach to Sustainable Development in Africa*. Addis Ababa: ECA.

Engerman, S.L., and K.L. Sokoloff. 2002. Factor Endowments, Inequality, and Paths of Development Among New World Economies. *Economia* 3 (2): 41–102.

Foody, G.M. 2003. Remote Sensing of Tropical Forest Environments: Towards the Monitoring of Environmental Resources for Sustainable Development. *International Journal of Remote Sensing* 24 (20): 4035–4046.

Giddings, B., B. Hopwood, and G. O'Brien. 2002. Environment, Economy and Society: Fitting Them Together into Sustainable Development. *Sustainable Development* 10 (4): 187–196.

Goos, M., A. Manning, and A. Salomons. 2014. Explaining Job Polarization: Routine-Biased Technological Change and Offshoring. *American Economic Review* 104 (8): 2509–2526.

Graham, C., and A. Felton. 2006. Inequality and Happiness: Insights from Latin America. *Journal of Economic Inequality* 4: 107–122.

Halter, D., M. Oechslin, and J. Zweimüller. 2014. Inequality and Growth: The Neglected Time Dimension. *Journal of Economic Growth* 19 (1): 81–104.

Hull, Z. 2008. Sustainable Development: Premises, Understanding and Prospects. *Sustainable Development* 16 (2): 73–80.

Jaumotte, F., S. Lall, and C. Papageorgiou. 2013. Rising Income Inequality: Technology, or Trade and Financial Globalization? *IMF Economic Review* 61 (2): 271–309.

Kelly, R. 2009. Science, Technology and Innovation and Inequality. *Journal of Eastern Caribbean Studies* 34 (3): 79–92.

Korotayev, A., and J. Zinkina. 2014. On the Structure of the Present-Day Convergence. *Campus-Wide Information Systems* 31 (2/3): 139–152.

Kuznets, S. 1955. Economic Growth and Income Inequality. *American Economic Review* 45 (1): 1–28.

Lee, D.J., and J.C. Son. 2016. Economic Growth and Income Inequality: Evidence from Dynamic Panel Investigation. *Global Economic Review* 45 (4): 331–358.

Lin, J.Y. 2012. *New Structural Economics: A Framework for Rethinking Development and Policy*. Washington, DC: World Bank Publications. Available from: http://siteresources.worldbank.org/DEC/Resources/84797-1104785060319/598886-1104951889260/NSE-Book.pdf. Accessed 20 Nov 2016.

Luke, T.W. 2005. Neither Sustainable Nor Development: Reconsidering Sustainability in Development. *Sustainable Development* 13 (4): 228–238.
Maddison, A. 2008. Shares of the Rich and the Rest in the World Economy: Income Divergence Between Nations, 1820–2030. *Asian Economic Policy Review* 3 (1): 67–82.
McMillan, M., and D. Rodrik. 2011. Globalization, Structural Change and Productivity Growth. In *Making Globalization Socially Sustainable*, ed. M. Bachetta and M. Jansen. Geneva: World Trade Organization Publications.
Meadows, D.H., D.L. Meadows, J. Randers, and W.W. Behrens III. 1972. *The Limits to Growth*. New York: Universe Books.
Milanovic, B. 2016. *Global Inequality: A New Approach for the Age of Globalization*. Cambridge: Harvard University Press.
Mokyr, J., C. Vickers, and N.L. Ziebarth. 2015. The History of Technological Anxiety and the Future of Economic Growth: Is This Time Different? *Journal of Economic Perspectives* 29 (3): 31–50.
Pearce, D.W., and G. Atkinson. 1993. Capital Theory and the Measurement of Sustainable Development: An Indicator of Weak Sustainability. *Ecological Economics* 8: 103–108.
Rawls, J. 1971. *A Theory of Justice*. Cambridge: Harvard University Press.
Rubil, I. 2015. Regional Heterogeneity in Growth and Inequality Elasticities of Poverty in Transition Countries. *Applied Economics Letters* 22 (6): 425–429.
Saboori, B., and J. Sulaiman. 2013. Environmental Degradation, Economic Growth and Energy Consumption: Evidence of the Environmental Kuznets Curve in Malaysia. *Energy Policy* 60: 892–905.
Schumpeter, J.A. 1939. *Business Cycles: A Theoretical, Historical, and Statistical Analysis of the Capitalist Process*. New York: McGraw-Hill.
———. 1994/1942. *Capitalism, Socialism and Democracy*. London: Routledge.
Spaiser, V., S. Ranganathan, R.B. Swain, and D.J.T. Sumpter. 2017. The Sustainable Development Oxymoron: Quantifying and Modelling the Incompatibility of Sustainable Development Goals. *International Journal of Sustainable Development and World Ecology* 24 (6): 457–470.
Stewart, F. 2014. Sustainability and Inequality. *Development* 57 (3–4): 344–361.
———. 2015. The Sustainable Development Goals: A Comment. *Journal of Global Ethics* 11 (3): 288–293.
Timmer, C.P. 2007. *A World Without Agriculture: The Structural Transformation in Historical Perspective*. Washington, DC: AEI Press.
Todaro, M.P. 2014. *Economic Development*, The Pearson Series in Economics. 12th ed. Philadelphia: Trans-Atlantic Publications.

UN (United Nations). 2012. *The Future We Want*. Resolution Adopted by the General Assembly on 27 July 2012. New York: UN. Available from: http://www.un.org/en/ga/search/view_doc.asp?symbol=%20A/RES/66/288. Accessed 18 Nov 2015.

UN. 2015. *Transforming Our World: The 2030 Agenda for Sustainable Development*, (A/RES/70/1). New York: UN. Available from: https://sustainabledevelopment.un.org/content/documents/21252030%20Agenda%20for%20Sustainable%20Development%20web.pdf. Accessed 10 Mar 2017.

UNCTAD (United Nations Conference on Trade and Development). 2012. *Economic Development in Africa Report 2012: Structural Transformation and Sustainable Development in Africa*. Geneva: UNCTAD.

UNCTAD. 2018. *UNCTADstat* [Online]. Available from: http://unctadstat.unctad.org. Accessed 5 Apr 2018.

UNSD (United Nations Statistics Division). 2016. *Millennium Development Goals Database* [Online]. Available from: http://mdgs.un.org/unsd/mdg/default.aspx. Accessed 11 Apr 2016.

Woodhouse, E., and D. Sarewitz. 2007. Science Policies for Reducing Societal Inequities. *Science and Public Policy* 34 (3): 139–150.

Zander, P., and H. Kächeleb. 1999. Modelling Multiple Objectives of Land Use for Sustainable Development. *Agricultural Systems* 59 (3): 311–325.

2

Reconstructing Modernisation Inclusive and Sustainable

Introduction

For a long period of time now, technology has been the main driving force for social changes. The first Industrial Revolution largely led to changes in the labour force through the evolutionary steam engine, mass production through the invention of electricity was realised in the second Industrial Revolution, while the third promoted a digital revolution through utilising computers and information technology, leading to influential and radical socio-economic changes. As a result, many Western societies have achieved modernisation by leading in these multiple technology-based Industrial Revolutions.

In addition, largely benefitting from this technology-driven form of modernisation, Western societies (re)colonised non-modern countries, and through their colony policies, they have plundered and exploited both physical and human resources to eventually grasp the title of *Advanced Countries*. Even in the process of modernisation, which embodied a selfish characteristic, their modernising experience must in itself be worthwhile to learn from. For this reason, the concept of *Modernisation* quickly became a mega developmental discourse, including Nurkse's theory of

S. J. Baek, *The Political Economy of Neo-modernisation*,
https://doi.org/10.1007/978-3-319-91394-0_2

Vicious Circle of Poverty (1953), Lewis's *Dual-Sector Model* (1954), Kuznets's theory of development (1955), Hirschman's *Unbalanced Growth* theory (1958), Rostow's *Take-Off* theory (1960), and many others.

In fact, how modernisation theory (hereinafter referred to as 'MT') became such a hegemonic discourse can be mainly attributed to two major drivers. The need for understanding a social scientific perspective on a developmental paradigm was first raised in the US when countries in Latin America, Asia, and Africa were liberated from imperialism and were faced with serious political instability and economic poverty. On the other hand, Western society, despite experiencing the end of imperialism, wanted to justify their intervention in developing countries by exporting a rational knowledge system and a belief in universal science (Acemoglu and Robinson 2012). The latter may also justify the claim of Prebisch's *Dependency* theory (1959). The following evolutionary development discourse that emerged from the controversy of this MT pointed out the limitations of these theories and expanded them to various fields: from the neo-classical growth theories (e.g., Solow 1956) which emphasise technology as an exogenous factor to the endogenous growth theories (e.g., Romer 1990) emphasising technology as an endogenous factor; under socio-political perspectives, collision between 'traditionality' and 'modernity', namely the *Clash of Civilisations* (Huntington 1996); the notion of *Multiple Modernities* (Eisenstadt 2002) where the term 'Western' does not necessarily have the same meaning as modern as there are various modernities among non-Western countries; and many others.

In addition to the fundamental criticisms against MT, I broadly attempt to identify the three major development paradigm shifts by summarising factors that have not been dealt with so far.

First, there has been a radical socio-economic change due to aggressive globalisation forces, especially during the past several decades (Koenig-Archibugi 2004). If globalisation can be considered a product of modernisation, it is difficult to erase the meaning that globalisation itself is a process of spreading ideas, artefacts, and institutions as a Western project (Giddens 1991). In other words, globalisation should be recognised not as a means to unify traditionality with modernity, but to their coexistence, as well as having a kind of unstable and possessive feature that even creates a trend of anti-modernism.

Stiglitz (2002) and Sachs (2006) pointed out that globalisation did not bring much economic benefit to the people of the developing world in the first place but rather widened the gap between its rich and poor. In fact, notwithstanding the significant socio-economic progress achieved by many developing countries for over a decade, a number of countervailing trends raise doubts about the sustainability of their development trajectory: their growth has been associated with limited employment opportunities, only marginal declines in poverty, and high levels of income inequality (Armah and Baek 2015, 2018). In this regard, without understanding the coexistence of traditionality and modernity through such globalisation, the reasons why some distorted and diversified developmental outcomes have been observed in African countries despite deeply embracing Western modernisation policies cannot be explained.

Second, while the focus of modernisation so far has been on a human-centred discussion of economic development, social integration, liberty, and peace, the aspirations for today and the future, along with strong globalisation driving forces, have begun to demand paradigm shifts to environmental development as a human-nature centred debate. In this regard, global politics introduced the concept of *Sustainable Development* (Hull 2008), which has become a democratic, globalised, competitive, and irreversible contemporary mega-narrative discourse (Castro 2004). In this context, the various developmental paths that have emerged in recent years cannot be explained without understanding how individual countries influence society's development through interactions with existing co-ordination mechanisms. In other words, when a new development idea such as sustainability is embedded within a society, there will be a mediating mechanism among the policy ideas of the main actors participating in the reform, the dominant political culture and social ideology, in order to determine the concrete contents and direction of the reform (Schmidt 2002).

Third, since the book *The Limits to Growth* by Meadows et al. (1972) offered caution on environmental crises such as resource depletion, which is one of the key challenges to sustainability, the importance of technological innovation has been re-emphasised (Myers and Simon 1994; Lomborg 2001; Kurzweil 2005). In fact, no development theory, including MT and mainstream economic growth theories, has ever denied the importance of technology. Because of the prevailing myth of technological progress, the side effects of technology have intentionally been neglected, even though they can lead to social conflicts, and especially to widening inequality (Cozzens 2007; Woodhouse and Sarewitz 2007).

In the forthcoming era, the convergence of technology and society will be accelerated by linking people with the *Internet of Things*, and the social impacts thereof is expected to be far greater than in the past (WEF 2015). Thus, without an objective understanding of the economic and social ripple effects of technological innovation and of detailed technological development patterns, it is very difficult to explain why those countries that are well known for technological innovations such as the US and South Korea have at times been so lacking in their recent economic performance and have even experienced social chaos.

The central objective of the chapter is thus to attempt to explore the three influential development paradigm shifts to reconstruct a country's modernisation process, the shifts that are aggressive globalisation forces, normative sustainable development ideas, and radical technological change. In so doing, a critical review of various MTs is imperative with the comparative application of the likely impact of recently emerging normative development ideas, particularly the notion of sustainability. The reconstruction effort is further grounded on technological innovation with a focus on the application of S-curve theory. Finally, a systematic assessment is undertaken on the theoretical integration of sustainability and radical technological change into the national framework where economic, social, environmental, and political dimensions are interacting with one another to shape new inequality dynamics.

Modernisation Theory Revisited

Critical Review of Modernisation Theory

In understanding the contemporary policy challenges and *constraints* that countries are facing, systematically addressing the theoretical foundation of MT shall be placed at the core of my arguments. In other words, critically examining its possibilities and limitations further associated with convergence discourse may provide theoretical and methodological implications that are more relevant to the context of the twenty-first century.

MT is a group of theories that explain the process of national development with a focus on the concept of modernity. According to Max Weber, a German political economist, modern society is interpreted by rational-, reasonable- and scientific thinking surrounding a society, which is in contrast to a premodern (traditional) society. As such, the dichotomy between traditional society and modern society, and the process of transition from traditionalilty to modernity, should be central to discussing what MT essentially focuses on.

The notion of modernity has, however, a very context-dependent nature in the sense that its time in history is perhaps assigned to the trisectional framework of *Premodern-Modern-Postmodern.* In this simple reasoning, 50 years ago, the end of the nineteenth century was regarded as modern, but now the period of global modernisation can be considered to be from the early to the mid-twentieth century. If such a trisectional substitution is applied at the national level, the timing of modernising could vary from country to country. Despite such ambiguity, MT became a hegemonic discourse from the point that a large number of nations were liberated from imperialism and were then faced with serious political instability and extreme poverty after the Second World War.

During the three or four decades after the 1950s, MT experienced a renaissance. The *Dual-Sector Model* (1954) became popular, convincing the intellectuals of the 1950s of its worth. In this model, agricultural and industrial sectors generally coexist in developing countries. These countries experienced rapid industrial development by benefiting from low-cost farming. Nonetheless, as the rural labour force continues to flow into the cities, certain challenges arise in the cities when the labour force from the rural area becomes exhausted—society passes through a turning point of labour force from the status of 'excess' to 'shortage'. This transitioning moment can be referred to as the *Lewisian Turning Point.* Here, there will be a structural discrepancy between the demand and supply of the labour market, which can be associated with a sharp rise in wages. As a result, a structure comprising high cost but low efficiency will become deeply embedded in society, which further leads to a slowdown in economic growth.

During a similar period, another influential thinker was Ragnar Nurkse, who introduced the theory of *Vicious Circle of Poverty* (1953). According to this, the lack of capacity for capital formation in underdeveloped countries leads to a higher poverty rate, which in turn can lead to challenges for effective capital formation. This bi-directional causality can be explored through three channels: high interdependence between income and capital, causality between poverty and health, and close linkages between poverty and education. As per his argument, the vicious circle of poverty would perpetuate the poverty level of developing countries, thus how to stem this vicious circle becomes a fundamental challenge for these countries.

In a study that attempted to address the challenges mentioned earlier, Paul Rosenstein-Rodan's *Big-Push Model* (1943) shall be revisited, which was later developed further by Murphy et al. (1989). This school of thought essentially asserts that strong effort in the early phase of development should be accompanied by a significant amount of investment in a rapid and timely manner so that a vicious circle can be stemmed.

Albert O. Hirschman (1958), who speaks in a similar way about development, argues that given the many economic sectors existing in a country, special efforts to foster a particular industry need to be made because such policy focus could create a leading industry, which will generate a chain reaction (domino effect) for other industries. In his argument, a positive circulating process would be repeated by 'lead versus pull' between industries. Thus, intensive competition between the leading and chasing sectors not only raises the level of competitiveness but can also naturally help stem the aforementioned vicious circle, which is dealt with by *Unbalanced Growth* theory.

Unlike various MTs that highlight the challenges and limitations of developing economies, Walt Whitman Rostow's *Take-Off* theory (1960) was derived from the modernisation experience of Western countries. He introduced the developmental process through five stages: traditional society, preconditions for take-off, take-off, drive to maturity, and the age of mass consumption. As his theory adopts the premise that the future of developing countries can be linked to the current state of developed economies, its extensibility when originally introduced was the most significant of its kind. Many developing countries rushed to find out where their economic status ranked among the five stages. Many also attempted to design and implement development strategies to jump to the next stage. In particular, such series of development were presented to explain the smooth transition process from a traditional society to an information society.

Since Rostow's theory caused a massive wave of academic attention, numerous follow-up studies have been conducted in a way that have attempted to present a link between various social sciences and political variables including governance (Apter 1965), democracy (Lipset 1959), psychology (McClelland 1967), and culture (Inkeles and Smith 1974). Meanwhile, under the economic pillar, neo-classical growth theory (e.g., Solow 1956) and endogenous growth theories (e.g., Romer 1990) were introduced, pointing out the alleged limitations of MT.

It seems clear that an extensive amount of research has been carried out in pursuit of one goal, namely understanding modernisation, which allowed MT to lead in an era of world development discourse. From the 1950s to 1960s, a need arose for a new paradigm for the socio-scientific understanding of underdeveloped countries, which was led largely by American scholars. The underdeveloped countries in question included those in Asia, Africa, and Latin America that gained their independence at various points after the Second World War. Since then, MT has evolved into an interdisciplinary theory that covers most sociological sciences including economics, sociology, political science, and psychology. Moreover, MT has been applied not only to academic debate but also to modernisation field projects.

In fact, MT has been closely linked to *Third World* development from the outset, substantially governed by the US and its foreign policy, which has had a profound impact on the modernisation processes of many developing countries. It is true that most of the modernisation-related research published in the 1950s and 1960s was led either directly or indirectly by a limited number of institutions that were funded by the US government. Among them, the Massachusetts Institute of Technology's Centre for International Studies is a foremost example, a type of military–industrial–academic complex (Leslie 1993). In the 1950s, under the leadership of the US Department of Defense, this institution assembled unprecedentedly diverse social scientists for the purpose of *Project Troy*, a top-secret project exclusively managed by James Killian, president of the Massachusetts Institute of Technology (Needell 1998). In the report, it was proposed that whips be used such as military power or propaganda to provide socio-economic incentives in order to prevent the expansion of communism and fascism, which implies that MT was also part of ideological competition.

Following the Cold War, the US appears to have built an empire that was not conquered by force (Hardt and Negri 2004). Indeed, it did not seek to build colonies nor to dominate other sovereign states. Instead, the apparent American empire has taken the initiative of the mega-development discourse to govern the world by institutionalising the free market and the value of democracy. In this regard, MT seems to have been a key driver of the world's hegemony for a century, in addition to its own theoretical values.

On the other hand, some theories that contradict these Western-centred modernisation discourses began to emerge. Werner Sombart, a German economist, proposed in the 1920s a new perspective, which criticised Eurocentrism by using the terms *Core* and *Periphery*. These terms had a profound effect on *Dependency* theory, further advanced by Raúl Prebisch (1959). According to this, core (i.e., industrial-based) nations are instructive to peripheral (i.e., agriculture-based) nations who are rather passive to the benefit of core nations. Through this concept of core versus periphery, he structurally identified a structural relationship between the two groups in the world's economic system. In it, Raúl Prebisch's argument was that these periphery countries (e.g., Latin American countries) may not escape severe poverty if maintaining an agrarian economic structure. This implies that coffee, bananas, sugar, and copper, which represent the core exported products for most Latin American countries, ironically prevented these countries from developing economically.

In this regard, it may be hard to deny that Western advanced economies had in the past exploited the natural and labour resources of many non-modern countries and that such exploitation had a great influence on their modernisation processes. As a result, countries that experienced colonisation have been institutionalised in either an inclusive or an extractive manner (Engerman and Sokoloff 2002). A number of colonised countries at that moment, when relatively inclusive institutions were embedded within society, were able to accept the values of modernity more easily, while most developing countries with relatively extractive institutions embedded in them faced socio-political conflicts in the process of accepting such values. In sum, it could be claimed that many Western nations modernised through extensive exploitation of colonial resources, while trying to share their modernising experiences with their colonised nations.

Epistemological Account

Having understood various MTs and their contradictory nature, it is worthwhile exploring the conceptual and epistemic aspects of MT. MT appears to share a core concept with the *Theory of Evolution* (Inkeles 1993). Just as human destiny is predetermined in evolutionary terms, MT seems to have an evolutionary implication that the transition to modernity from traditionality can be associated with a unidirectional process of change. Also, as the elements of biological organisms function independently in a large and structured framework, MT assumes that a harmonious system can be formed within the structured interdependence among major actors in society.

Due to these commonalities, MT can be interpreted to assume that all countries will eventually reach similar developmental outcomes from the processes of economic growth and democratisation. This is consistent with convergence development discourse. After starting to have a relationship with the West, developing countries may have pursued their developmental process in an inevitable and unstoppable fashion (Tiryakian 1991). On the other hand, MT played a central role in justifying Western intervention in developing countries via exporting a rational knowledge system and belief in universal science (Acemoglu and Robinson 2012). This therefore implies that the frame of thought is structured in a way that the present status of Western society (i.e., more capitalism and democracy internalised) can be the one in which the developing nations are supposed to aspire to in the future.

Second, MT also seems to adopt a *Functionalism* perspective (Inkeles 1993), which can be applied as a more detailed approach to the interdependence among actors of the social structural framework. In other words, this refers to a pattern of behaviour that occurs continuously through the interdependence of individual social factors, which will be defined as pattern variables, and that the collisions and harmonisation between these variables will ultimately lead to *Homeostatic* equilibrium. Such a process to reach equilibrium can be referred to as an endogenous process (of fluctuation, adaptation, and accommodation). In other words, a traditional society in which various pattern variables are interacting to shape some feature of endogenous process is naturally reaching a certain level of *Homeostatic* equilibrium, which means a stable process of development (e.g., industrialisation and urbanisation) by relaxing tension and conflict (Figueroa 2017).

This functionalist view is that traditional values are replaced with modern alternatives, or that modernity and traditionality are assumed to be mutually opposed (exclusive). From this perspective, it is assumed that adopting a Western-centred development model aiming at economic growth and democratised advancement is imperative for modernising aspirations through replacing traditionality with modernity. This is akin to stipulating that the right direction is to minimise traditional roles and values.

However, such a view can expose a limitation of MT because in developing economies, traditionality and modernity are neither asymmetric nor mutually exclusive. The majority of African countries have maintained, for instance, traditionality to emphasise the privileged class and their social status, while at the same time have accommodated universal values, such as capitalism and democracy. In this regard, modernity and traditionality coexist in many developing countries and, in fact, are compatible. It may therefore appear to be unreasonable to treat traditionality as an impediment to a country's development.

Notwithstanding the long-standing controversy of the traditionality versus modernity debate, the nub of such debate may have been hidden in the confrontation of the ideology of the Cold War. After the Cold War, when politics struggled to find the primary source of conflict (e.g., cultural and religious identities), the *Clash of Civilisations* emerged

(Huntington 1996). This normative discourse emphasised that there is not only a conflict between traditionality and modernity but that there is also a conflict among modernities themselves. These various types of conflict eventually lead to the formation of a national cultural identity. During today's post-Cold War era, conflicts arise between more comprehensive modern identities, which go beyond political and economic values. These complex conflicts are therefore a key factor reconstructing the world.

Using this line of reasoning, modernisation does not necessarily mean *Westernisation*, while there are also various forms of modernity among non-Western countries as well as even different modernisation paths among Westernised countries. The meta-debates on both *Westernisation* versus *non-Westernisation* and *Convergence* versus *non-Convergence* can thus be challenged from the viewpoint of evolution and functionalism as well as the concept of multiple modernities (Eisenstadt 2002).

Globalisation and Convergence

In the twenty-first century, the modernisation process has been underpinned by aggressive globalisation forces. It may be said that various modernisation paths along with competing development discourses have indeed been derived by accelerated processes of the globalisation phenomenon across countries.

During the 1970s and 1980s, the ideology of neo-liberal globalisation, mainly from the US, led the world development discourse, while various restructuring programmes were implemented around the 1990s, replacing the role of government with free market mechanisms. As a result, free trade, market openness, and deregulation were serving as social and political pressures for welfare retrenchment. For instance, states tend to reduce taxes and design deregulation policies in order to secure the business confidence of investors and capital. Moreover, because of the strong incentives to make the labour market more flexible, the role of welfare states was downgraded or converged to a liberal market economy (LME). Overall, neo-liberalism was recognised as a global standard that guaranteed economic efficiency as long as the globalisation of production and banking systems accelerated.

These policy trends may have resulted in the dismantling of local communities as multinational corporations became more powerful. When states hesitate to correct distorted resource allocation due to excessive global competition, the fairness and equity of the redistributional structure consequently deteriorate (Wallerstein 1996). In this sense, neo-liberal globalisation aimed at integrating the world into one market strongly influences the development path in many ways. The overall policy orientation towards a LME through deregulation, the liberalisation of banking and capital, flexible labour market, privatisation of public enterprises, and welfare retrenchment can be supportively associated with the convergence argument, which is further related to the intrinsic value of the MT.

On the one hand, strong resistance to globalisation forces has extensively continued. Critics of globalisation argue that although several advanced countries such as the US and Western European countries are exploiting powerless nations to accommodate their own globalisation, governments of these nations are actually proactive actors seeking to gain from entry into foreign markets.

In line with the critics, Joseph E. Stiglitz (2002) argues that globalisation has not brought much economic benefit to the people of the developing world but has rather widened income inequality. He provided evidence that in the 1990s, the world's total income grew by 2.5 per cent annually, but the number of people classified as poor across the world increased by 100 million. Moreover, Jeffrey D. Sachs (2006) also states that between 1980 and 2000, none of the rich nations in North America, Europe, and East Asia experienced economic decline, but many developing countries experienced negative growth in terms of PCGDP and the continuation of extreme poverty.

In sum, the rapid pace of globalisation has indeed increased competition in the consumer goods and services market, and has enabled capital flows without borders. In addition, as human resources are also shared across borders in the globalised world, some countries are benefiting from cheap labour forces thereby increasing their productivity while others are experiencing a widening inequality. It should be noted that globalisation is thus accompanied by the uncomfortable division of winners and losers (Koenig-Archibugi 2004).

As such, globalisation entails not only putting the arguments of convergence advocates underlining uniformity but also the notion of splitting winners and losers. In order to understand the essence of the pros and cons of globalisation along with the intensive debates on the *Convergence* versus *non-Convergence* debate, it is necessary to relate globalisation to a debate between modernisation and capitalism, which may make it possible to ascertain the limitations of convergence theory.

First, if globalisation is considered as a by-product of modernity, it would be difficult to rid itself of the essence of globalisation as a process of spreading ideas, artefacts, and institutions as a Western project (Giddens 1991). It is in this context that Huntington's *Clash of Civilisations* theory (1996) has plausibility. In other words, not only do traditionality and (post-)modernity coexist within the broad dimension of modernity but the tendency of anti-modernism also exists in modern society which is unstable and fluid under globalisation forces.

Although the postmodern era, in which mankind has been postulated as a postmodern stage, refers to the transformation to an information society, knowledge society, and green economies through globalisation, in stark reality such transformation is mostly a process under the system of capitalism. Due to this, globalisation contains the meaning of both globally restructured capitalism and a tendency for re(de)globalisation. Thus, developing countries like those in the semi-periphery or semi-core tend to be indwelled by unevenness (inequality in vertical hierarchy) rather than reciprocity equality (mutual benefit from the horizontal perspective) in the process of globalisation. The problem of globalisation can be seen as accelerating the endogenous irrationality previously raised by *Dependency* theory.

Although the great proliferation of neo-liberalism as a global standard seems untenable, globalisation seems instead to lead to the diversification of countries. As explained earlier, countries failing in terms of neo-liberal competition in the wave of globalisation undergo a huge social upheaval, evolve and overcome fierce clashes between traditionality and modernity in the process of modernisation, and also learn from the experience and know-how of winners in a globalising battle. Competition is becoming fierce here though, resulting in more and more losers.

Furthermore, the convergence phenomenon may not have come to us as predicted by the globalisationists, even among the most advanced countries that can be classified as winners in global competition. For instance, by looking at the development process of Western European countries largely regarded as welfare states, we may observe that the welfare system has been advanced more as an open economic system (Rieger and Leibfried 1998). Essentially, as governments try to open their economic systems in line with globalisation forces, they intend to stabilise the labour market by restricting wages in order to strengthen the competitiveness of domestic industries. It is often the case that more welfare systems are implemented to control problems such as unemployment and the gap between the rich and poor in order to pursue economic, social, and political stability (Garrett and Mitchell 2001).

Furthermore, as advanced welfare states have had to both focus on human capital development and expand industrial infrastructure, in order to increase economic productivity so as to attract more foreign investment, the role of government has been strengthened. Accordingly, the *Neo-liberal Convergence* advocates overemphasise the impact of globalisation pressure on the welfare state when taking into account the impacts of interdependence between the market and welfare system. Even if the argument of convergence can seem intuitively quite convincing, empirical evidence suggests that low wages, low tax rates, and deregulation may not be a recipe for capital investment but that rather the quality of labour forces and social overhead capital effective for production, market accessibility, and socio-political stability should all be ensured. Such capital investment dynamics appear to be not as straightforward as the arguments of convergence theorists may suggest.

Unlike the non-convergence phenomenon through globalisation between developed and developing countries (i.e., winner versus loser), the within-country aspect of the *Convergence* versus *non-Convergence* debate would be more complicated, as evidenced by the aforementioned diversified welfare regimes of developed countries. It is therefore necessary to discuss the three major channels of globalisations, namely trade, capital, and technological development.

First, through the globalisation of trade, it is hard to deny that advanced technology is one of the major catalysts for the modernising process, and effectively shared across the world, especially for developing countries with technological development having been a powerful driving force of economic development. As emphasised in *Stolper-Samuelson* theorem,[1] the revitalisation of international trade has indeed helped solve the problems of extreme poverty and inequality within developing countries (Jaumotte et al. 2013). As such, the globalisation of trade might have contributed to an escape from Nurkse's *Vicious Circle of Poverty* theory (1953). The negative aspects of the globalisation of trade have also, however, been strongly noted (Ravallion and Chen 2013). For instance, some insist that the increase in average national income does not automatically translate into an increase in the average living standard of households, and further that the alleviation of poverty is not always among the benefits of growth. Thus, the globalisation of trade has rather intensified this polarisation within countries.

The second point to look at here is associated with the globalisation of capital and banking. Indeed, today's global financial transactions are estimated to be about 50 times greater than the value of global trade (BIS 2017; WTO 2017), which clearly indicates that the impact of capital dynamics in a globalising market is enormous. As most capital usually comes from advanced countries, it could be helpful to understand the nature of the problem by highlighting a situation when huge amounts of capital are injected into developing countries. Once large capital flows into developing countries, it is usually associated with higher policy interest rates and a pegged exchange rate. Such policy responses could have a negative impact on the exporting industry to a certain extent. As such, the weakening of global competitiveness regarding the export industry in developing countries could be a key obstacle to sustainable economic development. Moreover, the labour force in trade-related industries could be absorbed into other industries or sectors that can be less productive, or it may lead to massive unemployment and social conflict. Generally speaking, when huge speculative capital from developed countries is invested as capital-intensive foreign direct investment in developing countries, it may be true that the market (and capital) mechanism strongly contributes to deepening the polarisation within the recipient country (Singh 2004). This may of course vary from country to country depending on the effectiveness of the financial and banking sectors and of poverty reduction strategies (Jaumotte et al. 2013).

Lastly, it can be said that the globalisation of technological innovation, accelerated by the diffusion of technology, has made a remarkable contribution to fostering national economic development. Such a contribution has been derived by promoting new industries and innovative products, and creating associated jobs (Mokyr et al. 2015). However, considering the intrinsic nature of technology, the outcome may not always be positive. In other words, the globalisation of technology could have an adverse impact on income inequality because it tends to increase the premium for skilled tasks in both developed and developing countries, thereby characterised by a skill-biased technological change (Jaumotte et al. 2013). In addition, globalising technology also has the nature of a routine-biased technological change, which could lead to the polarisation of workers along with the collapse of the middle class whose task areas are heavily repetitive (Goos et al. 2014).

The effect of the aforementioned technological innovation on society is only the tip of the iceberg. Technological advancements have indeed significantly transformed the structures of societies, industries, and economies (Freeman and Perez 1988), which combine to determine competitiveness in international trade and financial markets. As technological innovation has been the most influential aspect for the globalisation of trade and banking, and can therefore be seen as a kind of cross-cutting driver (e.g., to either accelerate or mitigate globalisation forces) of socio-economic change, it can be said that the socio-economic impact of technological development is taking place across all industries and sectors, so it may not be possible to analyse such a broad impact synthetically.

Narrative Analysis of Sustainability, Technology, and Inequality

Sustainability

Despite its greater influence over the world development discourse, MT has a number of limitations, particularly as it is based on a three-stage model (*Premodern-Modern-Postmodern*), which implies a dichotomy between traditionality and modernity. Partly because of this, today the term 'development' is sometimes substituted in for modernisation. Ultimately, it is rather too broadly used.

Development originally refers to the process of manifesting the potential of an object or organism to arrive in its natural and complete form (Esteva 1992). This concept of a strong biological implication has been applied to the *Monotheistic* advocates of Europe since the eighteenth century and thus has extended its application to social phenomena. Nisbet (1969) adequately explains it using a metaphor in which objects are perceived as organisms with biological development principles as development is applied to social phenomena.

Since the mid-twentieth century, development has been widely used by nations and societies and its meaning has mainly related to economic growth. As such, *Developmentalism*, mainly driven by economic growth, was actively accepted by countries, particularly developing ones who were enthusiastic about economic growth. The dominant elites in non-Western regions, such as in Asia, Latin America, and Africa, adopt developmentalism as a key strategy for their nation-building strategy, forming a single axis of consent to the global hegemony of developmental discourse (Ndlovu-Gatsheni 2012). In this context, developmentalism has been used primarily to refer to the ideology and strategy of national development, centred on industrialisation in *Third World* countries. As such, this has triggered a discussion on the so-called *Developmental State* (Evans 1995).

At some stage, the Western world created a new concept of modernity as part of its strategy to further expand its influence in developing countries. In fact, vague and ambiguous notions of modernity can be seen as a useful intervention tool for Western nations. As modernity tends to signify the direction that a society wants to pursue, it can function effectively in developing countries to deepen the idea of *Westernisation* or *Americanisation*. As a result, many developing nations around the world have recognised the Western hegemonic supremacy of modernity (Sikkink 1991). Thereafter, development and modernisation seem to have evolved into two separate branches although still sharing common values.

First, development began to branch off with a focus on the economic perspective. Economic growth means the process by which a country's economy produces more products than it did before. This is thus related to exploring the most efficient production methods in order to minimise the waste of resources while maximising the economic resources in the production system. The concept of development was then re-evolved from 'economic growth' to 'economic development' so to accommodate not only the growth of the economy but also the continuing enhancement of society. In this sense, the objective of economic development cannot simply be reduced or equated to that of economic growth. Rather, economic development is accompanied by economic growth and what kind of economic means are needed to achieve such a desirable social state, which implies a much broader and more comprehensive set of challenges in society.

According to Michael P. Todaro (2014), an American development economist, economic development is a concept that includes more than the material aspects of life, that is, it should be understood as a multidimensional process involving the continuation of economic growth, the reduction of inequality, the elimination of absolute poverty, social structure, public attitudes, and major changes in national institutions. In short, economic growth is a concept that focuses on the quantitative expansion of the economy, while economic development considers quantitative and qualitative aspects altogether. Nevertheless, given the fact that economic development indicators such as GDP and gross national income are widely used as criteria for determining the hierarchy of each country, we must admit that an economic growth-driven development process is still a global hegemonic trait.

As the concept of economic development began to pursue more than the material aspects of life, the development of society became more and more important and was directly involved in the shift in the development paradigm. Societal development can refer to the process of humans overcoming the structural change of a social system and also the process of taking control of the surrounding environment. In this process, the capacity of accommodating uncertainties related to the power shift, economic growth, social development, differentiation and specialisation, and adaptability to a new environment become core elements of economic development (Midgley 1995).

The concept of economic development may not, however, best justify Western intervention in developing countries. Therefore, many Western societies grew concerned about pursuing continuous intervention, eventually leading to the evolution of the concept of modernisation which acquired a meaning that integrates economic development into Western-centric development ideal. Modernisation today, therefore, constitutes a social environment in which activities of economic production are effectively integrated, and a process of re-defining the value system such as norms and culture by creating a social order that distributes the results of economic production evenly across society. However, such redefinition of the value system often involves internal conflicts and collisions, which again raises the issue of social integration in relation to institutional and cultural co-ordination. In this regard, society is confronted with social conflict, corruption, bad governance, and social movements that are associated with the dilemma of structural differentiation and integration during the process of modernisation, a process which is repeated (Diener et al. 1998).

In this sense, development and modernisation, which have evolved through complementing each other from the perspective of economic and social development, are faced with a great challenge. There has been a kind of human-centred discussion of economic development, social integration, and liberty and peace, but society of today and the future now demands another paradigm shift to a more human-nature-centred development.

Since Francis Bacon, a British Renaissance thinker, said that it is good to conquer and utilise nature to build civilisation, humanity has diligently sought knowledge based on a strong belief in the potential benefits of science and technology. In doing so, technological advancement has been exponential, allowing us to develop and utilise nature more efficiently. During the twentieth century and beyond, humanity has made tremendous efforts to move towards civilisation and development. However, due to the construction of civilisation, nature, the very foundation of civilisation around the world, is fundamentally collapsing.

Crucially, in this regard, nature relies on circulation. Indeed, the human biological system also circulates. That is, all natural bodies are driven by circulation. Even capitalism circulates to generate interest. Considering this *Principle of Circulation* with a focus on the earth's ecosystem, the ideal situation is that mankind enjoys modern values (capitalism) only with its by-product (interest generated from circulation).

With recent environmental challenges, humanity now needs to change its development paradigm and the concept of sustainable development has emerged as a result (Hull 2008). Sustainable development demands greater environmental considerations in addition to economic and social interests. Now sustainability, as sustainable development is also known, is a concept that is comparable to democracy and globalisation, which is everywhere (Castro 2004).

Conceptually speaking, sustainable development is a contested concept. From an economic view, Pearce et al. (1990) defined that 'sustainable development involves maximising the net benefits of economic development, subject to maintaining the services and quality of natural resources over time', while environmentalists put more emphasis on environmental conservation and protection. Furthermore, Conway and Barbier (1988) viewed it as a human-centred concept and focused on human capability and progress by arguing that 'sustainability is the ability to maintain productivity, whether of a field, farm or nation, in the face of stress or shock'. Although it is clear from the debates about sustainable development that there is no one single philosophy, people can agree to have their own interpretation of it (Giddings et al. 2002).

In fact, for a long time, social development and real democracy have been burdened by a strong desire for economic development. Similarly, there is also the possibility that such environmental considerations could still fall victim to an overly economic focus. Nonetheless, global politics today is exerting considerable effort to signal the seriousness of environmental issues, and to convince the public by presenting the logic that economic development and environmental protection can coexist. At the heart of this logic is *Ecological Modernisation*, which succeeds the discourse of Western modernisation and pursues the coexistence of environmental and economic development (McCauley 2006).

The ecological modernisation concept can, however, be less expandable than sustainable development. The term 'modernisation' inherently implies something that is Western-centred and used to legitimise and effect the transfer of Western experiences to developing countries. But environmental problems today are a common global challenge, and ecological modernising experiences of the West have not yet been sufficient to teach others about their environmental policy choices. In addition, many developing countries that are opposed to Western centrism tend to reject ecological modernisation, for fear of being dragged towards Western ideals. For this reason, ecological modernisation seems inadequate as a paradoxical discourse today to raise global awareness about environmental challenges.

Sustainable development is also a relatively self-directed concept, recognising that global environmental problems must be tackled by the developed and developing world together. In particular, developing countries are attracted by the opportunity to set their own policy directions in line with their own environmental and geographical conditions. Moreover, sustainable development may not recognise environmental development as a top priority but at least acknowledges that economic and social development are equal to environmental development, which has considerable implications for the interests of developing countries. Today, the hegemony of global economic development has already shifted from the global North (responsible for 40 per cent of global GDP) to the global South (nearly 60 per cent of GDP).[2] Such a shift may have also been reflected in the competing discourse of *Ecological Modernisation* versus *Sustainable Development.*

Against this backdrop, sustainable development can assist developing countries to utilise their environmental resources efficiently to contribute to their own modernisation processes, and to take responsibility for the environmental challenges with the overall aim of enhancing their international standing by making policy in a self-conscious manner. On the other hand, for advanced countries, economic development is already on track, but further growth may be difficult without the necessary capacity. Developed countries therefore ought to consider how they can sustain future growth by considering environmental, social, cultural, and political dimensions of sustainability in addition to economic dimensions. In this regard, for them, sustainability is not merely a desired goal of development, but rather a prerequisite for their competitive and continuous development (Langhelle 2000).

Technological Innovation

The notion of *Grow First and Clean Later* was prevalent in the 1950s and 1960s (Ophuls 1977). During that period, unprecedented economic growth led to a drop in the unemployment rate and to a material abundance in the West and in some communist countries as well. Almost all countries were convinced that there would be no problems regarding resource depletion or that any environmental crises would emerge in the future. Such optimistic views were mainly attributed to the blind faith of free markets and the perceived magic of technology. Many were so trusting of technology that any problem with depletion of resources could, they thought, be resolved through the development of alternative tech-based materials, and the free market that could control prices against demand-supply dynamics. Even Marxists believed that the advancement of science and technology could solve any problems regarding resource exhaustion.

Amid the earlier mentioned prevailing optimistic view of the world's development, raising ideas about global environmental crises that may occur in the next century was even hard to imagine. However, the Club of Rome[3] was concerned about the impact of rapid population growth and industrial growth based on a finite supply of resources. *The Limits to Growth* by Donella H. Meadows et al. (1972) also shared concerns about the predicament mankind faced in this sense. Immediately after the publication of its first edition in 1972, academics, regardless of being on the left or right wing of the political spectrum, voiced criticism. Later, the book received fierce criticism from entrepreneurs and even from advocates of capitalism.

But, over time, as the inequality problem and environmental challenges intensified, both people and organisations that used to have a negative view of the book were increasingly sharing its message. For example, some initiatives for the strengthening of environmental laws and the creation of cabinet-level environmental departments in various countries were launched. Sometime later, an updated revision of *The Limits to Growth* was published in 2004 (Meadows et al. 2004), which propelled environmental considerations back into the development discourse.

The main argument of the *Limitedness* advocates is that if the world's population continues to grow at the current pace, and if industrialisation, the pace of pollution, food production practices, and the use of resources continues unchanged, the planet will reach its limit for growth within the next 100 years. These advocates agree that technological advances may prolong this end point but warn that technology cannot be the fundamental solution. In particular, in the first edition of *The Limits to Growth*, Matthew Simmons noted that even decades after it was first released, the book's fundamental trend analysis is still valid, pointing out the widening inequality, and the level of environmental pollution and ozone depletion which are massively challenging issues today. In 2008, Graham Turner (2008)'s *A Comparison of The Limits to Growth with Thirty Years of Reality* concluded that changes in industrial production, food production, and pollution are all in line with the book's predictions of economic and social collapse in the twenty-first century.

Despite the *Limitedness* school of thought's great influence, the *Unlimitedness* advocates have also continuously tried to expand their influence by insisting that Earth will overcome its limitations (Jorgenson and Stiroh 2000; Hanson 2008). Such advocates are largely growth-oriented scholars, who argue that resource depletion and environmental degradation are secondary problems that can be resolved by technological advances (Myers and Simon 1994). These advocates further insist that the challenges of non-renewable resources and food shortages can also be addressed using alternative energy development through technology (Lomborg 2001). Most sensationally, futurist Ray Kurzweil (2005) popularised the idea of technological singularity by publishing *The Singularity Is Near*. In this book, he argues that through the law of accelerating returns, technology is progressing towards a singularity to create superintelligence. In this regard, this school of thought seems to believe that 'there is no impossibility' on the basis that technology will bring us more than we can currently imagine.

As far as the socio-economic system is concerned, it is necessary to pay attention to Schumpeter's *Creative Destruction* theory (1939) which systematically explains how technology contributes to the development of our economy and society. He focused on the *Creative Destruction* process, which breaks through circular flow, which is repeated every year on a constant basis through the notion of entrepreneurial spirit for innovation. In this vein, innovation is not just about creating new products or developing new production methods but also about a wide range of changes such as new material acquisition, new market exploration, and new organisation creation, in which the innovative efforts of entrepreneurs are leading the overall change of the economy and society. In this regard, he asked: 'how does capitalism create and destroy existing economic structures?' rather than 'in which ways does capitalism manage that structure'? This fundamental question leads us to *Disruptive Innovation* theory, later introduced by Bower and Christensen (1995) who attempted to refine innovation theories in a more empirical manner.

Briefly describing the concept of *Disruptive Innovation*, he first explained some observations that leading companies invest continuously to maintain their superior market position (e.g., by improving the quality of products) in order to try to meet shareholders' desired rate of returns (e.g., by setting higher prices). This type of innovation is regarded as *Sustaining Innovation*. In reality, however, the market sorely needs leading companies to provide higher performing products at lower prices. This mismatch could result in decreasing the correlation between customer loyalty and corporate profitability over time. Today, the market focuses on new innovations. Products that have been treated as cheap and poor and have not been given much market attention have improved their quality through continuous innovation efforts. At some point, the quality of goods with relatively lower prices matches with mainstream customers' needs. This represents the moment when market initiative moves from *Sustaining Innovation* to *Disruptive Innovation*. Christensen (1997) presented empirical evidence for this phenomenon and published *The Innovator's Dilemma* in 1997 (see Fig. 2.1).

From Schumpeter's *Creative Destruction* (1939) to Christensen's *Disruptive Innovation* (1997), the fundamental ideas can be associated

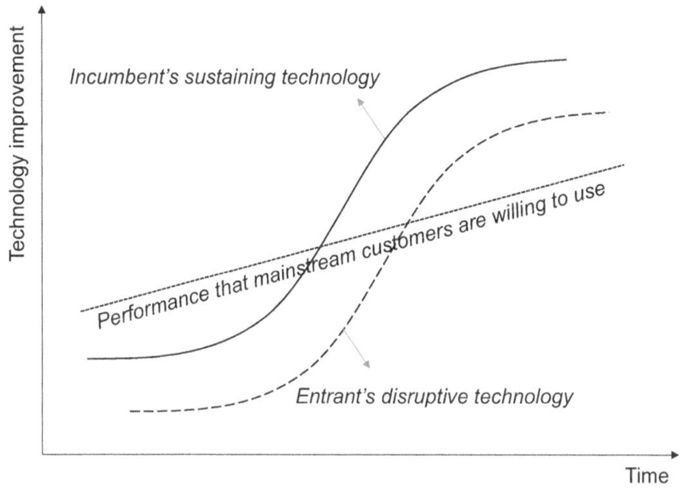

Fig. 2.1 Sustaining innovation versus disruptive innovation. (*Source*: Author's own elaboration on the basis of Christensen (1997))

with the S-curve pattern observed in technological progress. The proliferation of technology diffuses along with the shape of a typical S-curve: in the introductory phase, the improvement of the technological performance over time is slow; then it accelerates rapidly in the growth phase and then slows again in the maturity phase (Foster 1986; Schilling 2010).

On the other hand, the proliferation of multiple technologies spreads as competition between existing and new technologies spreads through the process of technology substitution (Chang and Baek 2010). In particular, technological discontinuity occurs in which the S-curves of a single technology are not connected in the replacement of existing and new technologies (Anderson and Tushman 1990). In this sense, Schilling (2010) argues that discontinuous technologies are based on entirely new knowledge while meeting market needs, and that the development of discontinuous technologies is much more advanced when new technologies are accelerated, or the development of existing technologies reaches a limit.

The importance of technology innovation cannot be overemphasised. It should however be kept in mind that blind faith could generate adverse effects as technology has a double-sided effect. Examining the light and shade of technology advancement in more detail in terms of how it is affecting inequality and polarising society can help us to understand the complexity of the transition from modernisation to sustainable development.

Firstly, science and technology contribute to the reduction of inequalities by strengthening people's knowledge and authority through education, scientific discovery, research, professions, and skills (Kelly 2009). Furthermore, R&D allows the prices of goods and services to become less expensive and more readily available to the wider public. This implies that the marginal benefit of increased accessibility to technology and products is often gleaned by the poorest among society (Woodhouse and Sarewitz 2007). This logic presents an almost entirely positive social impact from new scientific knowledge and technology and assumes that the benefits therefrom are automatically converted to benefit everyone (to varying degrees). In this sense, *Unlimitedness* advocates are stringent in their belief of continuous technological progress.

Until a new normative development discourse emerges with the notion of sustainability, in most countries, technological development is driven by an economic bias that is still valid (Cozzens 2007). This means that the main goal of a state is economic growth through improving its productivity and competitiveness, and social equity and distribution are still regarded as peripheral aims. In other words, the *Trickle-Down* effect is broadly assumed in a way that the fruits of total economic growth through technological innovation flow naturally to all members of society. This indeed implies that science and technological development are essentially based on utilitarian assumptions.

Today, however, this utilitarian way of thinking for technology advancement is challenged or often denied. Bourguignon and Morrisson (2002) studied the improvement of national income through several technological revolutions from the 1820s to the 1990s, and found that inequalities continued to widen between two groups: The US and Western European countries; and the rest of the world.

It is also necessary to examine the role of R&D for technological progress. Most R&Ds in terms of technological innovation have been led by the private sector. In understanding that the top priority of most corporations is to generate profits, there is a tendency for technology investments to be concentrated in wealthy potential customers rather than the poor who are not capable of creating profitable market demand. Besides, the demand for knowledge and innovation from the wealthy in rich countries should be very different from that of the poorest in developing countries. For this reason, unlike the notion of the *Trickle-Down* effect, the positive causality between technology and inequality is a plausible one.

Inequality

Over the past decade, many victims of the global financial crisis have been increasingly voicing criticism against mainstream neo-liberal discourse. Since the crisis, a number of countries have been striving to reshape their distributive justice while rethinking their economic development strategies in an inclusive fashion. It is thus no exaggeration to say that economic growth and inequality are forming the core of these development discourses. Such debate has been accompanied particularly with two *Nobel Prize Winners in Economics*, who have had an influential role in shaping such a narrative.

Simon Kuznets (1955), winner of the *Nobel Prize for Economics* in 1971 popularised the relationship between economic growth and inequality by hypothesising that its nexus is a kind of *Inverted U-Curve*. His line of reasoning can be interpreted in the sense that inequality in rural areas widens as economic resources are transferred from rural to urban areas in the early stage of development, but in the later stage of economic growth such deepened inequality challenges are mitigated due to industrialisation that substantially makes the rural-urban divide less significant. Starting with the concept of the *Inverted U-Curve*, critical debate on the *Inequality-Growth* nexus has experienced a golden era, with two schools of thought evolving therefrom.

On one hand, the discourse about the positive role of inequality has gained substantial public attention, arguing that inequality would induce intense competition that would eventually stimulate growth. Among the studies supportive of this stance, Rebelo (1991) and Deaton and Paxson (1997) insisted that inequality may contribute positively to economic activities so as to accelerate the developmental process on the basis that inequality itself can be a stimulus for people trying to secure a better education and/or job, and a source of motivation for entrepreneurs to invest more, thus ultimately creating new businesses and industries.

There have been, on the other hand, strong opponents of this argument. This school of thought believes that inequality could not provide many educational opportunities to the low-income segment of society, thereby preventing opportunities for social mobility and eventually hindering the development of human capital. It is therefore on this ground that widening inequality could seriously influence the dynamics of savings and consumption in a negative way, so inequality is regarded as an obstacle to economic growth (Feldstein and Horioka 1980; De La Croix and Doepke 2003).

Even though both schools of thought have been given due consideration, the strategic stance of the World Bank and the IMF, which indeed leads the world's neo-liberal development discourse, has tended to lean towards inequality as an *enabling* function. Nonetheless, reflective consideration on the positive hypothesis has recently been started to face much criticism, especially since the 2008 global financial crisis. *The Great Escape* by Angus Deaton (2013) has greatly popularised the idea of the negative function of inequality, and the author was later awarded the 2015 *Nobel Prize for Economics*.

In his book, Deaton agrees that inequality may be both a by-product of growth and an incentive for growth, but that a serious level of inequality could stifle growth, given his argument that the positive features could be erased when inequality comes from rent. Besides, a widened level of inequality tends to distort efficient allocation of resources, aggravate corruption, and advocate favouritism as the rich tend to be more enthusiastic about protecting their vested interests.

Such dichotomous discourse in understanding the *Inequality-Growth* nexus seems a bit overbearing, which has opened up the landscape for a significant amount of research on the coexistence of the two functions in one society. In this regard, Barro (2000) classified the world as rich and poor, and then analysed the changes in inequality versus economic growth in these two classifications of countries (rich and poor). Eventually, he found that the different stages of economic development could cause mixed causalities. In other words, there is a positive relationship in the rich group but a negative one in the poor. Very much consistent with Barros's study, Forbes (2000) argued that the impact of declining inequality in economic growth was positive for the early stage of economic growth but turned negative after a certain point (25 years later in the case of his studied countries).

Given this complexity, Benabou (1996) attempted to define inequality as an indispensable factor as well as a detriment to economic growth. In other words, the impact of inequality on economic growth is not in particular directly determined, nor is it in an absolutely positive or negative relationship. Hence, the direction of positive versus negative would depend on the economic situation in the relevant country (Barro 2000; Banerjee and Duflo 2003; Nemati and Raisi 2015). As such, the correlation between inequality and growth shall be by no means simple.

These theoretically different foundations and empirical results may have caused contrasting or sometimes contradicting assertions in explaining inequality growth dynamics. In other words, such an absence of consensus may be why the empirical evidence from the Kuznets process shows mixed results. For instance, Galbraith and Kum (2005) investigated the relationship and provided empirical evidence to support the notion that inequality is likely to reduce as the income level increases. In addition, they found some tendency for the reverse relationship for advanced nations, which is regarded as the *Augmented Kuznets Curve*. Similarly, in his book, Branko Milanovic (2016) claimed the existence of a *Second Kuznets Curve*, which is the current upswing of the inequality level over income for wealthy countries, mainly due to recent technological advances and the effects of globalisation. However, these two research findings are not comprehensive enough to be generalised in terms of the coverage of countries studied (the US, the UK, Canada, Finland, and some additional OECD countries). Also, their observations that appear to be the *Augmented Kuznets Curve* are based on the relationship between inequality and time period (not income level).

The causal relationship between inequality and growth varies depending on the country, region, and level of development. There is much complexity among social, economic, and environmental development that closely affects the duality feature of technological advancement (Baek 2017). Even if all of these variables were reflected in the correlation analysis between inequality and growth, it may still not be possible to generalise the relationship since the *Homeostatic* equilibrium generated through the endogenous process between these variables tends to evolve over time (Figueroa 2017). In this regard, a Kuznets curve can be largely context-dependent in terms of specific regions, countries, and the timing of the analysis.

Summary and Conclusions

Through a critical review of MT to understand the challenges and limitations of the current developmental landscape, we have derived theoretical implications from the point of view of globalisation, sustainability, and technological innovation dynamics, which can be highly relevant in the twenty-first century.

Contemporary concepts of modernisation with globalisation forces have taken social aspects seriously, and more recently environmental considerations have been strengthened in the policy syndrome. Thereafter, the recent shifting of meta-discourse from modernisation to sustainable development has been critically discussed. In addition, I have explained that technological innovation maintains a close relationship with development in the globalisation wave and even holds the key to win or lose the long-standing ideological debates on *Unlimitedness* versus *Limitedness* and especially with regard to the realisation of sustainability in the future. Despite technology playing such a positive role, it can also be a factor in widening the inequality situation within a country.

Since the complexity of the *Technology-Development-Inequality* nexus coexists within this grand discourse of modernisation, I have expressed concern that the development efforts to prepare for the twenty-first century may be much more complex than we have anticipated. For instance, Western society has pointed out the limitations of development. In fact, while several of these countries have achieved industrial development and have reduced poverty, these achievements have come at higher environmental and social costs. China, for instance, reduced poverty from over 60 per cent in 1990 to less than 10 per cent in 2010 lifting 566 million people out of extreme poverty; however, income inequality and greenhouse gas emissions increased substantially over the same period. For this reason, recent global development discourse, particularly with the notion of sustainability, calls for a new thinking on development.

Notes

1. *Stolper-Samuelson* theorem is a basic economic explanation for the relationship between trade liberalisation and income inequality. According to this, the expansion of trade opening in developing countries with an abundance of low-skilled labour leads to an increase in the wages of low-skilled workers, while lowering the wages of skilled labour in developing countries, which is likely to lead to a reduction in income inequality.
2. According to the World Bank's World Development Indicators (2017), Western-Advanced Nations (about 40 per cent of the global GDP): the US with 24.3 per cent; Germany with 4.5 per cent; the UK with 3.9 per cent; France with 3.2 per cent, Italy with 2.4 per cent; Spain with 1.6 per cent; the Netherlands with 1.0 per cent; and others with less than 1.0 per cent; and Non-Western-Advanced Nations (about 60 per cent of the global GDP): China with 14.8 per cent; India with 2.8 per cent; Brazil with 2.4 per cent; South Korea with 1.9 per cent; Mexico with 1.5 per cent; and others.
3. The Club of Rome is an organisation of individuals who share a common concern for the future of humanity and strive to make a difference.

References

Acemoglu, D., and J. Robinson. 2012. *Why Nations Fail: The Origins of Power, Prosperity, and Poverty*. New York: Crown Publishing Group.

Anderson, P., and M.L. Tushman. 1990. Technological Discontinuities and Dominant Designs: A Cyclical Model of Technological Change. *Administrative Science Quarterly* 35 (4): 604–633.

Apter, D.E. 1965. *The Politics of Modernization*. Chicago/London: University of Chicago Press.

Armah, B., and S.J. Baek. 2015. Can the SDGs Promote Structural Transformation in Africa? An Empirical Analysis. *Development* 58 (4): 473–491.

Armah, B., and S.J. Baek. 2018. Three Interventions to Foster Sustainable Transformation in Africa. *Journal of Social, Political, and Economic Studies* 43 (1–2): 3–25.

Baek, S.J. 2017. Is Rising Income Inequality Far from Inevitable During Structural Transformation? A Proposal for an Augmented Inequality Dynamics. *Journal of Economics and Political Economy* 4 (3): 224–237.

Banerjee, A.V., and E. Duflo. 2003. Inequality and Growth: What Can the Data Say? *Journal of Economic Growth* 8 (3): 267–299.

Barro, R.J. 2000. Inequality and Growth in a Panel of Countries. *Journal of Economic Growth* 5 (1): 5–32.

Benabou, R. 1996. Inequality and Growth. In *NBER Macroeconomics Annual*, ed. B.S. Bernanke and J. Rotemberg, vol. 11, 11–74. Cambridge: MIT Press.

BIS (Bank for International Settlements). 2017. *Triennial Central Bank Survey of Foreign Exchange and OTC Derivatives Markets in 2016*. Available from: https://www.bis.org/publ/rpfx16.htm. Assessed 22 Dec 2017.

Bourguignon, F., and C. Morrisson. 2002. Inequality Among World Citizens: 1820–1992. *American Economic Review* 92 (4): 727–744.

Bower, J.L., and C.M. Christensen. 1995. Disruptive Technologies Catching the Wave. *Harvard Business Review* 73 (1): 43–53.

Castro, C.J. 2004. Sustainable Development: Mainstream and Critical Perspectives. *Organization and Environment* 17 (2): 195–225.

Chang, Y.S., and S.J. Baek. 2010. Limit to Improvement: Myth or Reality? Empirical Analysis of Historical Improvement on Three Technologies Influential in the Evolution of Civilization. *Technological Forecasting and Social Change* 77 (5): 712–729.

Christensen, C.M. 1997. *The Innovator's Dilemma: When New Technologies Cause Great Firms to Fail*. Cambridge: Harvard Business School Press.

Conway, G.R., and E.B. Barbier. 1988. After the Green Revolution: Sustainable and Equitable Agricultural Development. *Futures* 20 (6): 651–670.

Cozzens, S.E. 2007. Distributive Justice in Science and Technology Policy. *Science and Public Policy* 34 (2): 85–94.

De La Croix, D., and M. Doepke. 2003. Inequality and Growth: Why Differential Fertility Matters. *American Economic Review* 93 (4): 1091–1113.

Deaton, A. 2013. *The Great Escape: Health, Wealth, and the Origins of Inequality*. Princeton: Princeton University Press.

Deaton, A., and C. Paxson. 1997. The Effects of Economic and Population Growth on National Saving and Inequality. *Demography* 34 (1): 97–114.

Diener, E., J.J. Sapyta, and E. Suh. 1998. Subjective Well-Being Is Essential to Well-Being. *Psychological Inquiry* 9 (1): 33–37.

Eisenstadt, S.N. 2002. Multiple Modernities. In *Multiple Modernities*, ed. S.N. Eisenstadt, 1–30. New Brunswick: Transaction.

Engerman, S.L., and K.L. Sokoloff. 2002. Factor Endowments, Inequality, and Paths of Development Among New World Economies. *Economia* 3 (2): 41–102.

Esteva, G. 1992. Development. In *The Development Dictionary: A Guide to Knowledge as Power*, ed. W. Sachs, 6–25. London: Zed Books.

Evans, P. 1995. *Embedded Autonomy: States and Industrial Transformation*. Princeton: Princeton University Press.

Feldstein, M., and C. Horioka. 1980. Domestic Saving and International Capital Flows. *Economic Journal* 90 (358): 314–329.

Figueroa, A. 2017. *Economics of the Anthropocene Age*. Basingstoke: Palgrave Macmillan.

Forbes, K. 2000. A Reassessment of the Relationship Between Inequality and Growth. *American Economic Review* 90 (4): 869–887.

Foster, R.N. 1986. *Innovation: The Attacker's Advantage*. New York: Summit.

Freeman, C., and C. Perez. 1988. Structural Crises of Adjustment, Business Cycles and Investment Behavior. In *Technical Change and Economic Theory*, ed. G. Dosi, 38–66. London: Francis Pinter.

Galbraith, J.K., and H. Kum. 2005. Estimating the Inequality of Household Incomes: A Statistical Approach to the Creation of a Dense and Consistent Global Data Set. *Review of Income and Wealth* 51 (1): 115–143.

Garrett, G., and D. Mitchell. 2001. Globalization, Government Spending and Taxation in the OECD. *European Journal of Political Research* 39 (2): 145–177.

Giddens, A. 1991. *The Consequences of Modernity*. Palo Alto: Stanford University Press.

Giddings, B., B. Hopwood, and G. O'Brien. 2002. Environment, Economy and Society: Fitting Them Together into Sustainable Development. *Sustainable Development* 10 (4): 187–196.

Goos, M., A. Manning, and A. Salomons. 2014. Explaining Job Polarization: Routine-Biased Technological Change and Offshoring. *American Economic Review* 104 (8): 2509–2526.

Hanson, R. 2008. Economics of the Singularity. *IEEE Spectrum* 45 (6): 37–42.

Hardt, M., and A. Negri. 2004. *Multitude: War and Democracy in the Age of Empire*. New York: Penguin.

Hirschman, A.O. 1958. *The Strategy of Economic Development*. New Haven: Yale University Press.

Hull, Z. 2008. Sustainable Development: Premises, Understanding and Prospects. *Sustainable Development* 16 (2): 73–80.

Huntington, S.P. 1996. *The Clash of Civilizations and the Remaking of World Order*. New York: Simon and Schuster.

Inkeles, A. 1993. Industrialization, Modernization and the Quality of Life. *International Journal of Comparative Sociology* 34 (1–2): 1–23.

Inkeles, A., and D.H. Smith. 1974. *Becoming Modern: Individual Change in Six Developing Countries*. Cambridge: Harvard University Press.

Jaumotte, F., S. Lall, and C. Papageorgiou. 2013. Rising Income Inequality: Technology, or Trade and Financial Globalization? *IMF Economic Review* 61 (2): 271–309.

Jorgenson, D.W., and K.J. Stiroh. 2000. Raising the Speed Limit: U.S. Economic Growth in the Information Age. *Brookings Papers on Economic Activity* 1: 125–235.

Kelly, R. 2009. Science, Technology and Innovation and Inequality. *Journal of Eastern Caribbean Studies* 34 (3): 79–92.

Koenig-Archibugi, M. 2004. Transnational Corporations and Public Accountability. *Government and Opposition* 39 (2): 234–259.

Kurzweil, R. 2005. *The Singularity Is Near: When Humans Transcend Biology*. New York: Penguin.

Kuznets, S. 1955. Economic Growth and Income Inequality. *American Economic Review* 45 (1): 1–28.

Langhelle, O. 2000. Sustainable Development and Social Justice: Expanding the Rawlsian Framework of Global Justice. *Environmental Values* 9 (3): 127–152.

Leslie, S.W. 1993. *The Cold War and American Science: The Military-Industrial-Academic Complex at MIT and Stanford*. New York: Columbia University Press.

Lewis, W.A. 1954. Economic Development with Unlimited Supplies of Labour. *The Manchester School* 22 (2): 139–191.

Lipset, S.M. 1959. Some Social Requisites of Democracy: Economic Development and Political Legitimacy. *American Political Science Review* 53 (1): 69–105.

Lomborg, B. 2001. *The Skeptical Environmentalist: Measuring the Real State of the World*. Cambridge: Cambridge University Press.

McCauley, D.J. 2006. Selling Out on Nature. *Nature* 443 (7107): 27–28.

McClelland, D. 1967. *The Achieving Society*. New York: Harper and Row.

Meadows, D.H., D.L. Meadows, J. Randers, and W.W. Behrens III. 1972. *The Limits to Growth*. New York: Universe Books.

Meadows, D.H., J. Randers, and D.L. Meadows. 2004. *Limits to Growth: The 30-Year Update*. Vermont: Chelsea Green.
Midgley, J. 1995. *Social Development: The Developmental Perspective in Social Welfare*. Thousand Oaks: Sage.
Milanovic, B. 2016. *Global Inequality: A New Approach for the Age of Globalization*. Cambridge: Harvard University Press.
Mokyr, J., C. Vickers, and N.L. Ziebarth. 2015. The History of Technological Anxiety and the Future of Economic Growth: Is This Time Different? *Journal of Economic Perspectives* 29 (3): 31–50.
Murphy, K.M., A. Shleifer, and R.W. Vishny. 1989. Industrialization and the Big Push. *Journal of Political Economy* 97 (5): 1003–1026.
Myers, N., and J. Simon. 1994. *Scarcity or Abundance: A Debate on the Environment*. New York: Norton.
Ndlovu-Gatsheni, S.J. 2012. Beyond the Equator There Are No Sins: Coloniality and Violence in Africa. *Journal of Developing Societies* 28 (4): 419–440.
Needell, A. 1998. Project Troy and the Cold War Annexation of the Social Science. In *Universities and Empire: Money and Politics in the Social Sciences During the Cold War*, ed. C. Simpson, 3–38. New York: The New Press.
Nemati, M., and G. Raisi. 2015. Economic Growth and Income Inequality in Developing Countries. *International Journal of Life Sciences* 9 (6): 79–82.
Nisbet, R.A. 1969. *Social Change and History: Aspects of the Western Theory of Development*. New York: Oxford University Press.
Nurkse, R. 1953. *Problems of Capital Formation in Underdeveloped Countries*. Oxford: Oxford University Press.
Ophuls, W. 1977. *Ecology and the Politics of Scarcity: Prologue to a Political Theory of the Steady State*. San Francisco: W. H. Freeman.
Pearce, D.W., E.B. Barbier, and A. Markandy. 1990. *Sustainable Development: Economics and Environment in the Third World*. London: Edward Elgar.
Prebisch, R. 1959. Commercial Policy in Under–Developed Countries. *American Economic Review* 49 (2): 251–273.
Ravallion, M., and S. Chen. 2013. A Proposal for Truly Global Poverty Measures. *Global Policy* 4 (3): 258–265.
Rebelo, S. 1991. Long-Run Policy Analysis and Long-Run Growth. *Journal of Political Economy* 99 (3): 500–521.
Rieger, E., and S. Leibfried. 1998. Welfare State Limits to Globalization. *Politics and Society* 26 (3): 363–390.
Romer, P.M. 1990. Endogenous Technological Change. *Journal of Political Economy* 98 (5): 71–102.

Rosenstein-Rodan, P. 1943. Problems of Industrialization of Eastern and South-Eastern Europe. *Economic Journal* 53 (210/211): 202–211.

Rostow, W.W. 1960. *The Stages of Economic Growth: A Non-Communistic Manifesto*. Cambridge: Cambridge University Press.

Sachs, J.D. 2006. *The End of Poverty: Economic Possibilities for Our Time*. New York: Penguin.

Schilling, M.A. 2010. *Strategic Management of Technological Innovation*. 3rd ed. New York: McGraw-Hill.

Schmidt, V.A. 2002. *The Futures of European Capitalism*. Oxford: Oxford University Press.

Schumpeter, J.A. 1939. *Business Cycles: A Theoretical, Historical, and Statistical Analysis of the Capitalist Process*. New York: McGraw-Hill.

Sikkink, K. 1991. *Ideas and Institutions: Developmentalism in Brazil and Argentina*. Ithaca: Cornell University Press.

Singh, A. 2004. *Multilateral Competition Policy and Economic Development: A Developing Country Perspective on the European Community Proposals*, UNCTAD Series on Issues in Competition Law and Policy. Available from: http://unctad.org/en/Docs/ditcclp200310_en.pdf. Accessed 11 Aug 2017.

Solow, R.M. 1956. A Contribution to the Theory of Economic Growth. *Quarterly Journal of Economics* 70 (1): 65–94.

Stiglitz, J.E. 2002. *Globalization and Its Discontents*. New York: W. W. Norton & Company.

Tiryakian, E.A. 1991. Modernisation: Exhumetur in Pace (Rethinking Macrosociology in the 1990s). *International Sociology* 6 (2): 165–180.

Todaro, M.P. 2014. *Economic Development*. 12th ed., The Pearson Series in Economics. Philadelphia: Trans-Atlantic Publications.

Turner, G.M. 2008. A Comparison of the Limits to Growth with 30 Years of Reality. *Global Environmental Change* 18 (3): 397–411.

Wallerstein, I. 1996. *Historical Capitalism with Capitalist Civilization*. London/New York: Verso.

WEF (World Economic Forum). 2015. *Industrial Internet of Things: Unleashing the Potential of Connected Products and Services*, REF 020315. Geneva: WEF. Available from: http://www3.weforum.org/docs/WEFUSA_IndustrialInternet_Report2015.pdf. Accessed 8 Sept 2017.

Woodhouse, E., and D. Sarewitz. 2007. Science Policies for Reducing Societal Inequities. *Science and Public Policy* 34 (3): 139–150.

World Bank. 2017. *World Development Indicators* [Online]. Available from: http://data.worldbank.org/datacatalog/world-development-indicators. Accessed 15 Feb 2017.

WTO (World Trade Organization). 2017. *World Trade Report 2017*. Available from: https://www.wto.org/english/res_e/publications_e/wtr17_e.htm. Assessed 22 Dec 2017.

3

Comparative Perspective on Modernisation and Institutionalism

Introduction

A newly emerged normative development idea of sustainability has been evolving, which was institutionalised by the UN in 2015, and which strongly reinforced, as an exogenous factor, all the responsible and accountable governments and other society actors within a country (i.e., top-down policy benchmark) to balance the three dimensions of sustainable development, including economic, social, and environmental aspects, namely the 2030 Agenda for Sustainable Development (i.e., SDGs) (UN 2015). But in the medium- to long-run, this normative development idea becomes endogenous in pursuit of development outcomes (i.e., bottom-up policy benchmark).

However, achieving the sustainable development process requires an integrated approach to development that simultaneously addresses these multidimensions of sustainability (Armah and Baek 2015). This implies breaking institutional silos, strengthening sectoral (i.e., horizontal) and sub-national (i.e., vertical) institutional and arrangement coordination within and among implementing entities (Armah and Baek 2018)—for example, economic growth strategies tend to be prioritised by ministries

S. J. Baek, *The Political Economy of Neo-modernisation*,
https://doi.org/10.1007/978-3-319-91394-0_3

of economy, finance, and trade while social development is likely pursued by ministries of labour and welfare, and environmental policies are primarily considered by ministries of environment, water, and fishery.

This endogenising process would be not only the case of normative development ideas but any case of external forces. Against the backdrop of an intensifying globalisation effect, the concept of technological innovation includes not only technology advancements (mostly in developed economies) but also its influence, to developing economies, on an increase in the overall technological possibilities for transforming labour, capital, information and communication, and natural resources into valued goods and services. Particularly, structural transformation can be achieved by being able to benefit from technological innovation in more advanced manufacture and services, rather than being left by the global division of labour to be primary commodity producers. Such technological innovation factor is mostly treated as exogenous in developing countries but becomes endogenous in the medium- to long-run. In this context, understanding the importance of how they are institutionalised (endogenised) into society well deserves careful analysis in exploring the central idea of this book.

Looking more closely at the process of institutionalisation from a more historical perspective, cultural heritage, migrations of people, and the legacy of colonisation have all shaped institutions in various countries differently. The influential role of European colonisation starting from the late fifteenth century is crucial in explaining the different institutional characteristics in different parts of the world. European, and particularly British, German, and Spanish, exploration and colonisation of Africa, Asia, Australia, Latin America, and North America impacted massively on the shaping of history and of the state of modern institutions in these places. Ideological wars between the promoters and opponents of slave trade and racial disaggregation, for example, led to the emergence of different interests and thus different types of institutions. Diseases, population density, and the discovery of natural resources played important roles in pushing colonising powers to specific areas (Rotberg 2003).

In much of Africa, Central America, the Caribbean and Asia, European powers set up extractive institutions that promoted their own interests as opposed to protecting property rights and establishing checks and balances on government (Acemoglu and Robinson 2012). The Europeans governed the 'new found lands' by institutionalising highly centralised systems to support the extraction of the abundant natural resources available. In particular, in the colonised nations of Africa and Latin America where natural resources were abundant and climatic conditions were favourable for growing crops but not for European settlement, the colonisers established extractive institutions to benefit their citizens back home (Acemoglu and Robinson 2012). In Latin America, where a large number of slaves were deployed, conditions were ripe for establishing plantations, which generated extreme inequality in terms of wealth, capital, and political power (Engerman and Sokoloff 2002).

As noted earlier in the text, these types of institutions are associated with limited growth and exclusion leading to large levels of inequality in these regions. On the other hand, in countries where the colonisers (i.e., mostly European settlers and their descendants) established institutions that promoted property rights for the majority, in the likes of Australia, Canada, New Zealand and the US, we see inclusive and sustained growth, underscored by innovation and technological advancement. In areas where colonisers settled in large numbers, the nature and structure of the institutions established were different. There was more focus on developing these areas for permanent settlement, rather than extractive institutions (Acemoglu and Robinson 2012). Protection of private property rights, particularly rights to land and trade under market institutions, was underscored. In addition, the relatively small size of farms in the northern parts of North America meant that more farmers were involved in agriculture, with low economies of scale in production (Engerman and Sokoloff 2002). The inclusive institutions established in these regions fostered a relatively fair (re)distribution of wealth, promoted innovation, and sustained economic growth.

Based on these broad regional categorisations, Africa, Latin America, and socialist groups that were shaped by strongly centralised and extractive institutions at their early stage of development are likely to experience relatively limited growth and retarded technological innovations (controlled innovations and talent in specific sectors), resulting in higher levels of inequality.

This historical analysis on the institutionalising process implies that a country tends to have its own adaptive capacity level for external forces, the capacity which is generated through the endogenous process between various institutions and actors (Baek 2017; Figueroa 2017). However, literature investigating such endogenising process within a country has received relatively less attention particularly in the context of normative force of sustainability. Therefore, this chapter is designed to explore why countries respond to external force (e.g., sustainability and technological innovation) differently and how the endogenous process affects such response. These two conceptual curiosities are addressed by in-depth theoretical analysis of competitive political economy and institutionalism in the process of development.

Varieties of Modernisation

This begins with a critical analysis of the grand discourse of comparative political economy. A mainstream comparative political economist argues that the various economic systems in each country are sustained based on the historic specificity of each country, the institutional and social cohesion of market mechanisms, and institutional complementarities (Esping-Andersen 1990; Hall and Soskice 2001). Accepting these arguments, for instance, even if the same technology is applied to society, developmental outcomes may differ as the endogenous processes in each society are different. Furthermore, if a negative effect of technology follows, the effect of the selective welfare policy of a government to minimise this negative effect also varies. When these experiences in a country are accumulated, individual countries arrive at their own unique mechanisms, which eventually lead to diverse economic systems rather than the convergence thereof.

The theory that has contributed most to systematising the core discourse of this school is the typology of *Welfare Regimes*, proposed by Esping-Andersen (1990), which categorised the world's various social systems into three types of welfare regimes, and included in the analysis de-commodification, social stratification, and state and market intervention. On this basis, Esping-Andersen argues that *Welfare Regimes* are classified into one of the following: liberal regime, conservative regime, or social democratic regime. In other words, in the triangular framework of the social system of state, market, and household, unique welfare systems are created, marked by path-dependency as each country tries to cope with the pressing demands of globalisation.

While Esping-Andersen's type of welfare state was entirely based on advanced European societies, James Ferguson (2015), who authored *Give a Man a Fish* focused primarily on developing countries. He found that developing countries have evolved with a particular emphasis on social assistance, without going through the Western welfare state model, which means that welfare states observed in developing economies are different from those of the West. Besides, he found that some countries in the southern hemisphere such as South Africa and those in South America, ahead of Western Europe, implemented a social assistance programme, a kind of basic income that provides cash unconditionally to the poor. He also noted that more than 40 per cent of the total population of South Africa received more than one subsidy, and that child allowances and pensions were actually effective in reducing poverty during periods of mass unemployment. In this sense, it has been suggested that a new type of welfare regime may work in many developing countries and various new types of welfare regimes could evolve in the future.

In this regard, academic research efforts have continuously been made to clarify the link between the study of the typology of the welfare state and the perspective of the production system. Hall and Soskice (2001) published *Varieties of Capitalism* (hereinafter referred to as 'VoC') which has been acknowledged as a seminal work on comparative political economy. Hall and Soskice classified advanced capitalism as liberal market economies (LMEs) and coordinated market economies (CMEs) by analysing various comparative systems such as wage relations, skill formation, corporate governance, and welfare system.

Their typology not only adapted a firm-centric approach that corporations are the decisive actors in a capitalist economy but also offered a relational view of firms, in that corporations have relationships with their employees internally and with parts suppliers, customers, entrepreneurs, trade unions, governments, and others. A corporation would be required to deal with various problematic coordination issues, including labour relations, vocational training and education, corporate governance, inter-firm relationships, and relationships with employees. As such, solving these coordination challenges through either market or non-market approaches is significant, in inducing various capitalist models.

According to their basis of distinction of the two groups, corporations in countries belonging to LMEs approach the coordination issues based on hierarchical organisation, competitive market system, and official contract. On the other hand, corporation in CMEs tends to address non-market approaches such as relational contract, networking supervision based on information exchange among insiders, and building cooperative relationships. In this context, the institutional infrastructure of each economic model formed in this process presents a formal and informal rule that actors generally follow, which can induce different corporate strategies that are best suited to each of their economic systems.

In addition, institutional complementarities exist between the sub-institutional domains of the economy, while countries with a certain type of coordination in one institutional domain tend to develop institutional practices that complement them in other domains. Thus, in the presence of institutional complementarity, different types of institutions are not randomly distributed in different countries, but rather complementary sets of institutions show different clusters depending on one country. Different combinations of these complementary institutions are key in establishing various capitalist models. From this point of view, Hall and Soskice use the US as a representative example of LMEs and use Germany for CMEs.

They further emphasised that these two distinct types of capitalist economies have generated comparative institutional advantages. LMEs have a comparative institutional advantage in areas that require rapid product development based on rapidly changing technology and research and development as LMEs tend to have an institutional arrangement of profit-based financial systems, flexible labour markets, and inter-firm market competitive relationships. Such areas that require radical innovation include biotechnology, semiconductors, software, communications, logistics, aviation, finance, and entertainment.

On the other hand, CMEs tend to have a comparative advantage in areas requiring gradual innovation for production, enhancement of competitiveness of products, and diversification of production methods that involve machinery, plant equipment, consumer durables, engines, and specialised transportation equipment. The CMEs have a competitive advantage here as they usually have a systematic arrangement such as the stable employment of workers, continuous skill-based learning systems based on enterprise (or industry) specific skills, and close production cooperation with sub-contractors.

These radical versus gradual innovations are in fact not new at all. As we have already seen through Christensen's *Disruptive Innovation* concept, the US (an LME) is making remarkable progress in the semiconductor industry, which requires rapid product development through *Disruptive Innovation*. On the other hand, in Germany (a CME), *Sustaining Innovation* in manufacturing is fostering globally competitive small- and medium-sized enterprises. Regarding the German-style innovation, Hermann Simon (2009) referred to this as *Hidden Champions* in that the country's medium-sized firms are becoming the reliable backbone of the national economic system.

In Western academia, theoretical debates and continuous follow-ups on VoC have been conducted extensively (e.g., Schmidt 2002; Amable 2003; Schröder 2013), with the latter having established itself as a modern capitalist theory in the era of globalisation and the information age; nonetheless, it still faces a lot of criticism.

First, it would be one-dimensional to understand capitalism as a dichotomy between LMEs and CMEs, leaving it subject to criticism that there could be various sub-groups within each of the two groups (Crouch 2005). For instance, for the distinction between LMEs and CMEs, Schmidt (2002) classified European countries into market capitalism, managed capitalism, and state capitalism. Kitschelt et al. (1999) and Amable (2003) insisted that there is an institutional differentiation within CMEs and several different sub-models. Schröder (2013) also echoed some of the criticism by attempting to unify VoC and Esping-Andersen's welfare regimes, claiming that there are three kinds of grouping, namely *United Typology*.[1]

It should also be noted that, like modernisation theory, VoC may reveal the limitation of an overtly Western-oriented analysis model. VoC theorists' typology covers only the advanced Western capitalist countries such as OECD members and does not cover all capitalist models in the developing world. Advocates of the dichotomous structure of LMEs versus CMEs tend to view Africa's least capitalist countries as a subtype of the CME grouping, which means that their analysis does not properly recognise the existence of varying economic models within African economies, as James Ferguson discovered and emphasised. This can be seen as a limitation stemming from Eurocentrism and can also be the fatal mistake of refuting the assumptions of *non-Convergence* advocates.

If this paradoxical self-logic refutation cannot be resolved, one may not be able to effectively respond to the criticism against VoC raised by *Convergence* theorists. Soederberg et al. (2005) argue that 'what we are witnessing today is the emergence of *Varieties of Neo-liberalism* rather than VoC, so it is only varieties within convergence trend'. They emphasised that there is a process called *Internalisation of Neo-liberal Globalisation* which includes complex adaptation processes, diverse policy experiments, and collapse/reconstruction of political alliances in each country.

Ironically, there is little between the two logics: one is that there is a substantial convergence of economic models in the ongoing trend of globalisation and neo-liberalism; and the other assumes convergence within CMEs when considering the national models of the poorest African countries as a subtype of the CMEs.

Taking these criticisms into consideration, the degree to which VoC is analysed seems to determine the number of typologies. In other words, VoC assumes the nation-state to be a unit of analysis for an economic model which forms a basis of institutional completeness. However, globalisation tendencies threaten nation-states according to the VoC typology. For example, multinational corporations mobilise resources and organise production systems in multiple countries and within distinctive economic models. Ascertaining what kind of capitalism these multinational corporations themselves are involved in could present a significant challenge to the VoC typology and could be used as a weapon for argument by *Convergence* theorists as well. Moreover, within the more complex and stratified international division of labour with globalisation of production and banking, the national model of capitalism may be viewed as institutionally incomplete (Crouch 2005).

To put it another way, the greater the system of capitalism as a higher concept, the closer to the viewpoint of *Convergence* theorists while the more the capitalist models are regarded as something that can vary by region or country, rather than as a global framework for moving the world, the more reliable the arguments of the various theorists will be. Accepting these controversies, for example, modernisation paths that have appeared in various ways can be interpreted as sub-types that appear in the mega-modernisation process, or that these various paths can be recognised as separate ones, which would really depend on the unit of analysis at the regional, national, or local level. Such controversies further depend on whether the concept of modernity is recognised as Western-centric or whether the modernity styles that are evolving should be viewed in a variety of ways as a sub-set of *Meta-modernity*.

Despite these fundamental limitations, the largest contribution of VoC is that it effectively explains the specific internal mechanisms that exist within society based on the actors and institutions that mediate its structure. Social systems, driven by institutions, are constantly being strengthened so as to eventually support the possibility of various economic systems, and this is understandable. It is, however, still assumed that the social internal system is systematised or ordered so that there is some sort of endogenous process that has already been predetermined. In other words, VoC emphasises that at some point each economic model maintains a constant *Homeostatic* equilibrium that either adapts to, or defends against, external economic pressures (Schmidt 2002) or, by suggesting institutional path-dependency as its main assumption, it tends to emphasise a historically fixed typification. Due to the limitations of this static analysis model, it is not well placed to explain why unexpected changes in the characteristics of economic models occur, or what causes such changes (Crouch 2005).

Evolutionary Institutions

Recognising the strengths and limitations of VoC, and incorporating them into the theories of development more systematically, we may be able to examine the adaptive capacity of society, which is a precondition for traditional society to fully integrate and maintain modern institutions.

When developing countries seek an ideal development model (i.e., future blueprints from the West), retracing the course of Western-centred development has become a central element of cultural infiltration of developing countries, and the destructive experiences of modernity that are constantly accumulated in traditional society (Vincent 1999). The collapse of the traditionality can be derived from the imbalance between modernising aspirations and the ways in which to achieve them, and manifests itself as an important issue of social change. The process of urbanisation is often seen as a significant social burden, where most of the population is concentrated in large cities, while rural areas are almost deserted. Populations of large cities that have lost their social productivity have caused social ills, which further hinder economic growth through the reduction in the ability to mobilise resources efficiently.

By further extending the aforementioned discussion, poor people in underdeveloped societies may not earn enough to increase their essential consumption, while there is also a pressure to save for future consumption or investment. Therefore, a situation may occur whereby people have to consume almost all of the resources they gain from production directly into their social lives and the education sector. As a result, as development progresses, the principle of *Selectivity* is applied differently to members of society, which could lead to the polarisation of capital accumulation, and finally to a drastic reduction in social solidarity. Relative deprivation thus continues to accumulate, and society faces a rather difficult situation. These cumulative fatigue symptoms in society fail to integrate the individual and institutional levels for self-development, which in turn hinders systematic stability in many economies.

Most Western societies have attempted to exert some influence by presenting themselves as a blueprint for development models in developing countries. Indeed, many developing countries have imported development models from Western industrialised societies, so they have a clear relationship with the West in relation to their future development strategies. In practice, developing countries have usually tried to achieve modernisation in a shorter period of time than Western societies have undergone, but there seems to be no universal standard by which national efforts to pursue such developmental processes can be measured (Scott 1998).

On the theoretical surface of development (modernisation), which has been expanded from the economic to the political and social level, since the intrinsic issue of the public participation of members of society and the mechanism of action for the emotional dimension of a given society is inherent, the balancing and integrating of the theoretical composition between traditionality and modernity needs to be emphasised. The point here is to focus on how to interact with the West while focusing on the inner traditionality rather than the external factors, which can be associated with the point that the theoretical framework should be used to illustrate specific cases rather than adopting concrete examples to illustrate one theory. It is therefore necessary to focus on various developmental processes and the interaction between sectoral elements through an analysis of multidimensions, which include political, economic, social, and cultural aspects, instead of emphasising a single element or a simplified description.

As mentioned earlier, despite the common pressures of globalisation, many countries respond differently from neo-liberalist claims, and they have not deviated so much from the fundamental framework of the existing *Welfare State* system that was formed long ago largely because of the path-dependent feature of institutional change. Institutions or their paths are formed increasingly by the mechanism of path-dependency, and this can be related to the interests of actors involved being interlinked. Positive feedback helps to influence such a mechanism (Pierson 2000).

However, there may be a limitation to the way in which this once-formed system implies a kind of deterministic view through path-dependency. It cannot, therefore, be argued that the policy of limited reforms in many countries simply depends on the conceptualisation of existing mechanical and deterministic path-dependence. Away from this, the concept of path-dependency in the future must be able to explain the dynamic process and mechanism of durability and change in institutional formation and change. To this end, Ebbinghaus (2009) introduced the concept of *Open Path-Dependence*, in which the branching pathways continue to select the best option under the structure of alternatives possibilities for historical critical conjunctures. Therefore, he argued that it is possible to form various paths, ranging from radical to gradual changes or to qualitative changes, depending on the level of choice in possible alternatives for each turning point.

Recently, a variety of institutionalist views offer theoretical innovations that transcend sectarian boundaries through learning effects that accept their theoretical and methodological weaknesses based on several empirical studies (Campbell and Pedersen 2001; Ebbinghaus 2009). Here, there are three major divisions: historical institutionalism, rational institutionalism, and sociological institutionalism. In addition, other points of views can be associated with factors that determine, in essence, increasing returns of scale, cohesion, positive feedback loop, and self-reinforcing. With regard historical institutionalism, power relations between major actors and institutions are seen as determining factors while rational institutionalists regard efficiency based on individual rationality. In addition, sociological institutionalism emphasises the legitimacy of institutions for members of society.

The procedural aspects of how the existing system changes and how a new path is formed are outlined as follows. In describing institutional change, the essential determining factor can be considered as a mutual relationship between institutions/structures and actors, which was raised earlier in the structuralist perspective. The institution/structure and actors are perceived to exist on the premise of each other (i.e., dualism), and not separately (Giddens 1979). Institutions are therefore not separate from human actors, and are forming and changing. Besides, the formation of institutions is not only reactionary. The formation and change of institutions is always mediated by human actions and ideas, while on the other hand restricting human actions and creating new institutions. This is especially so in the transition period when the existing system is collapsing and a new system is then created. Here, it is recognised that the strategic choice, ideas, and actions of major historical actors are all very important in given *constraints* (Campbell 2004).

With these views used to explain the formation and change of institutions, the importance of the concept of *Idea* starts with the assumption that a human can make a *Strategic Choice* as a rational actor. Although the concept of *Strategic Choice* was initiated by rational choice institutionalists, it also refers to the *Strategic Choices* and *Actions* of groups who try to change their power relations according to their interests. When using the concept of *Strategic Choice*, it is important to note that the creative choices of actors appear within the limitations of a given structure/institution, rather than in terms of the willingness of actors. In other words, the choice of these actors leads to the process of new *Path-Shaping* but within the context of long historical development.

In this perspective, the main actors (e.g., influential individuals) can through government policies and collective action create new paths according to the creation of Ideas and *Strategic Choices* for change. Hence, the Idea of major actors, followed by the formation of reform discourse, can play a mediating role for creating new institutional paths. According to ideologists, the formation of a new institution weakens the legitimacy of the existing system due to various reasons including uncertainty, important actors having to come up with new ideas for solving new problems, and new ideas that are appropriate and justifiably relative to other new ideas which will replace existing institutions, and, as a result, institutional change will take place.

In this context, Campbell (2004) attempted to systematise the mechanism of institutional change by focusing on the concept of Idea. The concept of Idea can be defined by taken-for-granted assumptions, which can be the background of decision making, as a belief system that is widely agreed to be desired by actors, such as worldviews, norms, identities, values, intellectual paradigms, culture, and interests. He then emphasised that the Idea functions as an important foreground to institutional change for problem solving, while at the same time it is also a background that limits actors' choices and decisions.

Campbell emphasised the Path-Shaping feature which is institutional path-dependence and its gradual institutional change. Such a feature can create new institutions by reassembling with institutional resources and principles that are available to them. So, in essence, it is a concept that explains the evolutionary process of institutional change in which new Path-Shaped institutions are similar to existing ones but evolving little by little. This process of change is explained through the introduction and application of the staged concepts of diffusion and the translation of Ideas. It should be kept in mind that Ideas are not the result of voluntary- and free-thinking by pure actors but rather through limiting the choice of actors in a historically formed institutional framework while carrying out specific roles and functions as means or mediators that enable institutional change.

Through a review of the concept of Ideas, the Path-Shaping dynamics are explored. In this process, the *Strategic Choice* of actors is important as the key mechanism of institutional change emphasises agent/actors' ideas and their strategic choices/decisions within the *constraints* of existing structures and institutions (Scharpf 2000). When the contradictions of institutions and institutional complementarities become a problem due to various factors such as external shocks and internal contradictions, existing power relations weaken or collapse and through the process of forming a new power relationship, *Strategic Ideas* are created for solving such problems; meanwhile, as competition between these Ideas becomes severe, institutional discourse is formed that can be justified in the society. This induced institutional Path-Shaping is again constrained by another new Path-Shaped institution.

To further develop the aforementioned discussion, according to the basic concept of path-dependency between institutions assumed in VoC, the developmental path of each country has already been predetermined due to their institutional patterns accumulated by experiencing the inevitable collision between *Traditionality* and *Modernity*. However, this predetermined assumption may not properly explain the unique and unexpected development outcomes of the last decades, such as the rapid development of the so-called Asian Tigers and the recent growth of Russia whose institutions have been radically transformed from those of the Soviet Union. Furthermore, it is unlikely to be able to correctly account for the rapid institutional change through innovations in the fourth Industrial Revolution era.

Thus, given the normative force of sustainability, which has been knocking on institutions and key actors in society, institutional Path-Shaping through reform discourse through normative ideas seems a more relevant assumption, and a more appropriate standard of analysis for drawing up the development process of a future society.

Considering the situation in the twenty-first century, where the notion of sustainability beyond the modernisation of past economic development heritage began to emerge, and recognising the nation-state as the most accepted unit of analysis, a configurational approach is taken to analysing the time and sequence of interactions over time on how structural reform discourse evolves. Adopting, beyond the main actor approach, multiple dimensions of analytical approaches on various institutional areas of development could not only contribute substantially to the debate on *Convergence* versus *non-Convergence* in the field of international development. More essentially, it may unfold a new horizon in the grand discourse of the complexity of the *Technology-Development-Inequality* nexus in the twenty-first century.

Concluding Remarks

By critically analysing VoC and its underlying assumptions that is path-dependency, I have emphasised that the reform discourse through ideas can be interpreted as the institutional Path-Shaping mechanism through the endogenous process of conflict and coordination between various institutional sectors in society. In other words, with the rapid acceleration of globalisation, and as modernisation progresses, the principle of *Selectivity* is applied differently to social members in terms of the process of accepting technological innovation so that social cohesion can be drastically reduced. However, if new ideas of sustainable development are accepted effectively within society, it may be possible to prevent to a certain extent the phenomenon of polarisation that might result from technological development. As a result, each country will be able to strategically embrace ideas that rapidly aid it through the process of globalisation, coupled with experiences gained from the past. Such ideas that drive the Path-Shaping capacity may determine whether technology-driven sustainable modernisation or technology-enforced unequal modernisation will be realised as well as whether already existing varieties of developmental paths will converge or further diverge in different fashions.

It can, however, be true that developing countries have an internally accumulated adaptation capacity (i.e., endogenous process), relatively lower than that of advanced countries (Huntington 1996; Rapley 2004). Because of this, developing economies have, over the last two decades, operated their own Path-Shaping institutions to find more appropriate development strategies to meet their real needs, rather than preparing against the possibility of social disruption that could result from imitating the paths of developed countries. Through these long-standing concerns, many developing countries have identified new development thinking (rather than modernisation), that is, *Structural Transformation* as a development priority in their national and regional development frameworks (Armah and Baek 2015). Similarly, the Common African Position on the post-2015 development agenda identified structural transformation as a priority for the successor global development agenda to the Millennium Development Goals (MDGs).

Notes

1. LMEs with liberal welfare states which are all English-speaking, relying upon market mechanisms to coordinate production and redistribution; CMEs with conservative welfare states which are continental European countries except for Japan, relying upon the state and civil society to coordinate production and redistribution but in ways that reinforce existing social inequalities such as occupational status; and CMEs with social-democratic welfare states which are all Nordic, being distinguished by a national coordination of production and welfare in ways that enforce a commitment to social equality. Each of three group's commonality is interpreted by a combination of culture and material circumstances, namely newly introduced concept of 'policy style'.

References

Acemoglu, D., and J. Robinson. 2012. *Why Nations Fail: The Origins of Power, Prosperity, and Poverty*. New York: Crown Publishing Group.

Amable, B. 2003. *The Diversity of Modern Capitalism*. Oxford: Oxford University Press.

Armah, B., and S.J. Baek. 2015. Can the SDGs Promote Structural Transformation in Africa? An Empirical Analysis. *Development* 58 (4): 473–491.

Armah, B., and S.J. Baek. 2018. Three Interventions to Foster Sustainable Transformation in Africa. *Journal of Social, Political, and Economic Studies* 43 (1–2): 3–25.

Baek, S.J. 2017. Is Rising Income Inequality Far from Inevitable During Structural Transformation? A Proposal for an Augmented Inequality Dynamics. *Journal of Economics and Political Economy* 4 (3): 224–237.

Campbell, J.L. 2004. *Institutional Change and Globalization*. New Jersey: Princeton University Press.

Campbell, J.L., and O.K. Pedersen, eds. 2001. *The Rise of Neoliberalism and Institutional Analysis*. New Jersey: Princeton University Press.

Crouch, C. 2005. Models of Capitalism. *New Political Economy* 10 (4): 439–456.

Ebbinghaus, B. 2009. Can Path Dependence Explain Institutional Change? Two Approaches Applied to Welfare State Reform. In *The Evolution of Path Dependence*, ed. L. Magnusson and J. Ottosson, 191–212. Cheltenham: Edward Elgar.

Engerman, S.L., and K.L. Sokoloff. 2002. Factor Endowments, Inequality, and Paths of Development Among New World Economies. *Economia* 3 (2): 41–102.

Esping-Andersen, G. 1990. *The Three Worlds of Welfare Capitalism*. Princeton: Princeton University Press.

Ferguson, J. 2015. *Give a Man a Fish: Reflections on the New Politics of Distribution*. Durham: Duke University Press.

Figueroa, A. 2017. *Economics of the Anthropocene Age*. Basingstoke: Palgrave Macmillan.

Giddens, A. 1979. *Central Problems in Social Theory: Action, Structure, and Contradiction in Social Analysis*. Berkeley: University of California Press.

Hall, P.A., and D. Soskice. 2001. *Varieties of Capitalism: The Institutional Foundations of Comparative Advantage*. Oxford: Oxford University Press.

Huntington, S.P. 1996. *The Clash of Civilizations and the Remaking of World Order*. New York: Simon and Schuster.

Kitschelt, H., P. Lange, G. Marks, and J.D. Stephens. 1999. *Continuity and Change in Contemporary Capitalism*. New York: Cambridge University Press.

Pierson, P. 2000. Increasing Returns, Path Dependence, and the Study of Politics. *American Political Science Review* 94 (2): 251–267.

Rapley, J. 2004. Development Studies and the Post-Development Critique. *Progress in Development Studies* 4 (4): 350–354.

Rotberg, R.I. 2003. Failed States, Collapsed States, Weak States: Causes and Indicators. In *State Failure and State Weakness in a Time of Terror*, ed. R.I. Rotberg, 5–10. Cambridge/Washington DC: Brookings Institution Press.

Scharpf, F.W. 2000. The Viability of Advanced Welfare States in the International Economy: Vulnerabilities and Options. *Journal of European Public Policy* 7 (2): 190–228.

Schmidt, V.A. 2002. *The Futures of European Capitalism*. Oxford: Oxford University Press.

Schröder, M. 2013. *Integrating Varieties of Capitalism and Welfare State Research: A United Typology of Capitalisms*. New York: Palgrave Macmillan.

Scott, J.C. 1998. *Seeing Like a State: How Certain Schemes to Improve the Human Condition Have Failed*. New Haven: Yale University Press.

Simon, H. 2009. *Hidden Champions of the 21st Century*. New York: Springer.

Soederberg, S., G. Menz, and P.G. Cerny. 2005. *International Political Economy: Internalizing Globalization: The Rise of Neoliberalism and the Decline of National Varieties of Capitalism*. Basingstoke: Palgrave Macmillan.

UN (United Nations). 2015. *Transforming Our World: The 2030 Agenda for Sustainable Development*. New York: UN, (A/RES/70/1). Available from: https://sustainabledevelopment.un.org/content/documents/21252030%20Agenda%20for%20Sustainable%20Development%20web.pdf. Accessed 10 Mar 2017.

Vincent, J.R. 1999. Net Accumulation of Timber Resources. *Review of Income and Wealth* 45 (2): 251–262.

Part II

Growth and Inequality Interaction in the Quest for Sustainable Development

4

Stylised Fact of the Changing *Inequality-Growth* Landscape

Introduction

In the policy arena, inequality and growth sit at the forefront of a state's agenda in many countries around the world largely because the raison d'être of the modern state is to succeed in attaining sustainable economic growth while reducing inequalities. The origin of the *Inequality-Growth* nexus can be traced back to Simon Kuznets (1955) who systematically discussed inequality in the course of modernisation. His hypothesis was that there is an inverse relationship between inequality and development, which became known as the *Inverted U-Curve*. Since then, investigation on the *Inequality-Growth* nexus has been of great interest not only to economists but also to the field of social science in general.

According to Kuznets's seminal work, at very low levels of income, inequality must also be low, particularly in rural areas, while a little bit higher in urban areas. However, on the whole, most live within a reasonable subsistence level. As the process of growth begins, the path to economic inequality is driven by changes in sectoral structure. Income inequality increases as people migrate from a relatively equal low-income rural sector towards a relatively unequal urban sector that offers higher wages. This migration is in line with the transition from a traditional

S. J. Baek, *The Political Economy of Neo-modernisation*,
https://doi.org/10.1007/978-3-319-91394-0_4

agricultural sector to a modern industrial one. Once the majority of the labour force is found in an industry or service sector in an urban area, then further movement results in a decrease in income inequality as the rural–urban divide becomes less influential. Therefore, the average per capita income increases further and income inequality within urban sector remains.

Such prime explanations have motivated a vast amount of empirical research over the past decades mainly looking for the *Inverted U-Curve* within and across countries. Various empirical studies found support for Kuznets's hypothesis (Papanek and Kyn 1986; Chang and Ram 2000), while more recent studies could not provide robust support for the hypothesis (Barro 2000; Herzer and Vollmer 2012). In particular, Barro (2000) somewhat agreed that the effect of income inequality is different contingent on the state of development. However, by using the panel data, the opposite of the *Inverted U-Curve* was found, in that inequality in poor nations (defined by PCGDP below US$2070) retards economic growth, while inequality in rich nations (defined by PCGDP above US$2070) stimulates growth.

The fundamental issues on whether or not to support Kuznets's hypothesis stem from two reasons. The first reason is the existence of two competing schools of thought on the relationship between income inequality and growth performance. Meanwhile, different views on the direction of causality running from inequality to growth or vice versa is another. Indeed, exploring either the positive or negative associationship between the two dimensions is critical when analysing whether or not countries may face trade-offs or synergies between inequality and growth. This should provide the basis for policymakers to target an effective (re) distribution of wealth to promote growth, *Trickle-Down* or *Trickle-Up*–oriented interventions or other policy options, which depend significantly on the causal orientations. These two issues are explored in more detail by reviewing past literatures in the next section.

Literature Review

One school of thought[1] holds that the positive relationship between inequality and growth predominates over society. According to this idea, income inequality provides incentives for more effort in order to gain more, which implies that inequality in a society is the outcome of the free choice of individuals according to their appetite for risk, or that distribution depends on the preferences of a society. As a result, inequality would encourage intense competition that would eventually stimulate growth. This school further insists that inequality may also contribute to an engagement in a wide range of economic activities on the basis that inequality itself can be not only a stimulus for people trying to secure a better education and job but also a source of motivation for entrepreneurs to invest more. According to this line of reasoning, this school of thought believes that a society as a whole may be better off with inequality than with equality.

A contrasting school of thought,[2] on the other hand, argues that the negative impact of inequality on growth activities inevitably predominates. Here, the view is that inequality cannot provide sufficient educational opportunities to the low-income segment of society, thereby preventing opportunities for social mobility and eventually hindering the development of human capital. In a more comprehensive manner, Wilkinson and Pickett (2010) investigated social problems in 23 countries with different levels of economic inequality and found that the more unequal the country was, the worse it performed in various dimensions of society, which substantially hinders economic activities. Meanwhile, Berg and Ostry (2013) also identified some negative consequences of income inequality through exploring the differences in the sustainability of economic growth between Asia and Latin America. They attributed such consequences to the difficulty in investing in education for the poor, which would likely result in a *Middle-Income Trap*. In particular, *The Great Escape* by Angus Deaton (2013) has greatly popularised the idea of the negative function of inequality, the author of which being awarded the *2015 Nobel Prize for Economics*.

These two competing views are based on a discussion of the effects of inequality on economic growth. I now turn the focus to the causation running from growth to inequality. A large body of literature[3] in the sphere of the reverse causal relationship supports the negative impact of economic growth on inequality by outlining empirical evidence that shows that negative growth deteriorates inequality mainly due to a market distortion. Likewise, the hypothesis that positive economic growth helps in the reduction of inequality has also been supported empirically by a variety of studies (e.g., Aghion and Bolton 1997; Kakwani and Pernia 2000; Ogun 2010). These studies claim in essence that growth provides the impetus for material wellbeing to actually mitigate the problem of inequality on the basis of an effective (re)distribution practice. This desirable phenomenon has been frequently cited by policy advisors who subscribe to the *Trickle-Down* effect (e.g., Basu and Mallick 2008; Akinci 2017).

The literature investigating a positive causality from growth to inequality has also received a fair amount of attention.[4] Rubin and Segal (2015) observed an upward trend in inequality by focusing on the US during the post-war period (1953–2008) and attributed its rampant inequality to a distorted economic structure where the top income groups receive a substantial portion of income from wealth and equity compensation, both of which are sensitive to economic growth. These authors claimed that the situation arose due to ineffective (re)distribution practices. Similarly, in his influential book, Branko Milanovic (2016) argued that the recent phenomenon of widening inequality in wealthy countries, especially in the US and the UK, could be attributed to radical technological advances and labour mobility, underpinned by the effects of globalisation. Moreover, he conceptualised such rising inequality as an N-shaped relationship between national income per person and inequality, which is also regarded as a *Second Kuznets Curve* or *Kuznets Waves*.

These four different theoretical foundations may have caused contrasting or sometimes contradicting assertions, which is why no consensus has yet emerged for the *Inequality-Growth* nexus in the historical process of development. Given the existence of the four aforementioned competing schools of thought, a vast amount of research and publications on this nexus have focused on the quantitative study of an individual or a very limited number of country case(s). Some studies[5] with extensive coverage of countries using a panel data approach have attempted to generalise but these have often been rejected because they present different results (mixed or non-linear) on the relationship and causality (Chang et al. 2018). There have been few attempts made to investigate two-way causality (i.e., bi-directional causal relation) on a large set of countries, which are challenged by technical problems and, more critically, complexities behind multiple causalities.

The early research here is classified into categories and summarised in Table 4.1.

Table 4.1 Previous literature on the relationship between inequality and growth

Relationship	Causality	Author(s)
Positive	Inequality to Growth	Friedman (1953), Rebelo (1991), Deaton and Paxson (1997), Galor and Tsiddon (1997), Partridge (1997), Li and Zou (1998), Forbes (2000), Bell and Freeman (2001)
	Growth to Inequality	Cutler and Katz (1991), Blank and Card (1993), Weriemmi and Ehrhart (2008), Rubin and Segal (2015)
Negative	Inequality to Growth	Stiglitz (1969), Feldstein and Horioka (1980), Alesina and Rodrik (1994), Persson and Tabellini (1994), Clarke (1995), Alesina and Perotti (1996), Benabou (1996), Benhabib and Rustichini (1996), Perotti (1996), De La Croix and Doepke (2003), Wilkinson and Pickett (2010), Razmi and Ashrafzadeh (2012), Berg and Ostry (2013), Lee and Son (2016)
	Growth to Inequality	Psacharopoulos et al. (1995), Aghion and Bolton (1997), Bruno, Ravallion and Squire (1998), De Janvry and Sadoulet (2000), Kakwani and Pernia (2000), Bourguignon (2003), Krongkaew and Kakwani (2003), Adams (2004), Heltberg (2004), Ravallion (2007), Basu and Mallick (2008), Ogun (2010), Ostry, Berg and Tsangarides (2014), Akinci (2017)
Mix (or Non-linear Relationship)		Deininger and Squire (1996); Barro (2000); Chen (2003); Galbraith and Kum (2005); Castelló-Climent (2010); Jihène and Ghazi (2013); Milanovic (2016); Baek (2017)

Source: Author's own elaboration

In addition to the causalities in question, much of the literature has not taken into account the distinction between the long- and short-run associationship which would have significantly different implications in explaining the inequality and growth dynamics. To be more specific, there are many transmission channels (beneficial or harmful) of interaction between inequality and economic performance. Meanwhile, the speed of transmission running from inequality to growth or vice versa can never be the same (Halter et al. 2014). For instance, some inequality as *enabling* mechanisms on growth tend to set in quickly—for example, aggregate saving (Kaldor 1956), thus stimulating R&D (Foellmi and Zweimüller 2006). On the other hand, some negative transmissions can be rather slow—for example, expensive fiscal policies (Alesina and Rodrik 1994), thus delaying human capital formation (Galor and Moav 2004). The field of research on the nexus with consideration of the time dimension together with multiple causalities has, in fact, received relatively little attention.

Besides, the consequences of inequality for growth performance may differ depending on the size of the economy (Lee and Son 2016), which could further vary by regional heterogeneity (Rubil 2015). Of the many empirical attempts, Galbraith and Kum (2005) have not only provided empirical evidence to support the notion that inequality is likely to reduce as the income level increases in general but they have also discovered a tendency of the reverse relationship for rich countries. As an example of 'within similar income group but different region', many developing countries in Latin America and Africa have experienced the process of development in an unexpectedly distorted direction in terms of inequality waves. Their distorted development trends can indeed be clearly distinguished from the development trajectories of the four Asian Tigers (i.e., Hong Kong, Singapore, South Korea, and Taiwan) who realised a radical growth while attaining a manageable level of inequality (Sachs and Warner 1997; Parente 2001). Although these regional (or income grouping) factors have been identified by some in the literature, no one has yet incorporated them together with the aforementioned issues in mind.

The *Inequality-Growth* nexus deserves careful analysis particularly to explain a historical process of development and growth pattern in general and providing basis of distinctive growth dynamics between rich and poor in particular. The focus of this chapter is therefore to explore empirically the growth and inequality interactions in long- versus short-run aspects based on bi-directional causality. From these analyses, I further attempt to examine possible distinctive characteristics in this domain by income groups or regional groups. These two methodological approaches will contribute to the existing literature in the field of the *Inequality-Growth* nexus where three critical research gaps, which have been making such nexus analysis complicated, can be identified: the complexity of causality (Chang et al. 2018), the neglected time dimension (Halter et al. 2014), and heterogeneities of the regional (Rubil 2015) and income groups (Lee and Son 2016).

Data and Empirical Regularities

The two variables included in this study are PCGDP and the Gini index (INE). First, PCGDP is one of the most widely used indicators for measuring progress towards economic development against inequality (Barro 2000; Frank 2009; Jihène and Ghazi 2013; Rubin and Segal 2015) and takes into account population size so as to provide useful information for country comparisons on the level and stage of development (Giddings et al. 2002). These annual data are sourced from the World Development Indicators of the World Bank (2016a).

As for an indicator to measure income inequality in income distribution, there are a number of inequality data sets available. Deininger and Squire (1996) developed and used new cross-country data sets on income and asset distribution to examine the interactions between inequality and growth. Then, the UN University-World Institute for Development Economics Research (UNU-WIDER) further advanced this data set by expanding the number of country-year observations of the Gini index. However, these data sets may not provide comparable information across countries due to differences in terms of the welfare definition employed and employment status, and the treatment of various other factors (Solt 2016a). Also, this data set tends to suffer from measurement error and some potential endogeneity problems (Subramanian and Kawachi 2004). In fact, the Luxembourg Income Study could be the only data source available which uses a uniform set of assumptions and definitions; however, the coverage of the data set is limited. In order to contribute towards making valid comparisons in inequality across countries and over time, the Standardised World Income Inequality Database (SWIID) was introduced in 2008, which was then expanded considerably since that time to maximise income inequality comparability for the broadest possible sample of countries and years (Solt 2016a). For this reason, the SWIID Version 5.1 (Solt 2016b) is used as a proxy indicator for inequality in this study.

Notwithstanding the advantageous, adopting the Solt SWIID data set would require a careful assessment of its suitability for purposes of the study. Solt gives estimates of 'market' and 'net' inequality, but it is worth remembering that 'market' measures could be distorted by the absence of non-market transfer incomes. The other measure of inequality is thus used for this study. However, it is true that Solt's estimates of 'net' income in developing countries may be distorted by applying imputations drawn from the effect of tax systems in developed countries that may not apply to the developing world. Further, inequality measures in sub-Saharan Africa can be high in part because a large share of households report no income at all, which raises a question of whether the concepts of 'household' and 'income' are the same as in other parts of the world.

Taking such *pros* and *cons* of its use into consideration, I count on its competitive advantageous over other data set as my empirical work can effectively be performed to meet the central purpose of this research.

To obtain a balanced data set, some incomplete country time series have been dropped. Thus, the number of countries covered is 175 but their time span varies: Lebanon (9 yrs: 1997–2005), Maldives (9 yrs: 2002–2010), Lithuania (10 yrs: 2004–2013), Djibouti (11 yrs: 1995–2005), Haiti (11 yrs: 1991–2001) … Japan (45 yrs: 1967–2011), Italy (47 yrs: 1967–2013), Sweden (54 yrs: 1960–2013), the UK (54 yrs: 1961–2014), and the US (55 yrs: 1960–2014).

Having formulated the data set, it is useful to perform a preliminary analysis of the data in order to observe some empirical regularities. In so doing, the scattergram is employed using *Cartersian* coordinates to display the INEs that determine the position on the vertical axis that corresponds to the position on the horizontal axis. As presented in Fig. 4.1, the scattergram analysis on the sample data sets demonstrates

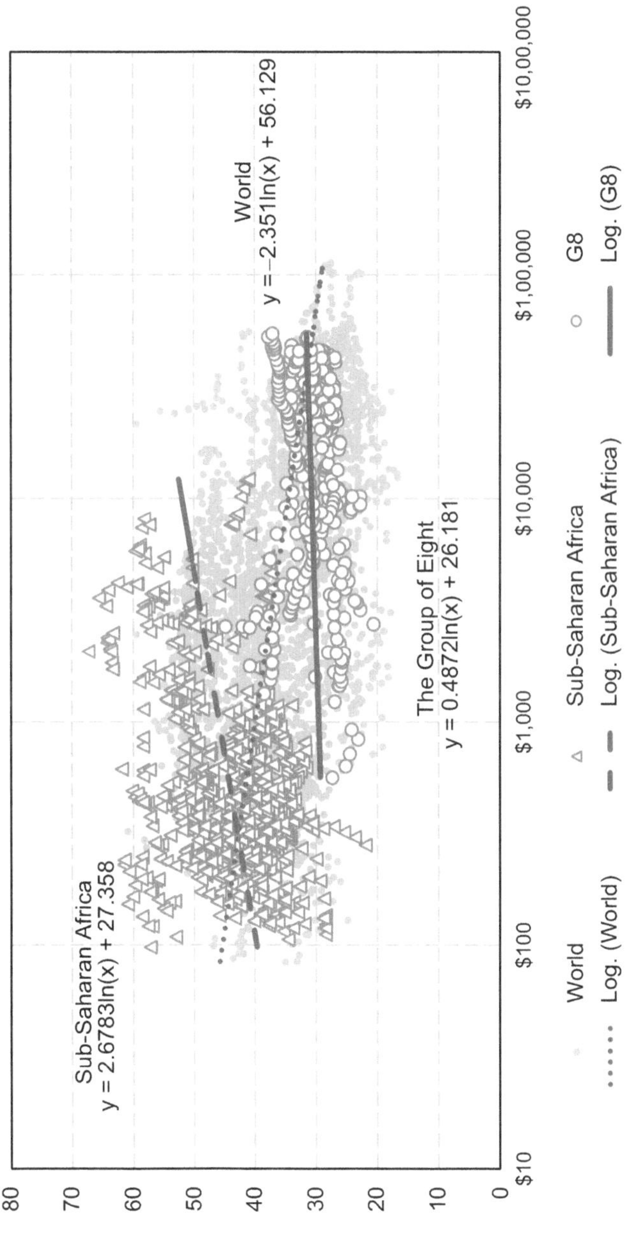

Fig. 4.1 Inequality-Income dynamics: Global, the Group of Eight and sub-Saharan Africa. (*Source*: Author's own elaboration on the basis of the World Bank's World Development Indicators (2016a) and the SWIID Version 5.1 (Solt, 2016b))

some empirical regularities—the negative association between inequality and growth in the group of 175 countries. Meanwhile, the coefficient of logarithmic equation for this group is approximately −2.35.

On the other hand, further analysis by country group provides additional regularities, which are behind the −2.35. Interestingly, 18 sub-Saharan African countries, widely considered the poorest group, appear to have a positive associationship with the logarithmic coefficient of 2.68. It can be inferred that higher growth is realised as income inequality increases and/or vice versa. Similarly, the Group of Eight (G8), the world's most highly industrialised economies, including Canada, France, Germany, Italy, Japan, Russia, the UK, and the US, is characterised by the positive relationship with the logarithmic coefficient of .49 although the magnitude of this coefficient is smaller than that of the poorest group.

Such empirical regularities observed from the preliminary analysis indeed provide very meaningful implications, that include, the world as a whole is approaching an environment where the more equal the income (re)distribution is, the greater economic growth is to be realised or vice versa. Notwithstanding this ideally negative associationship identified in global society as a whole, both the poorest and the richest groups of countries are likely to suffer from the undesirable circumstances with which societies become more unequal or unfair as their economies grow.

Hypothetically speaking, there is a trend behind a country's developmental process (or trajectory) under the *Inequality-Growth* nexus: as the process of growth begins from very low levels of income, inequality is likely to increase; after reaching a certain developmental stage, a country may transform into a society where its growth is translated into benefits for the poor; once shared broadly across society, an additional economic benefit has the tendency to be concentrated within the rich, which could aggravate an unequal income (re)distribution practice. Such regularities are extensively analysed in the remaining sections.

Econometric Modelling

The econometric method employed in this study to analyse the causality between income inequality and economic growth is the *Granger* causality model, utilised by *Eviews 8.0* statistical package. The *Granger* test was developed based on a major principle that presents a definition of cause and effect associationship between the two variables. Indeed, Granger (1969, 1988) showed that if two variables are individually integrated of order one and cointegrated, then there is the possibility of a causal relationship in at least one direction. Because of this definition, the *Granger* causality test has widely been used in research to test causality relationship and direction(s) between inequality and various dimensions of development (e.g., Assane and Grammy 2003; Frank 2009; Razmi and Ashrafzadeh 2012; Jihène and Ghazi 2013; Nuruddeen and Ibrahim 2014).

However, the application of the *Granger* causality test requires that all variables are time invariant (stationarity) (Assane and Grammy 2003). Since economic variables often contain unit root (possibly in existence due to a random process), this could invalidate statistical inferences and estimation processes in the econometric analysis. In order to check whether each variable is stationary or not, a unit root test should be performed. The *Augmented Dickey-Fuller* (ADF) (1979) is one of the most widely applied econometric methods to identify the order of the integration of each variable (Nuruddeen and Ibrahim 2014). According to the ADF assumption, if variables are found to be non-stationary at level but become stationary after the first differencing, they are considered reasonable to be used for econometric analysis. Two variables used in the study are converted into a natural logarithm to be tested via the ADF, which allows the results to be interpreted in elasticity and to avoid heteroscedasticity (Nuruddeen and Ibrahim 2014). The number of lags is in parentheses and selected according to the *Schwarz Information Criterion* (Assane and Grammy 2003). Also, as for robustness, the model containing both a constant and a trend is tested first since it is the least restricted. If the unit root is rejected, no further testing will take place, but if it is not rejected, then the model without a time trend will be tested (Sjö 2008). The unit root test is based on the estimation of an autoregressive model for Y_t and can be expressed as follows:

$$\Delta Y_t = X_0 + X_1\varphi + \alpha Y_{t-1} + \sum_{j=1}^{k} \omega_j \Delta Y_{t-j} + \varepsilon_t \qquad (4.1)$$

where Y_t, φ, and Δ confers a time series, a linear time trend and first difference operator, X_0 is a constant, k is respecting the optimum number of lags on the dependent variable, and ε_t is random error term.

If Y_t is found to be stationary, then a coevolution between the two variables may exist in the long-run. To investigate the existence of the long-run relationship (equilibrium), a cointegration test should be applied. The econometric framework used for a cointegration test is the *Johansen-Maximum Likelihood Cointegration* technique, which is best used to determine the number of cointegrating vector(s) (Johansen 1988; Razmi and Ashrafzadeh 2012; Jihène and Ghazi 2013). The *Johansen* method identifies the stationarity of the residue of two linear combinations. This method is based on the two likelihood ratio tests—trace test (λ_{trace}) and maximum eigenvalue test (λ_{max}). For the test, the order of the lag is determined according to the *Schwarz Information Criterion*, the *Hannan-Quinn Information Criterion*, and the *Akaike Information Criterion* (Assane and Grammy 2003), while the significance test is based on the test of MacKinnon-Haug-Michelis (1999). This multivariate cointegration test is expressed as:

$$Z_t = Kv_0 + K_1\Delta Z_{t-1} + K_2\Delta Z_{t-2} + \cdots\cdots + K_{p-1}\Delta Z_{t-p} + \Psi Z_{t-p} + \Phi_t \quad (4.2)$$

where Z_t represents a vector of variables that are integrated of order one, that is, $I(1)$; K represents a matrix of coefficients; Ψ represents a matrix of parameters; and Φ_t represents a vector of normally and independently distributed error term.

If no cointegration is found, then there is no evidence of the existence of long-run associationship between the two series. In this case, the *Granger* causality representation theorem in short-run can be tested based on the *Vector Autoregression* (VAR), which is explicitly specified as:

$$\Delta \text{PCGDP}_t = \Pi_1 + \sum_{i=1}^{n} \Pi_{i2}\, \Delta \text{PCGDP}_{t-i} + \sum_{j=1}^{m} \Pi_{j3}\, \Delta \text{INE}_{t-j} + \varepsilon_1 \quad (4.3)$$

$$\Delta \text{INE}_t = \Gamma_1 + \sum_{i=1}^{n} \Gamma_{i2} \Delta \text{INE}_{t-i} + \sum_{j=1}^{m} \Gamma_{j3} \Delta \text{PCGDP}_{t-j} + \varepsilon_2 \quad (4.4)$$

where PCGDP represents the per capita GDP, INE represents the income inequality index, Π_{ki} and Γ_{tj} ($k = 1, 2, 3$; $i = 1 \ldots n$; $j = 1 \ldots m$) are the coefficients, ε_1 and ε_2 are the error terms.

The causality test can be performed, based on the null hypothesis that there is no causal flow between inequality and economic growth (in both directions) as well as the alternative hypotheses, formally:

$$H_0 : \sum_{j=1}^{m} \Pi_3 = 0 \text{ and } \sum_{i=1}^{n} \Gamma_2 = 0$$

$$H_1 : \sum_{j=1}^{m} \Pi_3 \neq 0 \text{ and } \sum_{i=1}^{n} \Gamma_2 = 0, \text{ for unidirectional causality from INE to PCGDP}$$

$$H_1 : \sum_{j=1}^{m} \Pi_3 = 0 \text{ and } \sum_{i=1}^{n} \Gamma_2 \neq 0, \text{ for unidirectional causality from PCGDP to INE}$$

$$H_1 : \sum_{j=1}^{m} \Pi_3 \neq 0 \text{ and } \sum_{i=1}^{n} \Gamma_2 \neq 0, \text{ for bi-directional causality between PCGDP and INE}$$

On the other hand, if variables are found to be cointegrated, then there is evidence of the existence of a stable long-run equilibrium between the two variables. In other words, once the residue appears to be stationary, the *Granger* test under a *Vector Error Correction* (VEC) environment can be applied to test simultaneously the short- and long-run relationship. A major advantage of the VEC is that long-run effects are distinguished from short-run effects, with the specification of the long-run effect drawing on economic theory while allowing the short run to be in disequilibrium (Jansen 1996; Rocha 2006) and therefore both short- and long-run associationships can be identified. The model specification is presented as follows.

$$\Delta \text{PCGDP}_t = \Pi_1 + \sum_{i=1}^{n} \Pi_{2i} \, \Delta \text{PCGDP}_{t-i} + \sum_{j=1}^{m} \Pi_{3j} \, \Delta \text{INE}_{t-j} + EC^1_{t-j} + \mu_1 \quad (4.5)$$

$$\Delta \text{INE}_t = \Gamma_1 + \sum_{i=1}^{n} \Gamma_{2i} \Delta \text{INE}_{t-i} + \sum_{j=1}^{m} \Gamma_{3j} \Delta \text{PCGDP}_{t-j} + EC^2_{t-j} + \mu_2 \quad (4.6)$$

where Δ is a first difference operator, Π_{1}, Π_{2i}, Π_{3i}, Γ_{1}, Γ_{2j} and Γ_{3j} ($i = 1 \ldots n$ and $j = 1 \ldots m$) are the coefficients to the estimated, EC^1 and EC^2 are the lagged error correction term (ECT) derived from the long-run cointegrating relationship, and μ_1 and μ_2 are the error terms. For the VEC model (4.5), H_0: $\Pi_{3j} = 0$ for $j = 1 \ldots m$ is tested against H_1: $H_{3j} \neq 0$ at least one j, while for the VEC model (4.6), H_0: $\Gamma_{3j} = 0$ for $j = 1 \ldots n$ is tested against H_1: $\Gamma_{3j} \neq 0$ for at least one j.

In addition to the identification of the directions of causality, the signs of the associations between the two variables should also be investigated. The normalised cointegration coefficients in the long-run equilibrium (or associationship) through the VEC analysis convey useful information on the signs of the relationship between the two series although these estimates are not derived from structural equations. However, the short-run causality from the VEC and VAR specifications does not provide the signs.

Therefore, this study employs the *Impulse Response Functions* (IRF) to interpret the effect of a one-unit shock (change in one standard deviation) in one variable on the other (Pesaran and Shin 1998; Frank 2009). In other words, the IRF is systemised to identify the responsiveness of the dependent variables (endogenous variable) in the VAR and VEC system when a shock is put to the innovation (the error term: ε_1, ε_2, μ_1 and μ_2). But for calculating impulse responses, the ordering of the variables is important (Pesaran and Shin 1998). This study thus uses Cholesky (*dof* adjusted) ordering method as suggested by Doan (2007).

Modelling Results

Having described the methodologies in the preceding section, this section begins by interpreting the results of the ADF unit root test based on model (1). Out of 175 countries, 32 could not be tested due to an unbalanced sample observation between PCGDP and INE so these were dropped from the data sets. With 143 countries remaining, the ADF test results found that the PCGDP is non-stationary on their levels in 129 countries with the fact that in absolute terms, the computed values are smaller than the critical values at 5 per cent level; however, they become stationary after the first difference where the computed values are greater than the critical ones at 5 per cent level. As for INE, 117 countries show they are non-stationary on levels but become stationary after first differencing. In order to run the next econometric model, both inequality indicator and GDP indicator should pass the stationarity test. Therefore, an additional 33 countries are excluded from further econometric analyses. In other words, 110 pairs of the two variables are qualified and hence used for the next test, which is the *Johansen Cointegration* econometrics.[6] A summary of the ADF stationarity test results is presented in Appendix 1.

The two statistics for the *Johansen* Cointegration test in the VAR system are carried out as presented in Appendix 2. From the λ_{trace}, which is the value of the long-run test based on the trace of the stochastic matrix, 20 countries have no long-run cointegrating vector given the results where the estimated trace statistics are smaller than the critical value at 5 per cent level. On the other hand, the λ_{max} test, which is the test of long-run associationship based on the maximum eigenvalues of the stochastic matrix, indicates that the cointegrating vector is not found in 22 countries. Malta and Zimbabwe are the countries in which the λ_{trace} test identified one cointegrating vector, which is rejected by the λ_{max} test at 5 per cent level of significance. In this case, it is advisable to accept the result from the λ_{max} test especially for testing with small samples (Odhiambo 2005). Hence, the *Granger* causality investigation under the VEC environment is applied to 88 countries where the long-run relationship between inequality and growth is found, while the *Granger* test under the VAR system is employed for the remaining countries.

As presented, Appendix 3 summarises the test results of the *Granger* causality investigation under both the VEC and VAR environment. As explained, the countries where cointegration is detected are applied to the VEC model (5) and (6) as in these countries, the two variables are in long-run equilibrium state, and short-run dynamics of the variables are seen as fluctuations around this equilibrium. The VEC provides information on how such a short-run interacting system adjusts to converge to long-run state, which can be confirmed by the sign of the ECT. The ECT indicates the average speed of adjustment in the current period to disequilibrium in the previous period. In particular, the ECM should be statistically significant and negative sign. Otherwise, a cointegrating vector detected from the *Johansen* procedure does not actually exist in the VEC. Furthermore, the signs of the normalised cointegration coefficients (Coeff.) represent the signs of causality between the two variables: when the sign of the Coeff. is negative, the two variables are positively associated in long-run and vice versa.

Turning to error correction specifications in short-run, *Chi-Square Statistics* (χ^2 stat.) at 5 per cent level of significance show the direction of causation (either unidirectional or bi-directional causality). Similarly, for the countries where no cointegration is identified from the *Johansen* procedure, the *Granger* test under the VAR (3) and (4) is applied based on the test of the χ^2 stat. However, as discussed, the short-run *Granger* causality test does not provide information on the signs of the associations between the two variables, while the IRF test is applied to estimate the effect of a one standard deviation shock on the error terms. The results from the IRF (only those with statistical significance) are presented in Appendix 4.

For instance, in Argentina, long-run relationship from the *Johansen* Cointegration test shows that INE has a close relationship with PCGDP with the sign of the coefficient as expected.[7] It can thus be interpreted that for long-run results, as on average, an increase in INE by 1 per cent will decrease PCGDP by 7.89 per cent. But in the opposite direction that an increase in PCGDP by 1 per cent affects to decrease INE by .13 per cent is not statistically significant. Also, the speed of adjustment indicates that on average, there is a 168 per cent adjustment in the current period to disequilibrium in the previous period from short-run influence, which is bi-directional negative causality between PCGDP and INE. To sum up, Argentina appears to have negative long-run causality from INE to PCGDP with a normalised coefficient of 7.89 and also bi-directional negative causality in short-run, which contributes to long-run equilibrium with 168 per cent adjustment from disequilibrium in the previous period.

Overall, the long-run *Granger* causality appears to exist in 79 out of 88 countries analysed under the VEC system (a cointegrating vector was initially detected in nine countries, including Algeria, China, Latvia, Mauritania, Moldova, Norway, Peru, Philippines, and Sierra Leone but none of them are found to be statistically significant for the long-run). Moreover, there are 22 countries, in which the VAR applied appears to have the short-run *Granger* causality except Fiji and Zimbabwe where no coefficients are statistically significant. To sum up, 104 countries are found to have the *Granger* causality between inequality and growth at least in one direction: 79 countries with the long-run or both the long- and short-run causality, and 20 countries with only the short-run causality. In addition, all the identified causations in these 104 countries were analysed for the signs.

Framework of 11-Clusters Analysis

From a total of 104 countries identified for causations, it is useful for all these countries to be divided into sub-groups in order to derive some important implications. For this purpose, this study adopts the latest version of the *World Bank Country Classifications*. According to the classification, countries are defined, using the World Bank Atlas method (2016b) by the following: low-income economies with per capita gross national income (GNI) of US$1025 or less; lower middle-income economies with per capita GNI between US$1026 and US$4035; upper middle-income economies with per capita GNI between US$4036 and US$12,475; and high-income economies with per capita GNI of US$12,476 or more.

However, the application of these four sub-groups in this study may not be effective due to the incomparable and unbalanced number of country samples. Therefore, I modify this classification by combining the low and lower income to form one, namely *Developing Economies* with per capita GNI of US$4035 or less and renaming the upper middle income and the high income for the *Developed Economies* (with per capita GNI between US$4036 and US$12,475) and the *Advanced Economies* (with per capita GNI of US$12,476 or more), respectively. From this reclassification, the number of countries categorised into each of the three groups is fairly balanced for comparisons with 33 *Developing Economies*, 32 *Developed Economies* and 39 *Advanced Economies*.

Having classified the three levels of economic development, I needed to conceptualise the causality test results of the 104 countries with the 11 cluster (homogenous) groups.

First, *Group of Incentive* (GI) is identified as 13 countries where only unidirectional positive causality running from INE to PCGDP can be observed. These countries are characterised by inequality as *incentivising* economic growth. In other words, the higher (lower) inequality becomes, the greater (lesser) growth will be realised. Out of the 13 countries, 7 are categorised in the *Advanced Economies*, which infers that inequality as an incentive mechanism is a dominating factor in the GI cluster.

Second, *Group of Unequal income (re)distribution* (GU) is clustered with 22 countries. This cluster shows only unidirectional positive causality from PCGDP to INE and hence that growth benefits cannot effectively be shared with the poor segment mainly due to an unequal (ineffective) practice of income (re)distribution. In other words, the greater (lesser) growth is, the higher (lower) income inequality will be. Such unequal (re)distribution practice is mostly observed in both the *Advanced Economies* (11 countries) and the *Developing Economies* (10 countries).

Third, *Group of both Incentive and Unequal income (re)distribution* (GIU) is derived. This GIU cluster shows bi-directional positive causality between INE and PCGDP and consist of four countries: Austria, the Netherlands, Pakistan, and Tanzania where both an incentive mechanism and an unequal (re)distribution practice coexist. In this context, growth performance tends not to be radical because of its positive causality to income inequality while lowering inequality levels may not be targeted due to a strong demand for economic growth.

Fourth, *Group of Constraint* (GC) is clustered with 15 countries that can be characterised by only unidirectional negative causality running from INE to PCGDP. These 15 countries show that inequality as a constraint mechanism could hinder economic development. In other words, the higher (lower) inequality is, the lesser (greater) amount of economic performance will be achieved. Such a constraint mechanism tends to be observed in the *Developed Economies* (8 countries).

Fifth, *Group of Equal income (re)distribution* (GE) is established by unidirectional negative causality from PCGDP to INE. Like the GC, this cluster predominantly includes the *Developed Economies* (12 out of 18 countries) where the greater (lessor) the economic growth is, the lower (higher) the inequality will be in an economy.

Sixth, *Group of Constraint and Equal income (re)distribution* (GCE) shows bi-directional negative causality between INE and PCGDP. Such interaction is observed in two countries: Argentina and Burkina Faso. Based on the bi-directional negative relationship, they are not likely to implement policies targeting lower levels of inequality since too strong an intervention to lower inequality may also hinder economic performance.

Seventh, *Group of Incentive and Constraint* (GIC) consists of five countries: Canada, Chile, Senegal, Turkmenistan, and the UK. The GIC cluster is characterised by both unidirectional positive and negative causality from INE to PCGDP, meaning that inequality is acting not only as an incentive mechanism for economic growth but also as a constraint mechanism: Canada with mixed (both an incentive and a constraint) in short run, while an incentive mechanism in long run; Chile with a constraint mechanism in short run, while an incentive mechanism in long run; Senegal with an incentive mechanism in short run, while a constraint mechanism in long run; Turkmenistan with a constraint mechanism in short run, while an incentive mechanism in long run; and the UK with mixed in short run, while a constraint mechanism in long run. In this regard, these five countries tend to prioritise inequality mitigating policies rather than growth focus.

The eighth and the ninth cluster groups have substantial convergence regarding the *Inequality-Growth* dynamics: *Group of Incentive and Equal income (re)distribution* (GIE) is characterised by bi-directional causality between INE to PCGDP (positive from INE to PCGDP and negative from PCGDP to INE); and *Group of Constraint and Unequal income distribution* (GCU) is characterised by bi-directional causality between INE to PCGDP (negative from INE to PCGDP and positive from PCGDP to INE). The GIE and the GCU consist of nine countries and five countries, respectively, all of which are mostly found in the *Advanced Economies*. Both clusters can be characterised with circulating dynamics. For instance, in the GIE, as the income inequality level increases, greater economic performance is achieved, which in turn causes a lowering in inequality, which further hinders growth, which in turn causes greater levels of inequality. For this interaction, an adjustment mechanism may exist in preventing some marked deviation from the circulating dynamics.

Tenth, *Group of Unequal and Equal income (re)distribution* (GUE) includes seven countries which are characterised by both unidirectional positive and negative causation running from PCGDP to INE, meaning that countries which tend to have an economic structure that both equally and unequally (re)distribute its growth experience benefits across their societies. These seven countries are as follows: the Czech Republic and Honduras with unequal income (re)distribution in short run, while equal (re)distribution in long run; Hong Kong with equal income (re) distribution in short run, while unequal in long run; South Africa with mixed (both unequal and equal (re)distribution) in short run, while unequal (re)distribution in long run; Thailand with mixed (re)distribution in short run, while equal (re)distribution in long run; and the Kyrgyz Republic and Sierra Leone with mixed (re)distribution in short run only.

Eleventh, *Group of Complexity* (GX) is clustered with four countries, which include the countries where three characteristics or more (out of the four: incentive, constraint, unequal and equal distribution) are captured. Such complex causations are only observed in the *Developed Economies* (two countries) and the *Developing Economies* (two countries). Detailed analytical information for each of the 104 countries on causations is summarised in Appendix 5.

In analysing the 104 countries based on each of the three groups by income level, the *Developing Economies* are featured with predominantly positive relationship between inequality and growth, which accounts for 48.5 per cent (16 out of 33 countries). The next dominant feature for the *Developing Economies* is both positive and negative relations (10 countries), followed by negative relationship (7 countries). To be more specific for the positive relationship in 16 countries, an unequal income (re)distribution practice (GU) is the most common characteristic, which is observed in 10 countries, followed by inequality as an *Incentive* mechanism (GI) with four countries and both an incentive and an unequal (re) distribution feature (GIU) in two countries.

On the other hand, the *Developed Economies* appear to have a mostly negative associationship between the two variables. Its proportion is calculated for 65.6 per cent (21 out of 32 countries), followed by both positive and negative relationship (8 countries) and positive feature (3 countries). Out of 21 countries under the negative one, 12 countries show an equal income (re)distribution tendency (GE); eight countries feature inequality as a *Constraint* mechanism (GC); and only one country tends to show both a constraint and an equal (re)distribution (GCE).

Like the *Developing Economies*, the *Advanced Economies* also appear to be associated with the positive relationship between the two variables, which account for 51.3 per cent (20 out of 39 countries), followed by both positive and negative trend (12 countries) and negative tendency (7 countries). For the 20 countries that show the positive relationship, an unequal income (re)distribution practice (GU) is a dominant feature, which is observed in 11 countries. In addition, inequality as an *Incentive* mechanism (GI) and both an incentive and an unequal (re)distribution (GIU) are found in seven countries and two countries, respectively.

Based on income grouping analysis, it should be noted that the positive relations between inequality and growth predominates in the poorest group as well as the richest group, while the middle-income group tends to be associated with the negative relationship. Furthermore, the challenge that greatly belongs to the poorest and richest country group could be that these countries are unlikely to (re)distribute their growth benefits equally across society actors. In contrast, the middle-income group has a tendency for more equal (re)distribution across their societies. All these analysed results are summarised in Fig. 4.2.

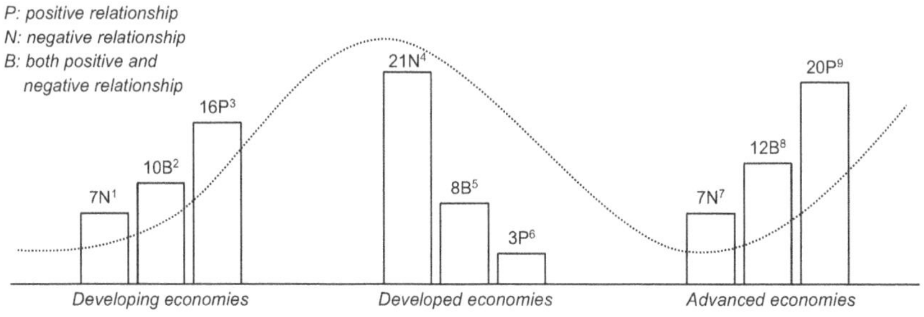

Fig. 4.2 Analysis of relationship and causality by income groups. (*Source*: Author's own elaboration on the basis of the modelling results)[8]

As for regional groups, my analysis is based on the *World Bank Regional Classifications* (2016c): Europe and Central Asia, sub-Saharan Africa, Latin America and the Caribbean, East Asia and Pacific, South Asia, North America, and Middle East and North Africa.

In the Middle East, North Africa, East Asia, and the Pacific, almost a half of the countries appear to feature both positive and negative associationship between inequality and growth. Meanwhile around 30 per cent of all the countries analysed in each of the two regions are characterised by the positive relationship while around 20 per cent by the negative one. South Asian countries tend to have a predominantly positive relationship, which accounts for 60 per cent of the sample countries in this region while Latin American and the Caribbean countries appear to show negative relationship with 52.9 per cent. Sub-Saharan African countries and European and Central Asian countries show a balanced result between positive and negative although the positive association is more observed with more than 40 per cent. In North America, only two countries belong to this group which are the US and Canada: The US is associated with positive relationship, particularly with regard its unequal income (re)distribution practice; meanwhile, Canada is likely to have both positive and negative relationship characterised by an incentive and a constraint mechanism. This regional gap analysis is reported in Fig. 4.3.

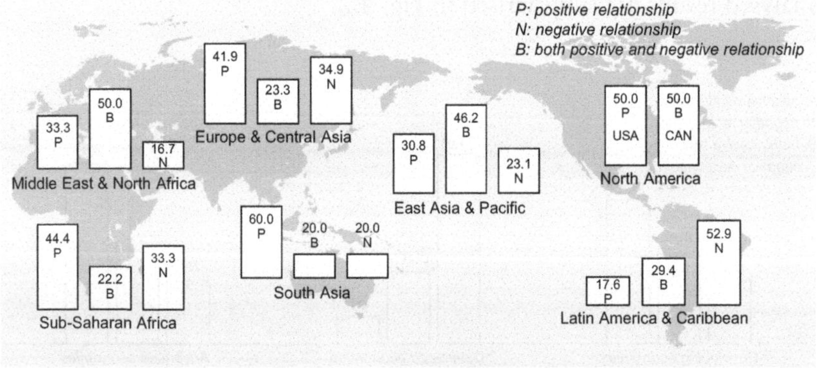

Fig. 4.3 Analysis of relationship and causality by regional groups. (*Source*: Author's own elaboration on the basis of the modelling results)[9]

Summary and Discussion

This study has explored empirically the long- and short-run relationship between inequality and development and their causality orientations, using a more recent and extensively larger data set of a 143-balanced country sample covering a period of 23.5 years on average. From the analyses, 79 countries (55 per cent of the total country sample) appear to have stable long-run equilibrium between inequality and growth by uni- and/or bi-directional causality. Additionally, 89 countries (62 per cent in total) are associated with at least short-run causations. Overall, countries are classified according to the combination of positive and negative one-way and two-way causal relations found to be statistically significant.

These results are noteworthy in that they provide both theoretical and empirical motivations to look into how growth affects the (re)distribution of income and vice versa. One of the most important insights gained from this causal-oriented investigation is that the Kuznets hypothesis is still valid for contemporary policy debate. More importantly, another upswing curve is identified in addition to the ups earlier then the downs later of the Kuznets curve, which implies some new inequality and growth dynamics that are likely to have existed for industrialised societies in the postmodern period from the 1960s to the present. The findings are also consistent with the theory of an N-shaped relation between national income per person and inequality: in other words, it confirms that for many high-income countries, growth is associated with rising inequality.

The N-shaped hypothesis is firmly supported by the cluster analysis by income groups: the *Developing Economies* are likely to be associated with a positive relationship between inequality and growth, mainly attributable to their ineffective (re)distribution practice of income and to the role of inequality functioning as an Incentive for growth; the *Developed Economies* appear to benefit from fair (re)distribution of income across their societies but to suffer from the detrimental role of inequality so as to hinder economic growth, commutatively resulting in the negative causal orientation; the *Advanced Economies* share a commonality of the inequality and growth dynamics observed in the *Developed Economies* in that growth benefits tend not to be effectively (re)distributed to the poor segments of society and that income inequality provides incentives for more effort to gain more, which are the two most common characteristics among the 11 clusters.

This study has further attempted to explore regional distinctive characteristics and found that the Middle East, North Africa, and East Asia feature a mixed relationship (both positive and negative); South Asia, Latin America, and the Caribbean have the opposite orientations of the positive and the negative, respectively; while sub-Saharan Africa, Europe, and Central Asia are associated with balanced characteristics between positive and negative relationship.

Our empirical findings on the existence of distinctive features of each of the income groups (Lee and Son 2016) and regional groups (Rubil 2015) can also be consistent with the competitive political economy discourse, particularly the typology of the *Varieties of Capitalism* (Hall and Soskice 2001) and *Welfare Regimes* (Esping-Andersen 1990).[10] Each of the regional groups in terms of production systems have been institutionalised by their own unique inequality feature of *enabling* and/or *restraining* growth. Similarly, the effectiveness of the (re)distribution system also implies that unique institutions have been shaped through features of institutional complementarities in each country or group of countries. Therefore, each of the countries' or regional groups' endogenous processes that shall be associated with inequality dynamics are likely to affect the process of continuous economic development, eventually directing some distinctive developmental paths (i.e., path-dependency) (Schmidt 2002; Amable 2003; Schröder 2013).[11]

Finally, there are several areas for future research, which can extend this study. Various causal links identified with the 11 clusters need to be explained in a more theoretical fashion by allowing for the possibility that these identified mechanisms are not mutually exclusive in explaining country patterns over time. Additionally, it should clearly be acknowledged that this study is largely empirical rather than atheoretical therefore further work would be needed to theorise about the possible explanations for what has been observed from this investigation. In addition to these, although this chapter contributes to the literature by presenting results of short- and long-run *Granger* causality tests in both directions, differentiating such short- and long-run causality should be further explored as a basis for explaining the complexity of nexus during upswings and downswings, which would provide more analytical information on country-specific trends over time (Halter et al. 2014: Chang et al. 2018).

On the technical side, there are a couple of issues I have encountered during the research, which I strongly believe that there would be areas for improvement if properly addressed. First, the limited number of observations for the *Granger* causality tests may have influenced the results, especially accounting for the high number of lags that would reduce the degree of freedom. Therefore, the robustness of the tests needs to be conducted because the results eventually influence the clusters identified. Moreover, the low number of observations is predominantly for developing countries (due to data unavailability), which would bias the sample and the results that allow to define the clusters. In this regard, my attempt to advance some interesting generalisations that highlight diversity of country evidence based on the methodology using the cointegration and error correction approach with *Granger* causality tests can be of great value-added interventions to ongoing scholarly communications in this field if such data challenges are substantially resolved.

In conclusion, the causal relationships between country-level economic development and inequality still remain a matter of considerable debate and policy interest. Clearly, identifying *Inequality-Growth* characteristics (i.e., the 11 clusters) that are embedded in society would be imperative before designing policy related to reducing income inequality and/or stimulating economic activities to promote inclusive and sustainable growth (Armah and Baek 2015, 2018). In this context, the methodology and results from this study can provide policymakers with a useful guideline on how to address the complexity (long-run versus short-run, positive versus negative associationship, unidirectional versus bi-directional causality, etc.) of the *Inequality-Growth* nexus and to prioritise policy intervention under the stylised dynamics.

Notes

1. For example, Friedman (1953), Rebelo (1991), Deaton and Paxson (1997), Galor and Tsiddon (1997), Partridge (1997), Li and Zou (1998), Forbes (2000), and Bell and Freeman (2001).
2. For example, Stiglitz (1969), Feldstein and Horioka (1980), Alesina and Rodrik (1994), Persson and Tabellini (1994), Clarke (1995), Alesina and Perotti (1996), Benabou (1996), Benhabib and Rustichini (1996), Perotti (1996), De La Croix and Doepke (2003), Wilkinson and Pickett (2010), Razmi and Ashrafzadeh (2012), Berg and Ostry (2013), and Lee and Son (2016).
3. For example, Psacharopoulos et al. (1995), Aghion and Bolton (1997), Bruno et al. 1998), De Janvry and Sadoulet (2000), Kakwani and Pernia (2000), Bourguignon (2003), Krongkaew and Kakwani (2003), Adams (Adams 2004), Heltberg (2004), Ravallion (2007), Basu and Mallick (2008), Ogun (2010), Ostry et al. 2014, and Akinci (2017).
4. For example, Cutler and Katz (1991), Blank and Card (1993), Weriemmi and Ehrhart (2008), and Rubin and Segal (2015).
5. For example, Deininger and Squire (1996), Barro (2000), Chen (2003), Galbraith and Kum (2005), Castelló-Climent (2010), and Jihène and Ghazi (2013).
6. The time period used between countries differ, which could provoke some econometric controversy with the extent to which *Granger* causality tests can be compared like-for-like (i.e., a structural break might appear in one series and not the other and hence affects the causality relationship differently in the results). Nonetheless, more than four-fifth of the studied countries have convergence of the period of 1985–2010. Furthermore, this econometric analysis covers 110 sample countries (out of 143 data available) that exceed 105 recommended number of the countries that can be appropriate to represent the world views (stylised facts) based on 5 per cent margin of error and 95 per cent confidence level so that the difference of the time period can be ignored.
7. Selection of ARG is unbiased. ARG is selected to describe the results simply because ARG is the first one both that is presented in Appendix 2 and that is confirmed with both long- and short-run associationships.
8. *Note*: 1 represents the 7 countries characterised for negative associationship between INE and PCGDP with GC cluster (KEN KHM TUN), GE cluster (MLI NPL TJK), and GCE cluster (BFA); 2 represents the 10

countries characterised for both positive and negative associationship between INE and PCGDP with GIE cluster (ARM LKA NGA), GUE cluster (HND KGZ SLE), GX cluster (EGY VNM), GIC cluster (SEN), and GCU cluster (LAO); 3 represents the 16 countries characterised for positive associationship between INE and PCGDP with GU cluster (BDI BGD GTM IDN IND MAR RWA UGA UKR ZMB), GI cluster (CIV GHA LSO UZB), and GIU cluster (PAK TZA); 4 represents the 21 countries characterised for negative associationship between INE and PCGDP with GE cluster (AGO CHN COL DOM MEX MNE MYS NAM RUS SRB TUR VEN), GC cluster (AZE BRA BWA ECU GEO KAZ PAN PRY), and GCE cluster (ARG); 5 represents the 8 countries characterised for both positive and negative associationship between INE and PCGDP with GCU cluster (JAM MKD), GUE cluster (THA ZAF), GX cluster (CRI PER), GIC cluster (TKM), and GIE cluster (JOR); 6 represents the 3 countries characterised for positive associationship between INE and PCGDP with GI cluster (BGR ROU) and GU cluster (BLR); 7 represents the 7 countries characterised for negative associationship between INE and PCGDP with GC cluster (FRA HUN IRL SWE) and GE cluster (HRV LVA SVK); 8 represents the 12 countries characterised for both positive and negative associationship between INE and PCGDP with GIE cluster (EST ISL KOR NZL PRT), GIC cluster (CAN CHL GBR), GCU cluster (ESP ISR), and GUE (CZE HKG); and 9 represents the 20 countries characterised for positive associationship between INE and PCGDP with GU cluster (AUS CHE CYP ITA JPN LUX MLT POL SGP TTO USA), GI cluster (BEL DEU DNK GRC NOR PRI SVN), and GIU cluster (AUT NLD)).

9. Six countries in Middle East and North Africa comprise GU (33.3 per cent), GC (16.7 per cent), GIE (16.7 per cent), GUC (16.7 per cent), and GX (16.7 per cent); 13 countries in East Asia and Pacific comprise GU (30.8 per cent), GC (77 per cent), GE (15.4 per cent), GIE (15.4 per cent), GUC (7.7 per cent), GUE (15.4 per cent), and GX (7.7 per cent); six countries in South Asia comprise GU (40 per cent), GIU (20 per cent), the (20 per cent), and GIE (20 per cent); 17 countries in Latin America and the Caribbean comprise GI (5.9 per cent), GU (11.8 per cent), GC (23.5 per cent), GE (23.5 per cent), GCE (5.9 per cent), GIC (59 per cent), GUC (5.9 per cent), GUE (5.9 per cent), and GX (11.8 per cent); 18 countries in sub-Saharan Africa comprise GI (16.7 per cent), GU (22.2 per cent), GIU (5.6 per cent), GC (11.1 per cent), GE

(16.7 per cent), GCE (5.6 per cent), GIC (5.6 per cent), GIE (5.6 per cent), and GUE (11.1 per cent); 43 countries in Europe and Central Asia comprise GI (20.9 per cent), GU (16.3 per cent), GIU (4.7 per cent), GC (16.3 per cent), GE (18.6 per cent), GIC (4.7 per cent), GIE (9.3 per cent), GUC (4.7 per cent), and GUE (4.7 per cent); two countries in North America comprise GU (50 per cent) and GIC (50 per cent)).

10. For a more detailed discussion, see Chap. 2.
11. For a more detailed discussion, see Chap. 3.

References

Adams, R., Jr. 2004. Economic Growth, Inequality and Poverty: Estimating the Growth Elasticity of Poverty. *World Development* 32 (12): 1989–2014.

Aghion, P., and P. Bolton. 1997. A Theory of Trickle-Down Growth and Development. *Review of Economic Studies* 64: 151–172.

Akinci, M. 2017. Inequality and Economic Growth: Trickle-down Effect Revisited. *Development Policy Review* 36 (S1): 1–24.

Alesina, A., and R. Perotti. 1996. Income Distribution, Political Instability, and Investment. *European Economic Review* 40 (6): 1203–1228.

Alesina, A., and D. Rodrik. 1994. Distribution Politics and Economic Growth. *Quarterly Journal of Economics* 109 (2): 465–490.

Amable, B. 2003. *The Diversity of Modern Capitalism*. Oxford: Oxford University Press.

Armah, B., and S.J. Baek. 2015. Can the SDGs Promote Structural Transformation in Africa? An Empirical Analysis. *Development* 58 (4): 1–19.

Armah, B., and S.J. Baek. 2018. Three Interventions to Foster Sustainable Transformation in Africa. *Journal of Social, Political, and Economic Studies* 43 (1–2): 3–25.

Assane, D., and A. Grammy. 2003. An Assessment of the Growth and Inequality Causality Relationship. *Applied Economics Letters* 10 (14): 871–873.

Baek, S.J. 2017. Is Rising Income Inequality Far from Inevitable During Structural Transformation? A Proposal for An Augmented Inequality Dynamics. *Journal of Economics and Political Economy* 4 (3): 224–237.

Barro, R.J. 2000. Inequality and Growth in a Panel of Countries. *Journal of Economic Growth* 5 (1): 5–32.

Basu, S., and S. Mallick. 2008. When Does Growth Trickle-Down to The Poor? The Indian Case. *Cambridge Journal of Economics* 32 (3): 461–477.

Bell, L., and R. Freeman. 2001. The Incentive for Working Hard: Explaining Hours Worked Differences in the US and Germany. *Labour Economics* 8 (2): 181–202.

Benabou, R. 1996. Inequality and Growth. In *NBER Macroeconomics Annual 11*, ed. B.S. Bernanke and J. Rotemberg, 11–74. Cambridge: MIT Press.

Benhabib, J., and A. Rustichini. 1996. Social Conflict and Growth. *Journal of Economic Growth* 1 (1): 125–142.

Berg, A.G., and J.D. Ostry. 2013. Inequality and Unsustainable Growth: Two Sides of the Same Coin? *International Organisations Research Journal* 8 (4): 77–99.

Blank, R.M., and D. Card. 1993. Poverty, Income Distribution, and Growth: Are They Still Connected? *Brookings Papers on Economic Activity* 2: 285–339.

Bourguignon, F. 2003. The Growth Elasticity of Poverty Reduction: Explaining Heterogeneity Across Countries and Time Periods. In *Inequality and Growth: Theory and Policy Implications*, ed. T. Eicher and S. Turnovsky. Cambridge: MIT Press.

Bruno, M., M. Ravallion, and L. Squire. 1998. Equity and Growth in Developing Countries: Old and New Perspectives on the Policy Issues. In *Income Distribution and High Growth*, ed. V. Tani and K.Y. Chu, 213–231. Cambridge: MIT Press.

Castelló-Climent, A. 2010. Inequality and Growth in Advanced Economies: An Empirical Investigation. *Journal of Economic Inequality* 8 (3): 293–321.

Chang, J.Y., and R. Ram. 2000. Level of Development, Rate of Economic Growth, and Income Inequality. *Economic Development and Cultural Change* 48 (4): 787–799.

Chang, S., R. Gupta, and S. Miller. 2018. Causality Between Per Capita Real GDP and Income Inequality in the U.S.: Evidence from a Wavelet Analysis. *Social Indicators Research* 135: 269–289.

Chen, B.L. 2003. An Inverted-U Relationship Between Inequality and Long-Run Growth. *Economics Letters* 78 (2): 205–212.

Clarke, G.R.G. 1995. More Evidence on Income Distribution and Growth. *Journal of Development Economics* 47: 403–427.

Cutler, D.M., and L.F. Katz. 1991. Macroeconomic Performance and the Disadvantaged. *Brookings Papers on Economic Activity* 2: 1–61.

De Janvry, A., and E. Sadoulet. 2000. Growth, Poverty, and Inequality in Latin America: A Causal Analysis, 1970–94. *Review of Income and Wealth* 46 (3): 267–287.

De La Croix, D., and M. Doepke. 2003. Inequality and Growth: Why Differential Fertility Matters. *American Economic Review* 93 (4): 1091–1113.

Deaton, A., and C. Paxson. 1997. The Effects of Economic and Population Growth on National Saving and Inequality. *Demography* 34 (1): 97–114.

Deaton, A. 2013. *The Great Escape: Health, Wealth, and the Origins of Inequality*. New Jersey: Princeton University Press.

Deininger, K., and L. Squire. 1996. A New Data Set Measuring Income Inequality. *World Bank Economic Review* 10 (3): 565–591.

Dickey, D.A., and W.A. Fuller. 1979. Distribution of the Estimators for Autoregressive Time Series with a Unit Root. *Journal of the American Statistical Association* 74 (366): 427–431.

Doan, T.A. 2007. *RATS User's Guide Version 7*. Evanston: Estima. Available from: http://digidownload.libero.it/rocco.mosconi/USER_gui_RATS.pdf. Accessed 23 Oct 2016.

Esping-Andersen, G. 1990. *The Three Worlds of Welfare Capitalism*. Princeton: Princeton University Press.

Feldstein, M., and C. Horioka. 1980. Domestic Saving and International Capital Flows. *Economic Journal* 90 (358): 314–329.

Foellmi, R., and J. Zweimüller. 2006. Income Distribution and Demand-Induced Innovations. *Review of Economic Studies* 73 (4): 941–960.

Forbes, K. 2000. A Reassessment of the Relationship Between Inequality and Growth. *American Economic Review* 90 (4): 869–897.

Frank, M.W. 2009. Inequality and Growth in the United States: Evidence from a New State-Level Panel of Income Inequality Measures. *Economic Inquiry* 47 (1): 55–68.

Friedman, M. 1953. Choice, Chance and the Personal Distribution of Income. *Journal of Political Economy* 61 (4): 277–290.

Galbraith, J.K., and H. Kum. 2005. Estimating the Inequality of Household Incomes: A Statistical Approach to the Creation of a Dense and Consistent Global Data Set. *Review of Income and Wealth* 51 (1): 115–143.

Galor, O., and O. Moav. 2004. From Physical to Human Capital Accumulation: Inequality and the Process of Development. *Review of Economic Studies* 71 (4): 1001–1026.

Galor, O., and D. Tsiddon. 1997. Technological Progress, Mobility, and Economic Growth. *American Economic Review* 87 (3): 363–382.

Giddings, B., B. Hopwood, and G. O'Brien. 2002. Environment, Economy and Society: Fitting Them Together into Sustainable Development. *Sustainable Development* 10 (4): 187–196.

Granger, C.W.J. 1969. Investigating Causal Relations by Econometric Models and Cross-Spectral Methods. *Econometrica* 37 (3): 424–438.

———. 1988. Some Recent Developments in a Concept of Causality. *Journal of Econometrics* 39 (1–2): 199–211.

Hall, P.A., and D. Soskice, eds. 2001. *Varieties of Capitalism: The Institutional Foundations of Comparative Advantage*. Oxford: Oxford University Press.

Halter, D., M. Oechslin, and J. Zweimüller. 2014. Inequality and Growth: The Neglected Time Dimension. *Journal of Economic Growth* 19 (1): 81–104.

Heltberg, R. 2004. The Growth Elasticity of Poverty. In *Growth, Inequality, and Poverty*, ed. A. Shorrocks and R. van der Hoeven. Oxford: Oxford University Press.

Herzer, D., and S. Vollmer. 2012. Inequality and Growth: Evidence from Panel Cointegration. *Journal of Economic Inequality* 10 (4): 489–503.

Jansen, W.J. 1996. Estimating Saving-Investment Correlations: Evidence for OECD Countries Based on an Error Correction Model. *Journal of International Money and Finance* 15 (3): 749–781.

Jihène, S., and B. Ghazi. 2013. The Causality between Income Inequality and Economic Growth: Empirical Evidence from the Middle East and North Africa Region. *Asian Economic and Financial Review* 3 (5): 668–682.

Johansen, S. 1988. Statistical Analysis of Cointegration Vectors. *Journal of Economic Dynamics and Control* 12 (2–3): 231–254.

Kakwani, N., and E.M. Pernia. 2000. What Is Pro-Poor Growth? *Asian Development Review* 18: 1–16.

Kaldor, N. 1956. Alternative Theories of Distribution. *Review of Economic Studies* 23 (2): 83–100.

Krongkaew, M., and N. Kakwani. 2003. The Growth–Equity Trade-Off in Modern Economic Development: The Case of Thailand. *Journal of Asian Economics* 14: 735–757.

Kuznets, S. 1955. Economic Growth and Income Inequality. *American Economic Review* 45 (1): 1–28.

Lee, D.J., and J.C. Son. 2016. Economic Growth and Income Inequality: Evidence from Dynamic Panel Investigation. *Global Economic Review* 45 (4): 331–358.

Li, H., and H. Zou. 1998. Income Inequality Is Not Harmful for Growth: Theory and Evidence. *Review of Development Economics* 2 (3): 318–334.

MacKinnon, J.G., A.A. Haug, and L. Michelis. 1999. Numerical Distribution Functions of Likelihood Ratio Tests for Cointegration. *Journal of Applied Econometrics* 14 (5): 563–577.

Milanovic, B. 2016. *Global Inequality: A New Approach for the Age of Globalization*. Cambridge: Harvard University Press.

Nuruddeen, T., and S.S. Ibrahim. 2014. An Empirical Study on the Relationship Between Poverty, Inequality and Economic Growth in Nigeria. *Journal of Economics and Sustainable Development* 5 (26): 20–24.

Odhiambo, N.M. 2005. Financial Liberalization and Financial Deepening: Evidence from Three Sub-Saharan African (SSA) Countries. *Savings and Development* 29 (suppl): 5–23.

Ogun, T. 2010. Infrastructure and Poverty Reduction: Implications for Urban Development in Nigeria. *Urban Forum* 21 (3): 249–266.

Ostry, J.D., A. Berg, and C.G. Tsangarides. 2014. *Redistribution, Inequality, and Growth*, IMF Staff Discussion Note 14/02. Available from: http://www.imf.org/external/pubs/ft/sdn/2014/sdn1402.pdf. Accessed 11 Nov 2016].

Papanek, G., and O. Kyn. 1986. The Effect on Income Distribution of Development, the Growth Rate, and Economic Strategy. *Journal of Development Economics* 23 (1): 55–65.

Parente, S. 2001. The Failure of Endogenous Growth. *Knowledge, Technology, and Policy* 13 (4): 49–58.

Partridge, M.D. 1997. Is Inequality Harmful for Growth. *American Economic Review* 87 (5): 1019–1032.

Perotti, R. 1996. Growth, Income Distribution and Democracy: What the Data Say. *Journal of Economic Growth* 1 (2): 149–187.

Persson, T., and G. Tabellini. 1994. Is Inequality Harmful for Growth? *American Economic Review* 84 (3): 600–621.

Pesaran, M.H., and Y. Shin. 1998. Generalized Impulse Response Analysis in Linear Multivariate Models. *Economic Letters* 58 (1): 17–29.

Psacharopoulos, G., S. Morley, A. Fiszbein, H. Lee, and W.C. Wood. 1995. Poverty and Income Inequality in Latin America During the 1980s. *Review of Income and Wealth* 41 (3): 245–264.

Ravallion, M. 2007. Inequality Is Bad for the Poor. In *Inequality and Poverty Re-examined*, ed. S. Jenkins and J. Micklewright, 37–61. Oxford: Oxford University Press.

Razmi, M., and S. Ashrafzadeh. 2012. Study of the Relationship Between Economic Growth and Income Distribution (A Case Study in Iran During the Period 1971–2007). *Economics and Finance Review* 2 (4): 24–31.

Rebelo, S. 1991. Long-Run Policy Analysis and Long-Run Growth. *Journal of Political Economy* 99 (3): 500–521.

Rocha, F. 2006. Are Error Correction Models the Best Alternative to Assess Capital Mobility in Developing Countries? *Economia Aplicada* 10 (2): 267–285.

Rubil, I. 2015. Regional Heterogeneity in Growth and Inequality Elasticities of Poverty in Transition Countries. *Applied Economics Letters* 22 (6): 425–429.

Rubin, A., and D. Segal. 2015. The Effects of Economic Growth on Income Inequality in the US. *Journal of Macroeconomics* 45(C: 258–273.

Sachs, J.D., and A.M. Warner. 1997. Fundamental Sources of Long-Run Growth. *American Economic Review* 87 (2): 184–188.

Schmidt, V.A. 2002. *The Futures of European Capitalism*. Oxford: Oxford University Press.

Schröder, M. 2013. *Integrating Varieties of Capitalism and Welfare State Research: A United Typology of Capitalisms*. New York: Palgrave Macmillan.

Sjö, B. 2008. *Testing for Unit Roots and Cointegration*. IEI: Linköping University. Available from: https://www.iei.liu.se/nek/ekonometrisk-teori-7-5-hp-730a07/labbar/1.233753/dfdistab7b.pdf. Accessed 13 Feb 2017.

Solt, F. 2016a. The Standardized World Income Inequality Database. *Social Science Quarterly* 97 (5): 1267−1281.

———. 2016b. *The Standardized World Income Inequality Database Version 5.1* [Online]. Available from: https://dataverse.harvard.edu/dataset.xhtml?persistentId=hdl:1902.1/11992. Accessed 10 Nov 2016.

Stiglitz, J.E. 1969. Distribution of Income and Wealth Among Individuals. *Econometrica* 37 (3): 382–397.

Subramanian, S.V., and I. Kawachi. 2004. Income Inequality and Health: What Have We Learned So Far? *Epidemiologic Reviews* 26 (1): 78−91.

Weriemmi, M., and C. Ehrhart. 2008. Inequality and Growth in a Context of Commercial Openness, Theoretical Analysis and Empirical Study: The Case of the Countries Around the Mediterranean Basin. *Journal of Social Management* 6 (2): 81–91.

Wilkinson, R., and K. Pickett. 2010. *The Spirit Level: Why Equality Is Better for Everyone*. London: Penguin.

World Bank. 2016a. *World Development Indicators* [Online]. Available from: http://data.worldbank.org/data-catalog/world-development-indicators. Accessed 19 Nov 2016.

———. 2016b. *World Bank Atlas Method – Detailed Methodology* [Online]. Available from: https://datahelpdesk.worldbank.org/knowledgebase/articles/378832-what-is-the-world-bank-atlas-method. Accessed 19 Nov 2016.

———. 2016c. *World Bank's List of Economies* [Online]. Available from: http://autism-insar.org/sites/default/files/docs/worldbanklist.pdf. Accessed 19 Nov 2016.

5

Theoretical Reshaping for the Augmented Inequality Dynamics

Introduction

For over a decade, impressive progress has been made on socio-economic development in the developing world, with more decent jobs being made available and more people lifted out of poverty. Economic growth has also been remarkable: the average rate of growth in low-income countries since 2000 (4.8 per cent) is far higher than the global rate (2.9 per cent) (World Bank 2017). These emerging economies are, thus, increasingly referred to as the new growth engine of the world as many have the tendency to grow more rapidly than high-income countries, thereby converging living standards between the two different worlds (Korotayev and Zinkina 2014).[1] Such *Catching-Up* or *Economic Convergence* across countries has been observed in the global development trajectory since the 1960s (Maddison 2008).

Such tremendous achievements can be attributed to several developing economies making an ongoing effort to structurally transform their economies in a competitive manner, which further influences the con-

S. J. Baek, *The Political Economy of Neo-modernisation*,
https://doi.org/10.1007/978-3-319-91394-0_5

cept of *Development* shifting into the one in a sustainable fashion (see Box 5.1). When it comes to questions of what drives structural change, there have been a number of hypotheses and models that contribute to explaining factors affecting the process of economic growth and development. Some of them provide a relatively clearer explanation on how economies evolve and how their structures shift while others are vague and sometimes fail to apply to the regional or country-specific development pathways. These competing (or complementing) arguments largely stem from definitional challenges and slightly different use of terminologies such as *Structural Change*, *Economic Transformation*, *Structural Shift*, and *Structural Adjustment*. However, in much of the literature on historical and modern economic growth and development, the most common terminology used to explain this concept is *Structural Transformation*.

At the very general view, structural transformation is regarded as a process characterised by a decline in the relative share of agricultural value-added and a rise in the modern industrial and service economy (Armah and Baek 2015, 2018). Based on this, sectoral shifts measured by the value-added contribution of industry as a percentage of GDP in low-income countries have been made with an increase of 1.6 percentage points from 19.5 per cent of GDP in 2000, while the world as a whole has registered a decrease of 3.0 percentage points over the same period (World Bank 2017). This transformation into the manufacturing sector in particular is the key enabler of growth in low-income countries, which was empirically studied by Szirmai (2012), who explored 67 developing countries in the period 1950–2005.

Unfortunately, however, structural transformation is not always accompanied by inclusive growth and prosperity for all; rather it often coexists with rising inequalities. With an average Gini coefficient over the period 2000–2009, greater income inequality is observed in Latin America with 52.2 per cent and Africa with 43.9 per cent. Almost all of the most unequal countries worldwide were in these two regions which are considered to be entirely part of the developing world. In particular, sub-Saharan Africa has the highest levels of inequality in the world, even after accounting for its initial level of development. Such high levels of inequality have been proven to be less poverty-reducing as the benefits of growth accrue to a few individuals, hence the *slower-than-anticipated* pace of poverty reduction in this sub-region, despite a decade of strong growth (ECA et al. 2016).

With this in mind, this chapter is motivated by a renewed emphasis by policymakers in developing countries in general and the poorest countries in particular to adopt a more inclusive growth trajectory during the course of structural transformation, which was set as the main theme of the 2017 UN high-level political forum on sustainable development, 'Eradicating Poverty and Promoting Prosperity in a Changing World' and of the 2017 Africa Regional Forum on Sustainable Development, 'Ensuring Inclusive and Sustainable Growth and Prosperity for All'.

Therefore, the central objective of this chapter is to examine first whether rising income inequality is a stylised fact for the process of structural transformation by revisiting classical accounts on inequality dynamics, modernisation, and endogenous growth, which has in fact been of great interest not only to the policy arena but also to the field of social science in general. A complex interacting nexus between development and inequality is then analysed by exploring social, political, and moral dimensions of inequality in addition to economics, in better explaining the impact of structural transformation on inequality and vice versa in the twenty-first-century context with reference to what can be learnt from the past experiences of countries in the previous centuries.

The rest of the chapter is organised as follows. In the following two sections, I extensively review the literature on the classical account related to the discussion on normative inequality and theories of modernisation and endogenous growth. Then, I attempt to theoretically integrate them for better explaining the complex nexus between development and inequality in the twenty-first-century context. My critical review allows me to introduce a proposal for *Augmented Inequality Dynamics* by systematising the endogenous process within a society in the following section. This chapter finally concludes with a summary and policy implications.

Box 5.1 'Shifting of normative development thinking' hypothesis

The notion of *Development* is widely used especially when it comes to the process by which a country produces more products by utilising economic resources in a more efficient manner. In the past, this concept has largely focused on economic dimensions (i.e., economic growth), but today it has become more and more complex so that social, environmental, political, and other dimensions are equally emphasised in the contemporary development discourse.

The ultimate goal of development can be grounded by creating monetary versus non-monetary value. From the instrumental perspective, there could also be a short-term approach to solving the problem at hand versus a long-term approach to target structural transformation for the future. As shown in Fig. 5.1, which is divided into quadrant bases, A and B groups of countries are mostly developing countries that are forced to focus on economic growth, while C and D groups are already developed countries mostly in Western Europe that have already achieved some material development and, as such, are now more likely to set goals for social and environmental development.

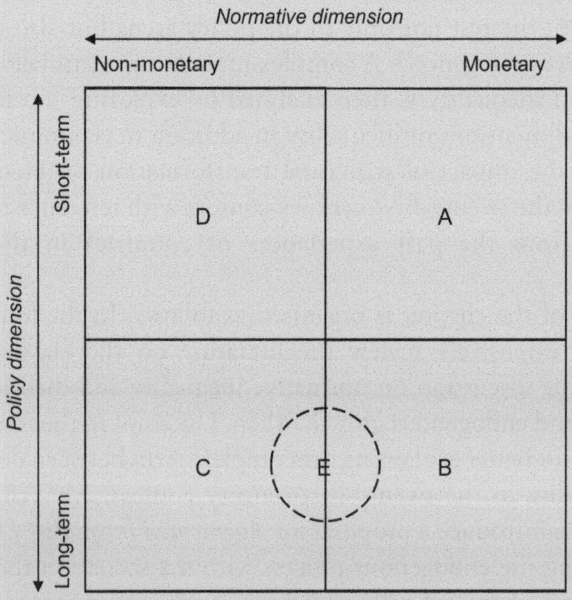

Fig. 5.1 Shifting development ideas. (*Source*: Author's own elaboration)

In more detail, the countries in group A have a desire for economic growth as described earlier. Here, most countries are receiving international development assistance, and donor countries tend to put more money into urgent challenges that need to be addressed immediately. Such aid actions could result in a much clearer and faster development impact for monitoring and evaluation, which is highly favourable for donors. That is why a large portion of development funds tend to be committed to short-term problem solving in many cases. Partly because of this, the countries in group A are exposed to various Western-style modern institutions, and, as a result, they have more decentralised and open decision-making systems, which may possibly slow down the process of designing and implementing a powerful long-term development planning framework.

Like group A, the countries in group B also have a strong desire for economic growth. These countries have relatively more resource endowments, such as natural resources and large populations that can be related to having greater potential for growth. However, largely due to their geopolitical or inherent advantages, they were colonised by the West, often resulting in being exploited by extractive institutions. Such an institutional feature is frequently associated with centralised and past decision-making systems. Besides, they are often not fully democratised and may rely on strong leadership driven by a small elite, and sometimes a dictatorship. As a result, group B is relatively easy in terms of formulating longer-term national planning compared to group A.

On the other hand, the group C countries have already achieved a considerable level of economic development, so their desire for social and environmental development is at a high level. The needs of their citizens are often reflected in national development policies through various institutions and coordination channels. Most of the countries have not, however, reached the level of strong democratisation, and, similar to the group B countries, these countries tend to have a presidential system in a centralised political institution, deeply embedded in society.

Lastly, countries in group D also tend to target non-monetary value creation because of their sufficient level of modernisation like those in group C. However, unlike those of group C, the power distribution is somewhat effective based on a parliamentary cabinet system, so the national development framework, particularly with regard to long-term planning, shall go through some processes of consultation and coordination, in order to reflect the voices of various stakeholders.

While the national development priorities of each group were dispersed in the past, their development focus is likely to converge into a group E.

Countries in groups A and B are frequently observing advanced countries today struggling with the side effects of economic growth such as widening inequality and climate change, forcing the world to rethink its development strategies. As such, the shifting of the development objectives of countries of groups A and B appears to be inevitable. Hence, civil society participates more in the policy decision-making processes in which environmental and social considerations are emphasised. Moreover, development partners are considering more and more the quality of development assistance in order to focus on long-term developmental impacts.

On the other hand, countries in groups C and D these days face some subdued economic growth challenges and, as a result, are now making significant efforts to look for new sources of growth. Such strong needs and efforts are balanced with civil society's voice for environmental development and wellbeing. There are also greater demands for transparency regarding decision-making processes and good governance. Thus, the shifting of policy response from economic to other dimensions becomes imperative, particularly considering the recent radical changes of socio-economic and environment dynamics.

In short, traditional development trajectories through which most advanced countries have passed can be 'from A to D' or 'from B to C'. In the past, both developmental paths have been likely to accomplish their development goals and priorities as targeted. In recent years, however, most developing countries that can currently be regarded as either 'A' or 'B' and that attempt to target their paths either 'from A to D' or 'from B to C' have realised that traditional styles of development strategies would not generate the expected outcomes any longer. Because of this, currently developing countries have started to look for alternative development strategies that might be more appropriate to their present circumstances. In this regard, the notion of structural transformation has recently been recognised as a means to achieve their long-standing development aspirations in a more sustainable and inclusive manner.

Normative Inequality for Development

Inequality as Enabling Versus Constraints to Development

The theoretical basis of investigation on the *Transformation-Inequality* nexus stems from two broad schools of thought. According to Friedman (1953), inequality provides incentives for more effort to gain more. Meanwhile, he further claimed that individuals have alternatives of actions not only with different rewards but also with different risks. Thus, individuals try to get the highest income possible, which implies that inequality in a society is the outcome of the free choice of individuals according to their risk appetite, and, as such, the redistribution practised depends on the preferences of a society. However, redistribution protecting individuals from taking responsibility for losses due to their own choices would lead to behaviour inspired by moral hazard, possibly making people overly carefree. Such practice is therefore to the detriment of people's efficiency and society as a whole.

This line of reasoning for the positive relationship between economic inequality and efficiency can also be associated with justice in that effort and skills translate into rewards. If the distribution of skills and effort in society were not equal, it would be unjust to reward (or treat) everybody equally. This straightforward intuition can be reflected by one famous sentence by John Smith: 'He who does not work, neither shall he eat.' In this sense, the role of the state in ensuring (re)distribution of incomes may need to be limited (Lissowska 2015). To sum up with this reasoning, a society as a whole may be better off with income inequality than with income equality.

At the very general view, I recognise the positive role of income inequality on efficiency (say, inequality as *enabling* or *incentivising*). However, this view overly focuses on the market economy and simple libertarian reasoning in analysing complex societies. In fact, rewards, which can be the basis of the incentive argument, could include not just money (income) but also several other forms (other than monetary value) such as moral value, trust, and good health. Furthermore, determining factors on economic inequality may not just relate to skills or effort but also inherited wealth, which further affects unequal opportunities to access education and social networks (Okun 1975). All these factors can be additional but significant ones that are likely to lower people's incentives for efficiency.

If the inequality as *incentivising* mechanism did not properly function in a society due to these reasons, then the not necessarily positive impact of income inequality on a society becomes a reality; and rather, its negative impact could inevitably dominate. Wilkinson and Pickett (2010) investigated 23 countries and concluded that income inequality may have negatively impacted in various indicators on social development dimensions, including health, education, social mobility, trust, and violence.

Furthermore, the research findings from Berg and Ostry (2013) are consistent with this negative associationship, in that a decrease in income inequality may be hindered from sustainable growth, with the use of the two regional comparative studies (i.e., Asia and Latin America).

Whether the *incentivising* or *constraints* mechanism comes into play is a field of study essential for policymakers, social leaders, politicians, and other important society stakeholders. Some countries or societies are likely to evolve and develop with very high levels of income inequality (e.g., Brazil, Bostwana, and China), while others are somewhat stagnating with very low levels of income inequality (e.g., Japan, Norway, and Sweden). Due to the multidimensional nature of inequality, there is no single consensus on the role of economic inequality. Nonetheless, both negative and positive advocates regarding the role of inequality have provided a significant amount of research evidence to support each of their positions: positive role (e.g., Li and Zou 1998; Forbes 2000; Chen 2003; Voitchovsky 2005) and negative role (e.g., Alesina and Rodrik 1994; Thorbecke and Charumilind 2002; Easterly 2007; Lissowska 2015).

Inequality in the Context of Rawls's Difference Principle

How then could economic inequality stimulate efficiency or impede socio-economic development? In other words, why is inequality considered good in some countries but bad in others? The theory of *Distributive Justice*, pioneered by John Rawls four decades ago greatly answers to these questions.[2] In his book, *A Theory of Justice* (1971), he argued that people not only tend to favour an equality in terms of income and wealth distribution but also tend to permit certain levels of inequality. To put it another way, what his argument implies is that up to a certain level of income inequality, a society would not be damaged by moral, trust, monetary, or other values, which is the so-called Difference Principle.[3]

According to the *Difference Principle*, inequality in a society is acceptable only if such inequality is to the advantage of the least advantaged members of a society. This is in line with the inequality as *incentivising* mechanism and by combining both arguments this mechanism is likely to function up to a situation where the least advantaged members of a society are motivated and incentivised to engage in social and economic activities—as income inequality increases, the rich get richer but share more with the poor. For instance, if a higher salary is paid to a medical doctor (than to a grocery clerk), this salary disparity may help increase access to healthcare service for all (even for the poor such as a grocery clerk). However, if the level of inequality exceeds an acceptable (tolerable) range, a kind of threshold where inequality ceases to become *enabling*, and become *disabling*, the *incentivising* mechanism is not likely to function but is rather likely to be detrimental to socio-economic development.

This principle is, although subject to much criticism particularly Rawls's account of the least advantaged and its lack of representativeness (Weatherford 1983; Schaller 1998), very influential on the moral guidance for societal processes and structures that fundamentally affect different redistribution practices of societal benefits or burdens, which ultimately cause more or less income equality in any given society. In particular, (re)distribution practices (or tax reforms) underpinned by political reaction with regard to the *Difference Principle* may, however, sometimes be limited. Anthony Downs (1957) proposed the idea that politicians are likely to be not too far from voters at the philosophical centre, otherwise they would be outside the so-called *Median Voter* theorem. Simply speaking, political parties of the left may tend to shift their policy stance towards the right and vice versa in order to win an election. Therefore, each party's political stance becomes increasingly similar (converging) so that taxing policies and income (re)distribution practices may not significantly differ among them. This shifting tendency to either the left or the right could minimise the likely impact of the (re)distribution of income.

Overall, inequality can enable or constrain the course of a developmental society, which tends to shape the level of inequality. In turn, the economic inequality level (increases or decreases) influences the societal distribution practice of income and wealth. However, it is important to note that there is a certain limit of inequality that determines whether inequality is functioning as *incentivising* or as *constraints* for the development of a country.

Transformational Process and Inequality

Linking Modernisation Theory to Inequality Dynamics

Further discussions directly linking transformational process with inequality dynamics are worthwhile, in addition to previous normative ones. Such attempts can be traced back to an era in which the theory of modernisation dominated around the 1950s, during which a vast number of development ideas were theorised to generalise development processes where we can derive some implications on inequality patterns. Although various modernisation theories have already been discussed in Chap. 2, four selected theories that have had a substantial impact on development discourse with the particular focus on inequality dynamics are reviewed in turn below—that is, Lewis's *Dual-Sector Model*, Nurkse's theory of *Vicious Circle of Poverty*, *Unbalanced Growth* doctrine, and Kuznets's *Inverted U-Curve*.

In the 1950s, there is little doubt that William Arthur Lewis's *Dual-Sector Model* (1954) substantially led the idea of modernisation. According to the model, agricultural and industrial sectors generally coexist in developing countries. Relatively inexpensive rural manpower begins to be transferred to the industrial sector, which leads to progressive industrial development. As the rural labour force is transmitted to a city that provides better wages, a country as a whole tends to achieve industrial development. The national inequality problem can therefore be alleviated to some extent as wage growth in the industrial sector, which absorbs the excessive labour force, slows down. As the trend of urban migration continues, it will then come to a point when the labour force of the rural area transmitted to the city is exhausted, which is referred to as the *Lewisian Turning Point*. When such a turning point is reached, there is a mismatch between the demand for, and supply of, labour, which eventually leads to a sharp rise in wages. This can therefore be associated with a widening inequality.

The theory of *Vicious Circle of Poverty* is also relevant in this regard, as introduced by Ragnar Nurkse (1953). He attempts to explain, using supply-demand dynamics, why underdeveloped countries cannot keep up with continuous economic development. In terms of the supply side, the least developed countries (LDCs) in general lack the capacity for capital accumulation and productivity, which can be negatively associated with real income dynamics. As a result, lower levels of savings are inevitable, impeding capital formation and ultimately leading to capital shortages. On the other hand, purchasing power may not be sufficient due to lower real income, which can make such markets unattractive from a firm's point of view. Since firms are less motivated to make additional investments in this market, this leads to lower levels of investment demand and ultimately lower incomes.

This theory does not, however, seem to directly address inequality within a country. Nevertheless, considering the perspective of inequality at the national level, we may get the following implications. Poor people have relatively little money to save, so there is little room for education to improve productivity. Real income is therefore also low. It can also be true that there are few incentives for customised welfare benefits for the poor who have a relatively lower rate of socio-political participation. In the end, the gap between the poor and the rich continues will increase or stay at a certain level. Although this analogy may be a little exaggerated and overly simplistic, many studies have indeed argued that welfare programmes for the poor such as those introducing equal access to education and raising minimum wages may be effective in solving the inequality challenges (De Gregorio and Lee 2002).

Next, it is worth exploring the *Unbalanced Growth* doctrine—for example, Albert Hirschman (1958), Walt W. Rostow (1960), Marcus Fleming (1955). This school of thought stresses the need to invest in strategic sectors of the economy rather than simultaneously investing in all sectors. This is because sectors other than strategic ones would automatically be developing themselves through, namely, the *Linkages* effect. In short, when a certain sector is first developed, other sectors try to catch up, which could lead to positive competition between these industrial sectors, resulting in *Unbalanced Growth*.

Among the various theoretical models of *Unbalanced Growth*, Rostow's *Take-Off* theory (1960) is reviewed here. Walt W. Rostow clarified the developmental process for the modernisation of the country in a clear, step-by-step manner. This model concerns a country that passes through a predefined set of five stages as follows.

At the initial stage, a society signifies a primitive society having no access to modern technology. As such, the society is driven by a highly intensive labour force, mostly in agriculture and other primary industries. Meanwhile, a significant portion of the country's wealth is likely to be invested in non-productive activities such as religion or military (*Traditional Society*).

A development process begins with some innovative activities by the rich, which is likely to boost economic productivity. This rising trend of productivity is supported by income shifting from a feudal society to a productive society. Also, some external forces or demands for primary goods or raw materials can initiate structural change to a country's economy. Further investment in infrastructure and technology is then made for the country to move up to the next stage (*Preconditions for Take-Off*).

Rapid growth is generated by one or two manufacturing sectors where most workers are employed. This stage is further characterised by an increase in urbanisation, industrialisation, and technological breakthrough that is significantly supported by continuous investment which comes with changes in income (re)distribution. A substantial portion of income is entrusted with a capitalist who can re-invest it to increase the rate of capital formation, which could further be promoted by government policies (fiscal measures) and institutions (banking institutions and capital markets) (*Take-off*).

Modern technology is applied widely across almost all industries and sectors, including entrepreneurial development so that an economy can start to diversify into new innovative and more productive areas, so that the country can eventually be self-sustained for future growth. This self-sustainability is accompanied by intensified industrial development (shift from heavy engineering to machine tool and electrical equipment) with further capital formation (*Drive to Maturity*).

An industrial base dominates the whole economy with some transition from industrial production to consumer goods and services such as an increase in real incomes, which could lead a society to be concerned with normative consumption practices (by the people) regarding high-value goods and services. At this stage, a society pays more attention to social welfare and security than on economic value. An increasing trend in income inequality may thus be relaxed (*Age of High Mass Consumption*).

With regard to the inequality dimension, first, in the *Traditional Society*, the agricultural sector is at the centre where the core of political power is generally dominated by those classes who own or control land. As such, there may be a high level of inequality throughout the country, but there is no great social resistance. These inequality dynamics do not change much in the *Preconditions for Take-Off* stage. But when society reaches the *Take-Off* stage, this could change. On the basis of technological development, the economic fundamentals are shifted from the agricultural to the industrial sector. As we have already analysed in the *Dual-Sector Model*, the challenges of inequality can be mitigated slightly as labour supply migrates to the industrial sector. Thereafter, when entering the *Drive to Maturity* phase, modern technology becomes widespread throughout the industries. Such influential technological development could cause capital to be concentrated in technology-intensive industries in which simple labour tasks tend to be replaced by technology automation and even skilled labour can also be substituted if it fails to adapt quickly to new technologies. The final stage, the *Age of High Mass Consumption*, represents the shift from technology-labour-intensive industries, which have so far led the national development drives, to technology-service-centric industries so that the inequality dynamics of the previous stage may further be intensified or maintained at a certain level.

Finally, Kuznets curve (1955, 1973), which is the most direct and clear analysis of the relationship between inequality and economic growth, needs to be discussed. Kuznets's hypothesis was that there is an inverse relationship between inequality and economic transformational process, which became known as the *Inverted U-Curve*. In particular, Kuznets argued that structural transformation is inevitably associated with and is a pervasive phenomenon (one of the six stylised facts) of modern economic development.

Kuznets's prime explanation can be described as follows. At very low levels of income, inequality must also be low, particularly in rural areas while a little bit higher in urban ones, whereby most live on a fairly even amount of subsistence. As the process of growth begins (as an economy industrialises), the path to economic inequality is driven by changes in sectoral structure. Income inequality increases as people migrate from a relatively equal low-income rural sector towards a relatively unequal urban sector which offers higher wages. This migration is in line with the transition from a traditional agricultural sector to a modern industrial one. Once the majority of the labour force is found in an industry or service sector in an urban area, further transfers result in a decrease in income inequality as the rural-urban divide becomes less influential. Therefore, the average per capita income increases further and income inequality within the urban sector remains.

Each of the various modernisation theories discussed so far have attempted to inherit the developmental discourse through their own assumptions and approaches. Nonetheless, these influential theories of modernisation have clear both divergence and convergence areas. The Kuznets process emphasises the importance of economic growth in generating demographic transitions that affects changes in income (re)distribution and inequality, while the other models focus on the role of technological advances in some leading industries and sectors as an essential driver for other sectors and eventually for a country as a whole. Moreover, Kuznets described the ups and downs of income inequality as being inevitable for the development process, while the others did not regard it as 'inevitable' *per se* but rather put a different emphasis on income (re)distribution practice as one of the 'conditions' necessary for a society to move up to the next stage and especially Nurkse's theory warned that inequality level continuously becomes aggravated in the developing economies context.

On the other hand, there is also a significant area of convergence among these theories. They are likely to argue that almost all societies naturally pass through certain steps (or stages) of transformation, starting from traditional economies that focus on subsistence farming to modern and advanced ones through industrialisation. In fact, this central idea of process is common in other growth theories. Hence, inequality dynamics are evolving "Ups and Downs" in the transformation from agricultural sector to the high productivity industrial sectors such as manufacturing, chemical, services, and high-tech.

With some optimistic views on the *Transformation-Inequality* nexus in mind, policymakers (or politicians) in middle- or low-income countries since the start of their long-term processes of transformation have often subscribed to the notion that "Rising income inequality is inevitable during economic transformation" in order to justify their economic policies that could inevitably widen the levels of inequality in the short term. In addition, many further emphasise to the public the belief that their proposed transformation policies would eventually achieve an egalitarian form of income (re)distribution practice in the long term (Korzeniewicz and Moran 2005).

Limitations of the Classical Account on Transformation-Inequality Nexus

Notwithstanding their influential explanations of a country's modernisation (transformation) process, the theories remain the subject of controversy when it comes to the concept of inequality in the developing countries context. There are three broad interconnected contexts to be reconsidered.

First, a majority of these theories and models are rather *Eurocentric*. Their exclusive focus on Western society tends to exclude the relevance of local or regional configurations of society and inequality, particularly in the global South where most poor nations are located. The results from thesis transformation frameworks were mainly derived from the historical geography of the global West. As their processes are a mechanism for explaining a long-term transformation process, most of the past literature may have had to focus on rich countries that have already accumulated a significant amount of historical records that can provide enough analytical evidence on their development trajectories (Itagaki 1963): the UK (Polak and Williamson 1991), Germany (Dumke 1991), Sweden (Soderberg 1991), Australia (Thomas 1991) and the US (Ram 1991).[4]

Due to the exclusive focus on the West, these theories may not therefore be effectively applied to the context of the developing world. Bah (2011) found that many developing countries are following processes that are very different from the path of structural transformation experienced by most developed countries in the past. In Africa, the continent has been yet to structurally transform largely due to the fact that the majority of economies therein are still agricultural based. In particular, industrial sectors have been very slow to grow while transfers of labour and other resources have mainly taken place between agricultural and service sectors (thus somewhat leap-frogged a stage from a traditional society to the age of high mass consumption), whose share in GDP is high at the relatively lower income level, which is not the case for some Asian countries who followed the path of the Western model. In sum, the two theories may have framed the processes without seriously taking into consideration any distinctive regional or national characteristics of inequality.

Second, these theories do not seriously take into account initial socio-economic condition, which could have considerable implications on the transformational process and inequality dynamics. Their assumption that all countries automatically start at the same stage and with the same capability to progress further through the stages of economic transformation may not be agreeable. This is because some countries could start their development, for instance, from the second stage of Rostow's theory, thanks to an abundance of natural resources, while others start from the bottom. Moreover, some countries neighbour others that have a high productivity of labour that can easily be transferrable, while the others are geographically disadvantaged, thus significantly limiting access to this type of labour force. In this regard, most modernisation theories may not properly reflect the inequality evolution, as embedded in the transformation process.

As a result, empirical evidence on the Kuznets process has shown mixed results, which further depend significantly on data (choice of variables) and methodological approaches (Herzer and Vollmer 2012). An *Inverted U-Curve* relationship largely depends on other characteristics (such as political and social institutions and cultural heritage) rather than economic perspective (income level) only (Kaelble and Thomas 1991). Without serious considerations of the other factors embedded in a particular society, one cannot determine accurately what level of inequality is accrued from any particular amount of structural transformation (Temple and Ying 2014).

To be more specific, Kuznets's argument was largely favourable to Latin American countries, which were at an intermediate stage of development, thus questioning its validity for different groups of countries (Milanovic 2000). Latin America is a region with historically high levels of inequality due to its colonial past, which was likely to have generated extreme inequalities of wealth, capital, and political power (Engerman and Sokoloff 2002). In this regard, these theories did not take into consideration the fact that countries have different histories and therefore distinctive institutional features, thus structuring different configurations of inequality.

In the twenty-first century, capitalism followed by institutions based on the Western model attempted somewhat to standardise institutions across the world. However, by doing so, they could still never erase history. As a result, the question of validity and various factors affecting inequality dynamics in a certain country or region started to arise (e.g., extractive institutions due to European colonisation in the case of Latin America). Their theoretical and empirical experiences of emerging and developed countries point to potential trade-offs between growth and inequality, while some countries have managed to reduce poverty and inequality through structurally transforming.

Next, both processes rarely consider human capital aspect, which is one of the main approaches used to explain different individuals' salary profiles over time (Mincer 1974). This human capital idea takes into consideration individual educational choices based on a constrained maximisation process in that the return on investing in education is greater than any alternative financial investment. Therefore, the earning profile of a person critically depends on the amount they have invested in education that essentially determines their wages. In this regard, this rational choice regarding education by individuals matters significantly, which includes such decisions made by house-workers, students, pensioners, and informal workers who are largely absent from their frameworks of modernisation process. Rather, they are broadly based on a relationship between capitalists and workers, and major and minor industrial sectors. Additionally, individual rational choices for investing in education can substantially be affected by the characteristics of the individual (e.g., family background, birthplace, and inherited wealth), which often go beyond an individual's mere abilities (Guidetti 2014). For instance, unequal opportunities for schooling may hinder a country's transformation process by lowering people's incentives for efficiency and changing the baseline scenario for the distribution of resources (Brunori et al. 2013).

Technology in Growth Theory

Through exploring the key development theories, we may be able to better understand the development challenges and opportunities in the context of inequality dynamics for developing countries. However, a particular attempt needs to further be made in mainstreaming a technology factor into the *Transformation-Inequality* nexus.

As a representative of neo-classical growth theory, Robert M. Solow's theory (1956) distinguishes the effects of changes in labour, capital, and technology from the economic growth process. His empirical results show that PCGDP in the US doubled from the early 1900s to the 1950s, and that the effect of capital on such growth was only 12.5 per cent, while the remaining 87.5 per cent was derived from technological advances. However, the most controversial issue in his theory can be associated with the productivity improvement from technological advances that are partly contributing to growth; however, such technological advances are assumed to be exogenous variables that are not closely related to economic factors. His model assumes technological development to be an exogenous variable, which inevitably leads to the conclusion that long-term equilibrium growth shall also exogenously be determined outside the economic system.

In reconstructing the neo-classical growth view on exogeneity, theories of Arrow (1962) and Romer (1990) can be most influential in this regard. The former (Arrow) emphasises that, unlike products that cannot generally be used with other people (i.e., rivalry), knowledge does not compete with that of others (i.e., non-rival). However, the fact that knowledge is regarded as such does not mean that everyone can use it. Some knowledge can be excludable (i.e., excludability), depending on the legal and institutional apparatuses. Therefore, most of the knowledge accumulated by R&D is highly likely to be excludable, thus tending to ensure high returns. Knowledge accumulation is also done in everyday life through the *Learning by Doing* mechanism, and thus technological advances based on these learning effects are caused by internal dynamic factors within the economy.

Similarly, the latter theory (Romer) attempts to demonstrate that there is a mechanism to achieve technological progress within the economy without an exogenous technological impact. This is especially so, as it sees productivity enhancement as the most critical determinant of economic growth rather than capital accumulation. He argues that endogenous technological progress can be achieved in an economy through investment in technology, knowledge accumulation, and technology diffusion. Therefore, this theory emphasises the role of R&D as an essential economic activity that can bring technological progress. It further asserts that in countries where the level of technology advancement is poor, technology adoption or spillover from the outside can also be a major driver for growth. But it would be impossible for these countries to connect such technology to their economic growth without the underlying human capital. Human capital shall be accumulated so that existing knowledge can effectively be utilised properly. It can further create new knowledge and innovation based on existing knowledge, *enabling* productivity improvement, which is a clear stimulus for growth.

These two endogenous growth theories that were designed to overcome the limitations of the neo-classical view on economic growth also have their own limitations. For instance, although endogenous growth theory emphasises the importance of knowledge and human capital, in reality, it is difficult for low-income countries to accumulate knowledge and human capital. Thus, as Nurkse points out, underdeveloped countries experience various difficulties in escaping from the poverty circle, and the income gap between low- and high-income countries is widening. For this reason, it is increasingly difficult for LDCs to catch up with advanced ones through the accumulation of knowledge and the improvement of human capital.

Nonetheless, during the past several decades, we have observed that the poorest economies have jumped up to the rank of advanced countries, especially with regard to the four Asian Tigers (i.e., Hong Kong, Singapore, South Korea, and Taiwan) (Sachs and Warner 1997; Parente 2001). In addition, endogenous growth theory assumes several key factors such as endogenous mechanisms of human capital improvement and of technological innovation, and even the relationship between them is also limited by multiple assumptions. Due to this, empirical grounds may be hard to prove.

Towards Better Accounts

As emphasised previously, it is essential to internalise Rawls's normative theory to understand inequality dynamics properly in terms of the *enabling* and *constraints* discourse. We have also explored various development theories in an attempt to integrate them into inequality dynamics. In so doing, it shall be noted that they could explain to some extent the contemporary inequality dynamics, but more critically contain a series of limitations. We have further examined the importance of technology and human capital as endogenous variables by looking at the two most influential endogenous growth theories.

Let us take a closer look at the three key points of integrating key arguments that can be valid for the twenty-first-century context of inequality dynamics. First, as emphasised in Rawls's account, multidimensional concepts of inequality, including moral, political, and social characteristics, shall be discussed first in an attempt at such integration.

Understanding inequality dynamics in a society is imperative because it can have a significant influence on whether a country's developmental aspirations are to be achieved or not. But it would not be possible without an effective consideration of Rawls's account. In other words, the dual roles of inequality (as *incentivising* and/or *constraints*) can significantly affect the ease of states' transformation among developmental actors." In this regard, there might be little point in comparing the inequality levels between countries since each country has its own moral and justifiable degree of inequality. When the level of inequality exceeds a tolerable level (according the *Difference Principle*), which is naturally set by a society and should be different from one society to another, inequality as *constraints* mechanism rather than inequality as *incentivising* comes into play. On the other hand, when the level of inequality is somewhere within an acceptable range by the majority of the public, both mechanisms interact with each other, along with other socio-political dimensions, to guide a country's developmental path and its potential capacity to do so.

Synthetically, the accumulated experience of being incentivised and restrained by inequality configuration in a society shapes a country's own inequality dynamics. Therefore, once inequality dynamics are formed as historically driven systems of social, economic, and political relations that frame the regulation and coordination mechanism that governs a society, the dynamics can be so durable in a way that they structurally transform themselves (depending on past interacting trajectories and decisions made).[5]

My aforementioned points of criticism cannot however invalidate their inspired analysis of the long-term modernisation and economic growth process. Rather, the criticisms should be used not just to draw some insights from this developmental process, but also to recast, without ruling out the relevance of critically the reviewed theories and models, some original arguments by shifting the analytical focus from 'on transformation' to 'on income inequality', and particularly on the inequality dynamics within a society.

Second, the most insightful learning point across the review of modernisation and endogenous growth theories is associated with the essential role of technology advancements. Without this consideration, we may miss the essence of this inequality dynamics of the day where I firmly believe that information and technology is playing as a cross-cutting catalyst of change for developmental processes and as an accelerator for the contemporary inequality dynamics.

We are now living in a society in which information (knowledge) can spread within minutes all across the world. Thus, diffusion of information (knowledge) has firmly entered the arena of policymaking. This diffusion can help an individual country prepare its strategy or policy that is relevant to its present socio-economic conditions so as to avoid repeating past mistakes made by other (advanced) countries. For instance, some Asian countries such as South Korea and Taiwan have followed similar patterns that were contrary to the Kuznets curve. In other words, a number of East Asian countries have experienced rapid economic transformation (substantially through the export sectors development, which benefitted from globalisation) without a significant increase in inequality, which is often described as the so-called Growth with Equity. Stiglitz (1996) explained this growth pattern in the equity phenomenon by referring to the *Re-investment* mechanism applied, in that these countries immediately re-invested initial benefits from rapid growth into land reform, universal education, and equally distributed industrial policies, all of which in turn support high rates of growth to the benefit of society as a whole.[6]

Besides, Birdsall et al. (1997) explained that productivity and income in rural areas were rising much faster than expected, thereby deviating from the conventional process of demographic transition between rural and urban areas. This is largely due to knowledge or technology shared radically across countries, cities, and towns. For instance, information and communications technology (ICT) has substantially contributed to modern farming methods. Farmers can now benefit from using technology for wireless local area networking around fields to track crop production, smart management, standardisation management, and sales through e-commerce. Farmers have even started to spray pesticides through *Drone* technology, which makes it possible to identify seeds and pests for crop growth in a very efficient manner. In this regard, it should be noted that technological innovations have dramatically improved agricultural productivity, leading to the development of agricultural industries and the boosting of the economic capacity of farming households. Such technology and its radical diffusions imply that the inequality dynamics (*Up* and *Down*) through the sectoral shift, emphasised in modernisation theory, may no longer be valid for modern society in the twenty-first century.

As already discussed in my argument regarding the exclusive focus on the West, some African countries have experienced sectoral shifts only favourable to the services sectors, while the value-added contribution of industry, particularly manufacturing, to GDP has declined since 2001, which is sometimes described as *De-industrialisation* (Armah and Baek 2015), and may contradict those of modernisation process. Even the structural transformation process of skipping manufacturing development and going directly into the services sectors undertaken by most Caribbean countries can refute them further. Such a non-traditional modernisation pathway may be attributable to the knowledge transfers. Many developing countries appear to have adopted policy mechanisms that were developed and used by governments in middle-income countries, according to Palma (2011). This similarity (what it calls, *Knowledge Sharing* and *Lessons Learned* mechanism) often deviates from the traditional process, which is indeed consistent with Arrow (1962)'s central argument on the role of knowledge. Even still, such a knowledge sharing mechanism may help reduce the process of economic transformation with some countries opting to leap-frog a stage, possibly benefitting from technology transfer (say, green technology). In this case, the pattern (movement between the stages) of development by a country would be very different from the conventional pathway, which would further alter the inequality evolution.

Third, in line with the second point, such diffusion dynamics are deeply associated with a notion of globalisation. In other words, discussion on inequality and development in present day may not be useful if we are largely focusing on internal/domestic factors, not external ones. Nowadays, for instance, one of the most serious challenges in Europe is the issue of migration to the West, which significantly pushes or even creates transnational structures for greater inequality (Guidetti 2014). External factors created by globalisation are vital with regard to inequality challenges (Mills 2009). In general, globalisation is underpinned by a worldwide evolution of increases in human capital mobility, consumer goods, and services. Giddens (1991) pointed out that globalisation can intensify societal relations, which link distant localities so that some economic events in rural areas can significantly be influenced by events that occur in other areas or even in other countries. This implies reciprocal influences of certain tendencies and simultaneously a localisation or urbanisation process, which could deviate from the traditional pathway proposed by previously discussed eminent scholars. In this regard, no analytical view on inequality would be intelligible without taking into consideration the impact of globalisation and its impact to knowledge and information.

A new international division of labour derived by globalisation could reject traditional arguments on the transition of market and labour. In other words, this new division of labour can be profoundly affected by certain external forces and internal demands for outsourcing and offshoring activities that favour international trade and technology-based manufacturing businesses. For instance, if multinational companies in advanced countries keep highly skilled and innovative practices (from the introduction of new technologies) with few high-income workers, which implies a reduction of jobs and wages for low-skilled workers, factory workers' jobs can be outsourced (or relatively easily replaced with technology-related jobs) with lower-salary workers in developing countries. On the other hand, this outsourcing practice for developing countries could be rather positive since such a practice can provide more employment opportunities and narrow the gap between highly skilled and low-skilled wage earners while favouring the process of industrialisation (Mills 2009).

Recent perspectives by the new structural economics school of thought provide useful insights in this debate (Lin 2012). Motivated by long-standing and continuous quest for how economic growth is to be sustainable (in response to the failure of structural policies for stimulating growth in a sustainable manner), the new structural economists have attempted to explore the nature of economic growth and the determinants of economic structure and its evolution in the process of economic transformation. They particularly focused on exploring the factors of distinguishing the Western-driven economic development in the long time ago from the Asian development pathways, largely by China, India, South Korea, Singapore, and Taiwan, realised relatively in more recent periods where these Asian countries were not following the dominant neo-liberal market approach. From extensive empirical investigations, the new-structuralists concluded that the economic structure to promote industrial development is the most critical for economic growth in a way that factor endowments determine such structure. While factor endowments are given in the short term, they can be altered in the long term through upgrading and technological innovation.

This school of thought's theoretical foundation of analysis stemmed from taking advantage of: old structural approach to explain countries' structural differences caused by market failures; and structural market-based approach to economic development nested in neo-classical doctrine. Attempting to bring together best from these two development thinkings, the new-structuralists introduced new term, *New Structural Economics*. Their main stance is that they would agree with a pivotal role of the market mechanism for resource allocation and in facilitating upgrading of factor endowments, but at the same time active role of governments is more important. It is because government is essential in 'coordinating investments for industrial upgrading, technology advancements and diversification and in compensating for externalities generated by first movers in the dynamic growth process' (Lin 2012). It is also necessary for governments to compensate upgrading inheres with large externalities to corporate's transaction costs.

Proposal for the Augmented Inequality Dynamics

In an attempt to build new framework of inequality dynamics in the twenty-first century, our discussion on inequality dynamics should start with the basic idea that the degree of past inequality itself significantly affects the current level. This is somewhat related to previous arguments on human capital aspect (Mincer 1974; Brunori et al. 2013; Guidetti 2014). If we can recall our review on the models of modernisation and economic growth, income inequality is very low at the very beginning of the premodern stage. With a very low level of income inequality, people would have no reason to complain. Rather, people would tend to perceive their society as *Fair*, gradually making them believe that they are living in a society that is not unjust. As a result of this perceived *Fairness* accumulated, people ultimately become to believe that individual skills and effort are fairly rewarded and that they all have the right to enjoy the fruits thereof (Alesina and Angeletos 2005). In other words, people tend to believe that economic inequality and poverty is due to an irrational choice or lack of effort in a society where inequality functions as *incentivising* people to put in more effort in order to receive the benefits therefrom. As a result, more creative and innovative activities therefore come into play.

This debate is deeply related to the psychosocial theory of the effects of inequality on social cohesion. According to this psychosocial perspective, negative consequences such as a lower level of social trust, higher crime rates, poor levels of health services, and a lower rate of democratic participation tend to severely undermine social cohesion, especially in a society with a high level of inequality (Elgar and Aitken 2011). Psychological consequences of inequality have also been presented by Neckerman and Torche (2007) who argue that in a society with higher levels of inequality, the relative deprivation of low-income individuals can be severe. Moreover, Wilkinson and Pickett (2010) explain that gaps in social status are observed more in societies with higher inequality and, as a result, emotions of *Stress* and *Unrelatedness* arise. People also want to spend time with people of a similar social status (McPherson et al. 2001). The more socially stratified a society is, the further people from different classes are kept from one another, and, as a result, the degree of interaction between the social groups is lower. In particular, this may cause the poor to think that they are helpless and could hinder their civic participation. The poor may thus believe that their opinions are not heard and reflected in the political system at all (Uslaner and Brown 2005).

Politics also react to this phenomenon (mainly to obtain more votes; *Median Voter* theorem, discussed in the previous section) by trying to sustain the *incentivising* mechanism thus ensuring the limited effect of the (re)distribution of incomes in a society. In other words, *Inequality-Growth* nexus would tend to depend on the ability of the poor to vote for redistributive policies (Alesina and Rodrik 1994). Otherwise, a society would be considered *Unjust* as pointed out by Friedman (1953), who implies that politics choose low-redistribution practices and low-taxing policies (Alesina and Angeletos 2005). This is typically an ideology favourable to republican in the US (or conservatives in the UK). Such social consciousness forces the public to tend to accept lowering taxes, regardless of their social status and income levels, which is likely to go against the narrowing inequality levels in a society in a situation where the Difference Principle is applied. Alesina and Rodrik (1994) argue that the negative impacts of inequality on growth appear through the political dimension. They focus on how the government and politics react and cope with the amplification of social conflicts due to socio-economic inequalities, and on whether the widening of socio-economic inequality can be resolved, which largely depends on how the national political system responds (see Box 5.2).

Box 5.2 Inequality is all about politics and human capital, not market economy

In the inequality research, understanding the arguments by Thomas Piketty (2013) is imperative. Although Piketty uses economic inequality as a theme of research, his main focus is not on inequality itself, but the polarisation of the degree of inequality. This is the reason why his discussion is on the basis of the decimal multiplier, which shows how much income and capital are concentrated in the highest hierarchy, instead of the Gini coefficient that shows the overall degree of income inequality in the country.

Piketty compares the rate of return on capital with the rate of economic growth. His analysis of data spanning 200 years shows that the rate of return on capital has largely been at around 4–6 per cent, while the rate of economic growth has been only 1–2 per cent except some special cases. The top ten richest in terms of deposits, securities, and real estate actually own most of the country's capital, while people that are ranked in the bottom half are unlikely to have any capital or are even in debt. The middle class, between the top ten and the bottom half, also lack capital other than their homes.

More importantly, the top ten do not just earn capital gains, but they are more likely to find better jobs and thus earn higher incomes. Therefore, their capital income is more likely to be saved than consumed. That is why the vicious circle of polarisation could have a significant impact on economic growth. The rich are likely to save without spending even if their income doubles, while the poor are more likely to spend for consumption. Therefore, when the same money is put into the pockets of the poor rather than the rich, the market contribution would be greater. Furthermore, when economic polarisation becomes severe, society tends to question the legitimacy of the state, decreasing the feeling of *Belonging* and eventually making social integration difficult to achieve.

To solve these social problems, Piketty argues that the capital to income ratio should be reduced, either to 'increase income' or 'reduce capital'. First of all, an increase in income means economic growth; however, today many countries are suffering from having an ageing population, which leads to weakening the driver for increasing productivity. The remaining option to lower the capital to income ratio is either that the state to 'collect taxes on capital income' or 'tax on capital itself'. A typical example is when Franklin Roosevelt, president of the US, introduced a highly progressive tax rate of nearly 80 per cent for the highest income bracket, which only very few people fell under. The tax collected by such radical reforms was invested in public spending and as a result strengthened social welfare, and effectively contributed to promoting the working class to the middle class, which ultimately reduced inequality.

Piketty attributes this successful case of Roosevelt not to the self-regulating system of the capitalist market economy but to the equality created by non-economic factors, especially political dimensions. In other words, he emphasises that polarisation is not a question of the economy but of politics. He explains that politics is not merely a question of justice but a matter of the consensus of many. The power to derive consensus is not about the number of people or their physical strength but their power to negotiate and persuade. The former power comes from the power one has while the latter comes from the power of knowing. Workers who have only a labour force which can be replaced by machines or computer tools at any time are unable to exercise their power of negotiation no matter how many there are. Therefore, in order to change this, more need to acquire a high-quality education to acquire advanced knowledge and skills, while, according to Piketty, persuasive power can also be educated. The power of persuasion that drives voluntary concessions from the top ten, that is, attitudes towards democracy and fairness and the ability to draw consensus around community values are things that should be educated, and not considered as things that naturally occur.

Additionally, relatively big enterprises that benefit most from low tax rates could save proportionally more of their earnings than small- and medium-sized enterprises, according to post-Keynesians (Kaldor 1956). Such savings could accelerate more investment that is beneficial to economic transformation overall (Lissowska 2015). This process can be interpreted as favourable for the overall value of savings and eventually for society and its transformation as a whole. However, due to some externally negative impacts such as the pessimistic outlook for the global economy, the trend of outsourcing practices and others discussed previously, it is likely that big enterprises will retain vast amounts of working capital for short- and mid-term operations and giant investments and decide not to invest. Such decision-making would result in greater income inequality.

The role (size) of the government can also be limited partly due to the downsizing of the role of income (re)distribution between profits and wages, which further influences the limited redistribution practices. A society may experience soaring inequality because subsidies or other forms of social support from the government to the poor are restrained, while the rich save more than they invest, eventually accumulating more wealth. Even the salary gap of workers between large-sized firms and small- and medium-sized firms becomes larger thereby burdening the poor mainly because of limited redistribution practices. This is the moment when the poor start to complain about income inequality and feel a sense of discontent about such circumstances.

Furthermore, in the case of being severely affected by external forces, this taxing practice would not effectively lead to private sector development, which is required for economic transformation process. As a result, more discontent in social system prevails, followed by greater Unfairness perceived by the public. Eventually, the political system will suffer significantly from very strong pressure to distribute income more evenly, especially to the poorer segments of society. On the other hand, if the investment conditions are favourable to big enterprises and the economic structures that are underpinned by effective institutions, a society would experience the *Trickle-Down* effect of capital (Aghion and Bolton 1997) so that small- and medium-sized enterprises gain more profits, which would then be shared with their employees.

In cases where the level of inequality reaches or even exceeds a tolerable range, a society is likely to transform into believing that their negative situation is largely due to luck, birth, and a lack of connections rather than skills and effort (Alesina and Angeletos 2005). From this moment on, the *incentivising* mechanism rarely works, and a society may be considered Unjust and Unfair, which can incur society's mistrust in politics, and particularly in the rich. As the perception of unfairness in a society grows, there is a trend that liberals (or the democrat party in the US) would be more likely to come to power whose preferable doctrine is to raise tax rates for redistribution practice. On the other hand, this social consciousness may limit not just the *incentivising* mechanism but also the creative and innovative activities and the overall benefits from savings and investment activities. For instance, when a society relies more on human capital (mostly in cases of countries with an advanced stage of development), the poor who feel discontent and distrust due to high levels of economic inequality may hesitate in investing in education and other socio-economic engagements, which may significantly hinder transformational process (Lissowska 2015).

So far, I have attempted to explain and systematise the endogenous process within a society, underpinned by social, economic, political, and moral dimensions of inequality (see Fig. 5.2). In other words, inequality dynamics are very interconnected with the multidimensions of society's development (Figueroa 2017). This proposal can help explain how inequality can incentivise or restrain the process of various societal interactions by itself going Up and Down. These Ups and Downs were partially described in the Kuznets process but not in detail (only it Ups in

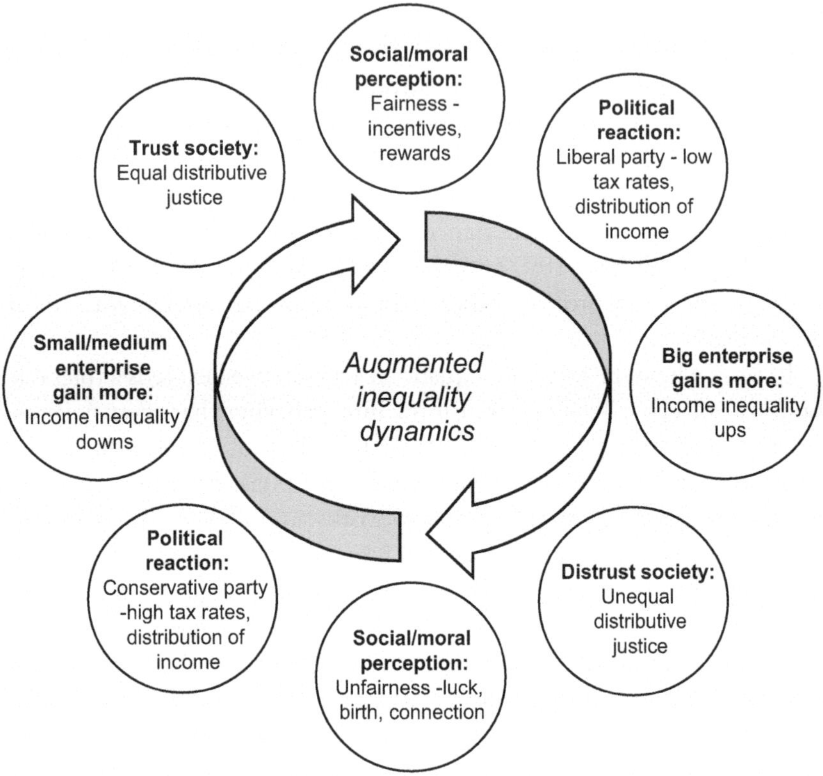

Fig. 5.2 Conceptual framework for augmented inequality dynamics. (*Source*: Author's own elaboration)

the initial stage and then the Downs in the later stage of development, mainly due to the demographic transition). Unlike the Kuznets process, the centre of my argument is that income inequality is not only a necessary but rather an essential element to be linked with other dimensions by increasing and decreasing continuously. In addition to conceptualising the endogenous process itself, these inequality dynamics can have further important implications that can explain the reasons for the differences in levels of inequality and transformational process across countries. In other words, some deviation from any part of the inequality dynamics in this internal mechanism would cause a different inequality dynamics process, which can be the basis of explaining a country's particular situation and comparing it with others. Therefore, the inequality dynamics can be so evolutionary in a way that they structurally transform themselves by interacting with various dimensions and institutions to shape their own pathway depending on both past interacting patterns and strategic decisions made. This is what it calls Augmented Inequality Dynamics.

In sum, each factor functioning in the endogenous process within the societal system plays a pivotal role in determining the degree of inequality and level of economic transformation. The factors further determine the pathway of transformation in a country, whether its economy grows with high or low inequality or stagnates with high or low inequality—growing economies with low levels of income inequality are ideal but rarely observed in the real world. This is why a large amount of empirical evidence has appeared to often reject the Kuznets curve and other models or the myth that advanced countries are likely to have lower levels of income inequality while developing countries higher levels of income inequality. It is therefore imperative to systematise how the Augmented Inequality Dynamics consistently entail the social, political, moral, and economic dimensions of a society, thereby shaping its own inequality dynamics (see Fig. 5.2).

Concluding Remarks

The inequality mechanism as both *enabling* and *constraints* exists in society. As for the *enabling* mechanism, economic inequality can provide incentives for more effort to gain more. Thus, individuals would try to get the highest income possible, which implies that inequality in a society is an outcome of the free choice of individuals according to their risk appetite. If the inequality as *incentivising* mechanism were not properly functioning, then its positive impact to society may be called into question. Rather, it would more likely act as *constraints* to developmental activities. However, it should be noted there is a certain limit of inequality that determines whether it is functioning as *incentivising* or *constraints* for the development of a country, which is the aforementioned *Difference Principle.*

Furthermore, it is found that most of the influential development theories in the contemporary social sciences may not properly be able to explain inequality dynamics in the twenty-first century, given the multidimensional characteristics with which inequality is inherently concerned.

From my critical review and linking development theories and normative theories of inequality dynamics, the dual roles of inequality can significantly affect the determination of states' structurally transforming. As time goes by, experiences of the dual roles of inequality and societal interactions for inequality, Up and Down are accumulated in a society, which shape its own inequality dynamics. In other words, once the inequality dynamics are formed as historically driven systems of social, economic, and political relations that frame the regulation and coordination mechanism that governs a society, the dynamics can be so evolutionary in a way that they structurally transform themselves by interacting with various dimensions and institutions to shape their own pathway.

With the new concept of the Augmented Inequality Dynamics, policymakers should try to maintain a manageable and acceptable level of inequality by taking seriously into consideration the idea of the *Difference Principle* in a more analytical way while pursuing inclusive growth and transformation. This sustainable policy would not only minimise the detrimental impact of inequality as *constraints* but also maximise the favourable utilisation of inequality as *enabling* more productive and inclusive processes—for example, growth with equity through re-investing institutional reforms (Perotti 1996), human capital (Galor and Zeira 1993), social and environmental dimensions of a society. This is because a society can benefit from economic growth, development, industrialization, and structural transformation in a way that people's lives become more convenient, productive, and efficient, supported in particular by tremendous technological advancement and diffusion of information and knowledge (Chang and Baek 2010). In this hypothetical world, the *incentivising* role would be the predominant one over any other role of inequality in the sense that economic capital does not only enable people to buy something, it also gives them a greater sense of prestige that is socially most valuable than capital for a leading position in society.

Finally, there are several areas of future research, which can extend or improve this study. First, the augmented inequality dynamics can be empirically grounded and nationally or locally sensitive. Furthermore, it would be extremely valuable to measure empirically the level or range of tolerating income inequality (in line with the Difference Principle) in an individual country, in order to determine exactly what level of income inequality is at any particular level of economic transformation and to inform policymakers how to manage this in a sustainable manner. On the theoretical side, a more comprehensive study to identify broad areas of convergence for the inequality and growth dynamics will be of great value: more recent endogenous growth theories and new structural economics. In sum, these suggested areas of research can contribute to the continuous and persistent global effects to resolve the dilemma of inequality, equality, and growth.

Acknowledgements This chapter is restructured and rewritten based on a journal article, 'Is Rising Income Inequality Far from Inevitable During Structural Transformation? A Proposal for An Augmented Inequality Dynamics', published in *Journal of Economics and Political Economy*, Vol. 4, Issue 3, pp. 22–35, 2017. Copyright for this article is retained by Seung-Jin Baek, with first publication rights granted to the journal.

Notes

1. This was based on general assumption that income growth leads to improved living standards. However, there were a number of exceptions against convergence of living standards between the two different worlds. For instance, despite income growth in India, living standards (as measured, e.g., by access to toilet or cooking fuel) have not improved much, cf. the differences between income and multidimensional measures of poverty.
2. Although Rawls comes down to us mainly as a philosopher rather than as a student of either economic growth or structural transformation, and his framework is not somewhat 'developmental', his argument by difference principle would fundamentally be applicable to and still be valid to contemporary development discourses, particularly for inequality debate.
3. Each person is to have an equal right to the most extensive basic liberty compatible with a similar liberty for others.
4. Particularly, Piketty and Saez (2003) criticised the Kuznets hypothesis, 'Inequality reduces in mature stage of development' by putting a compelling case of the US where the level of inequality has grown in recent decades.
5. For a more detailed discussion, see Chaps. 3 and 4.

6. What is interesting is that his re-investment argument was already identified as one of the conditions in the take-off stage of Rostow's model. Although Rostow did not explicitly spell out the inequality problem in the take-off stage, he might have implied that continuous capital formation not only promotes re-investment activities but also causes the concentration of capital into the hands of a few capitalists, which could intensify inequality. Importantly, the problem of Rostow's model here lies in his assumption that re-investing boosts economically focused industrial activities, and not the likes of institutional reforms or education as proposed by Joseph E. Stiglitz.

References

Aghion, P., and P. Bolton. 1997. A Theory of Trickle-Down Growth and Development. *Review of Economic Studies* 64: 151–172.

Alesina, A., and G.M. Angeletos. 2005. Fairness and Redistribution. *American Economic Review* 95 (4): 960–980.

Alesina, A., and D. Rodrik. 1994. Distribution Politics and Economic Growth. *Quarterly Journal of Economics* 109 (2): 465–490.

Armah, B., and S.J. Baek. 2015. Can the SDGs Promote Structural Transformation in Africa? An Empirical Analysis. *Development* 58 (4): 1–19.

Armah, B., and S.J. Baek. 2018. Three Interventions to Foster Sustainable Transformation in Africa. *Journal of Social, Political, and Economic Studies* 43 (1–2): 3–25.

Arrow, K.J. 1962. The Economic Implications of Learning by Doing. *Review of Economic Studies* 29 (3): 155–173.

Bah, E.M. 2011. Structural Transformation Paths Across Countries. *Emerging Markets Finance and Trade* 47 (2): 5–19.

Berg, A.G., and J.D. Ostry. 2013. Inequality and Unsustainable Growth: Two Sides of the Same Coin? *International Organisations Research Journal* 8 (4): 77–99.

Birdsall, N., D. Ross, and R. Sabot. 1997. Education, Growth and Inequality. In *Pathways to Growth: Comparing East Asia and Latin America*, ed. N. Birdsall and F. Jaspersen, 93–127. Washington, DC: Inter-American Development Bank.

Brunori, P., F. Ferreira, and V. Peragine. 2013. *Inequality of Opportunity, Income Inequality and Economic Mobility: Some International Comparisons*, World Bank Policy Research Working Paper, (WPS6304). Available from: http://

documents.worldbank.org/curated/en/864781468331737091/pdf/wps6304.pdf. Accessed 5 Dec 2016.
Chang, Y.S., and S.J. Baek. 2010. Limit to Improvement: Myth or Reality? Empirical Analysis of Historical Improvement on Three Technologies Influential in the Evolution of Civilization. *Technological Forecasting and Social Change* 77 (5): 712–729.
Chen, B.L. 2003. An Inverted-U Relationship Between Inequality and Long-Run Growth. *Economics Letters* 78 (2): 205–212.
De Gregorio, J., and J.W. Lee. 2002. Education and Income Inequality: New Evidence from Cross-Country Data. *Review of Income and Wealth* 48 (3): 395–416.
Downs, A. 1957. An Economic Theory of Political Action in a Democracy. *Journal of Political Economy* 65 (2): 135–150.
Dumke, R. 1991. Income Inequality and Industrialization in Germany, 1850–1913: The Kuznets Hypothesis Re-examined. In *Income Distribution in Historical Perspective*, ed. Y.S. Brenner, H. Kaelble, and M. Thomas, 117–148. Cambridge/Paris: Cambridge University Press.
Easterly, W. 2007. Inequality Does Cause Underdevelopment: Insights from a New Instrument. *Journal of Development Economics* 84: 755–776.
ECA (Economic Commission for Africa), AUC (African Union Commission), AfDB (African Development Bank), and UNDP (United Nations Development Programme). 2016. *MDGs to Agenda 2063/SDGs Transition Report 2016: Towards an Integrated and Coherent Approach to Sustainable Development in Africa*. Addis Ababa: ECA. Available from: https://www.uneca.org/sites/default/files/PublicationFiles/mdg-sdg-report-2016_en_revfeb2017.pdf. Accessed 11 Apr 2016.
Elgar, F.J., and N. Aitken. 2011. Income Inequality, Trust and Homicide in 33 Countries. *European Journal of Public Health* 21 (2): 241–246.
Engerman, S.L., and K.L. Sokoloff. 2002. Factor Endowments, Inequality, and Paths of Development Among New World Economies. *Economia* 3 (2): 41–102.
Figueroa, A. 2017. *Economics of the Anthropocene Age*. Basingstoke: Palgrave Macmillan.
Fleming, M. 1955. External Economies and the Doctrine of Balanced Growth. *Economic Journal* 65: 241–256.
Forbes, K. 2000. A Reassessment of the Relationship Between Inequality and Growth. *American Economic Review* 90 (4): 869–897.
Friedman, M. 1953. Choice, Chance and the Personal Distribution of Income. *Journal of Political Economy* 61 (4): 277–290.

Galor, O., and J. Zeira. 1993. Income Distribution and Macroeconomics. *Review of Economics Studies* 60 (1): 35–52.

Giddens, A. 1991. *The Consequences of Modernity*. Stanford: Stanford University Press.

Guidetti, R. 2014. Theoretical Approaches to Inequality in Economics and Sociology. *Transcience* 5 (1): 1–15.

Herzer, D., and S. Vollmer. 2012. Inequality and Growth: Evidence from Panel Cointegration. *Journal of Economic Inequality* 10 (4): 489–503.

Hirschman, A.O. 1958. *The Strategy of Economic Development*. New Haven: Yale University Press.

Itagaki, Y. 1963. Criticism of Rostow's Stage Approach: The Concepts of Stage, System and Type. *Developing Economies* 1 (1): 1–17.

Kaelble, H., and M. Thomas. 1991. *Income Distribution in Historical Perspective*. Cambridge: Cambridge University Press.

Kaldor, N. 1956. Alternative Theories of Distribution. *Review of Economic Studies* 23: 83–100.

Korotayev, A., and J. Zinkina. 2014. On the Structure of the Present-Day Convergence. *Campus-Wide Information Systems* 31 (2/3): 139–152.

Korzeniewicz, R.P., and T.P. Moran. 2005. Theorizing the Relationship Between Inequality and Economic Growth. *Theory and Society* 34: 277–316.

Kuznets, S. 1955. Economic Growth and Income Inequality. *American Economic Review* 45 (1): 1–28.

———. 1973. Modern Economic Growth: Findings and Reflections. *American Economic Review* 63 (3): 247–258.

Lewis, W.A. 1954. Economic Development with Unlimited Supplies of Labour. *Manchester School* 22 (2): 139–191.

Li, H., and H. Zou. 1998. Income Inequality Is Not Harmful for Growth: Theory and Evidence. *Review of Development Economics* 2 (3): 318–334.

Lin, J.Y. 2012. *New Structural Economics: A Framework for Rethinking Development and Policy*. Washington, DC: World Bank Publications. Available from: http://siteresources.worldbank.org/DEC/Resources/84797-1104785060319/598886-1104951889260/NSE-Book.pdf. Accessed 20 Nov 2016.

Lissowska, M. 2015. Is Poverty and Inequality Actually Good for Growth? *Forum for Social Economics* 44 (2): 133–158.

Maddison, A. 2008. Shares of the Rich and the Rest in the World Economy: Income Divergence Between Nations, 1820–2030. *Asian Economic Policy Review* 3 (1): 67–82.

McPherson, M., L. Smith-Lovin, and J.M. Cook. 2001. Birds of A Feather: Homophily in Social Networks. *Annual Review of Sociology* 27: 415–444.

Milanovic, B. 2000. Determinants of Cross-Country Income Inequality: An 'Augmented' Kuznets Hypothesis. In *Equality, Participation, Transition: Essays in the Honor of Branko Horvat*, ed. M. Uvalic and V. Franicevic, 48–79. London: St. Martin's.

Mills, M. 2009. Globalization and Inequality. *European Sociological Review* 25 (1): 1–8.

Mincer, J. 1974. *Schooling, Experience, and Earnings*. New York: Columbia University Press.

Neckerman, K.M., and F. Torche. 2007. Inequality: Causes and Consequences. *Annual Review of Sociology* 33: 335–357.

Nurkse, R. 1953. *Problems of Capital Formation in Underdeveloped Countries*. Oxford: Oxford University Press.

Okun, A.M. 1975. *Equality and Efficiency: The Big Tradeoff*. Washington, DC: The Brookings Institution.

Palma, J.G. 2011. Homogeneous Middles vs. Heterogeneous Tails, and the End of the 'Inverted-U: It's All About the Share of the Rich. *Development and Change* 42 (1): 87–153.

Parente, S. 2001. The Failure of Endogenous Growth. *Knowledge, Technology and Policy* 13 (4): 49–58.

Perotti, R. 1996. Growth, Income Distribution and Democracy: What the Data Say. *Journal of Economic Growth* 1 (2): 149–187.

Piketty, T. 2013. *Capital in the Twenty-First Century*. Cambridge: Harvard University Press.

Piketty, T., and E. Saez. 2003. Income Inequality in the United States, 1913–1998. *Quarterly Journal of Economics* 118 (1): 1–39.

Polak, B., and J.G. Williamson. 1991. *Poverty, Policy and Industrialization: Lessons from the Distant Past*. Background Paper for the 1990 World Development Report, (WPS645). Available from: http://documents.worldbank.org/curated/en/879321468739136427/pdf/multi0page.pdf. Accessed 9 Jan 2017.

Ram, R. 1991. Kuznets's *Inverted U-Hypothesis*: Evidence from a Highly Developed Country. *Southern Economic Journal* 57: 1112–1123.

Rawls, J. 1971. *A Theory of Justice*. Cambridge: Harvard University Press.

Romer, P.M. 1990. Endogenous Technological Change. *Journal of Political Economy* 98 (5): 71–102.

Rostow, W.W. 1960. *The Stages of Economic Growth: A Non-Communistic Manifesto*. Cambridge: Cambridge University Press.

Sachs, J.D., and A.M. Warner. 1997. Fundamental Sources of Long-Run Growth. *American Economic Review* 87 (2): 184–188.

Schaller, W.E. 1998. Rawls, the Difference Principle, and Economic Inequality. *Philosophical Quarterly* 79: 368–391.

Soderberg, J. 1991. Wage Differentials in Sweden, 1725–1950. In *Income Distribution in Historical Perspective*, ed. Y.S. Brenner, H. Kaelble, and M. Thomas, 76–95. Cambridge/Paris: Cambridge University Press.

Solow, R.M. 1956. A Contribution to the Theory of Economic Growth. *Quarterly Journal of Economics* 70 (1): 65–94.

Stiglitz, J.E. 1996. Some Lessons from the East Asian Miracle. *World Bank Research Observer* 11 (2): 151–177.

Szirmai, A. 2012. Industrialisation as an Engine of Growth in Developing Countries, 1950–2005. *Structural Change and Economic Dynamics* 23 (4): 406–420.

Temple, J., and H. Ying. 2014. *Life During Structural Transformation*, Discussion Paper 14/650. Bristol: University of Bristol. Available from: http://www.efm.bris.ac.uk/economics/working_papers/pdffiles/dp14650.pdf. Accessed 14 Jan 2017.

Thomas, M. 1991. The Evolution of Inequality in Australia in the Nineteenth Century. In *Income Distribution in Historical Perspective*, ed. Y.S. Brenner, H. Kaelble, and M. Thomas, 149–173. Cambridge/Paris: Cambridge University Press.

Thorbecke, E., and C. Charumilind. 2002. Economic Inequality and Its Socioeconomic Impact. *World Development* 30: 1477–1495.

Uslaner, E.M., and M. Brown. 2005. Inequality, Trust, and Civic Engagement. *American Politics Research* 33 (6): 868–894.

Voitchovsky, S. 2005. Does the Profile of Income Inequality Matter for Economic Growth? Distinguishing Between the Effects of Inequality in Different Parts of the Income Distribution. *Journal of Economic Growth* 10 (3): 273–296.

Weatherford, R.C. 1983. Defining the Least Advantaged. *Philosophical Quarterly* 33 (130): 63–69.

Wilkinson, R., and K. Pickett. 2010. *The Spirit Level: Why Equality Is Better for Everyone*. London: Penguin.

World Bank. 2017. *World Development Indicators* [Online]. Available from: http://data.worldbank.org/data-catalog/world-development-indicators. Accessed 22 Feb 2017.

6

Can the SDGs Promote Structural Transformation in Africa? An Empirical Analysis

Introduction

For over a decade, remarkable progress has been made on socio-economic development in Africa, with more children enrolled in primary school and steep declines in child and maternal deaths. The incidence of extreme poverty has also declined albeit marginally. The positive trends in Africa's social indicators have been underpinned by real GDP growth which averaged approximately 5 per cent over the last decade and contributed to significant progress towards achieving the MDGs. Indeed, despite its relatively low initial conditions, the continent was close to attaining three of the eight MDGs: Goal 2 on achieving universal primary education, Goal 3 on promoting gender equality and empowering women, and Goal 6 on combating HIV/AIDS and other diseases (ECA et al. 2015).

The continent's socio-economic advancement has been supported by strong international support and prudent macroeconomic management. Over the period 1990–2014, Africa accounted for 42 per cent on average of total net official development assistance (ODA) disbursements, which also makes the continent the largest recipient of the ODA.[1] Africa is also benefitting from the World Trade Organization Bali

S. J. Baek, *The Political Economy of Neo-modernisation*,
https://doi.org/10.1007/978-3-319-91394-0_6

packages, particularly the Trade Facilitation Agreement, which seeks to promote global trade competitiveness. Furthermore, the Heavily Indebted Poor Countries (HIPC) initiative and Multilateral Debt Relief Initiative (MDRI) have contributed to significant reductions in Africa's debt burden.[2]

However, trends in Africa's socio-economic and environmental indicators raise questions about the sustainability of the continent's development trajectory. At the socio-economic level, Africa's growth performance has not been associated with significant reductions in poverty or improvements in employment opportunities. The growth elasticity of poverty in sub-Saharan Africa is estimated at −1.57 which is relatively low compared to other regions (ECA and AUC 2012). As a result, the continent has the highest rate of extreme poverty globally and the number of poor people is on the rise despite reductions in the poverty rate. Extreme poverty (measured at US$1.25 per day) in sub-Saharan Africa declined 14.2 per cent over a 22-year period from 56.9 per cent in 1990 to 42.8 per cent in 2012. Meanwhile, the number of poor people swelled to 389 million from 280 million during the same period owing to Africa's rapid population growth (UNSD 2015) (see Fig. 6.1).

A contributing factor to the limited impact of growth on poverty is the fact that Africa's growth acceleration has not been associated with commensurate increases in decent employment opportunities. Approximately 70 per cent of jobs in sub-Saharan Africa are classified as vulnerable, and youth unemployment rates are high averaging 21.08 per cent. Unemployment and underemployment are particularly pervasive among the youth. Using the International Labour Organization's Trends Econometric Models (2014), the overall unemployment rate in 2015 for the continent was 12 per cent, with the highest unemployment rates in Southern Africa (18.0 per cent), North Africa (13.9 per cent), East Africa (10.7 per cent), West Africa (9.1 per cent), and Central Africa (8.2 per cent). Unemployment rates for female youth are also significantly higher than those for their male counterparts in all regions, highlighting gender disparities in access to jobs. The disconnect between growth and employment creation can be attributed to the primary commodity

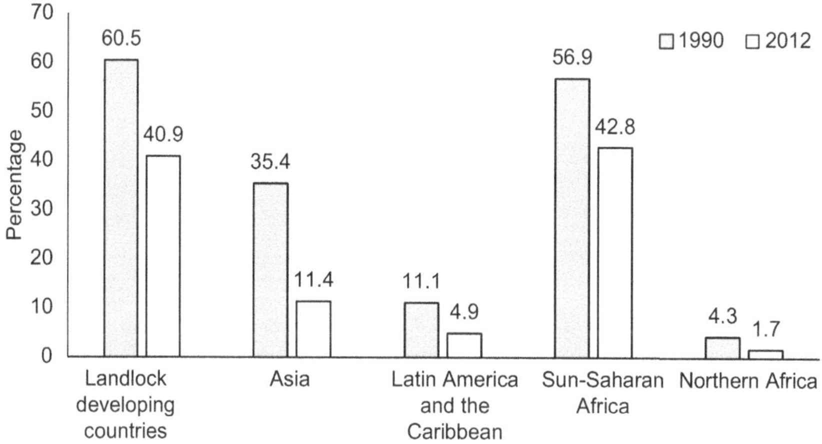

Fig. 6.1 Proportion of population living below US$1.25 purchasing power parity per day. (*Source*: Author's own elaboration on the basis of the World Bank's World Development Indicators (2016))

dependence of most African economies which renders them vulnerable to external shocks and undermines prospects for job creation through value-addition and beneficiation. Indeed, the continent has experienced a prolonged era of *De-industrialisation* as evidenced by the stagnation in manufacturing value-added at around 11 per cent of GDP. On the other hand, service sector value-added has increased substantially over the same period (see Fig. 6.2).

The lack of inclusive growth has contributed to social exclusion in Africa. Social exclusion is manifested by high-income inequalities and disparities in access to social services. With an average Gini coefficient of 43.9 per cent over the period 2000–2009, income inequality in Africa is second only to Latin America which averaged 52.2 per cent over the same period (AfDB 2012). In 2010, six out of the ten most unequal countries worldwide were in Africa; the Southern Africa sub-region exhibits the highest level of inequality in the continent (Armah et al. 2014).

Disparities in access to social services are also pervasive and pronounced in the rural areas. Births to women in the top quintile group are nearly

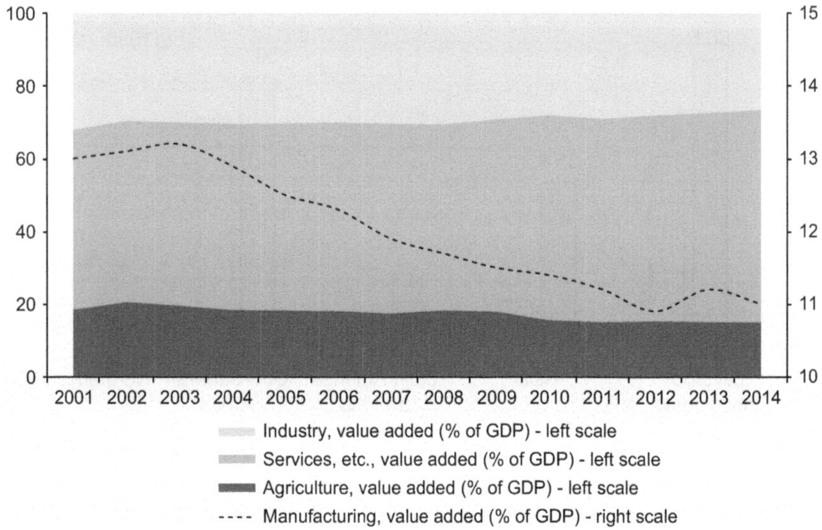

Fig. 6.2 Sectoral contributions to GDP in Africa. (*Source*: Author's own elaboration on the basis of the World Bank's World Development Indicators (2016))

three times more likely to be attended by a trained professional than births to women in the poorest quintile. Ninety per cent of women living in urban areas have at least one antenatal care visit during pregnancy, compared to 71 per cent of women living in rural areas (ECA et al. 2011). Children and adolescents from the poorest households are at least three times more likely to be out of school than children from the richest households; and the poorest households are more than twice as likely to be stunted as children from the richest households, leading to further school dropouts (ECA et al. 2014).

At the level of the environment, the continent has the lowest greenhouse gas emissions globally; however, like the rest of the world, the continent's emissions are closely related to the rate of PCGDP growth (see Figs. 6.3 and 6.4). Carbon dioxide emissions peaked with per capita income growth in 2004, declined sharply in 2005, and resumed an upward trend thereafter. These trends suggest that at current rates, Africa's greenhouse gas emissions will rise with accelerated growth.

In effect, on current trends, Africa's growth performance can be described as unsustainable due to excessive dependence on a narrow range of primary commodities, high levels of income inequality and disparities

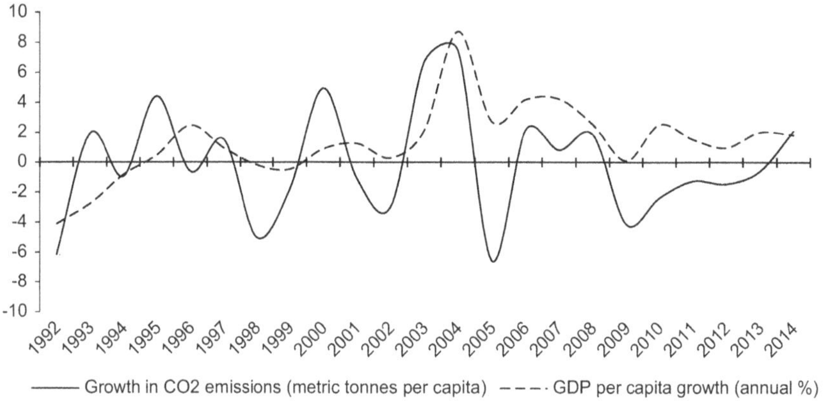

Fig. 6.3 Carbon dioxide emissions and PCGDP growth trends in sub-Saharan Africa. (*Source*: Author's own elaboration on the basis of the World Bank's World Development Indicators (2016))

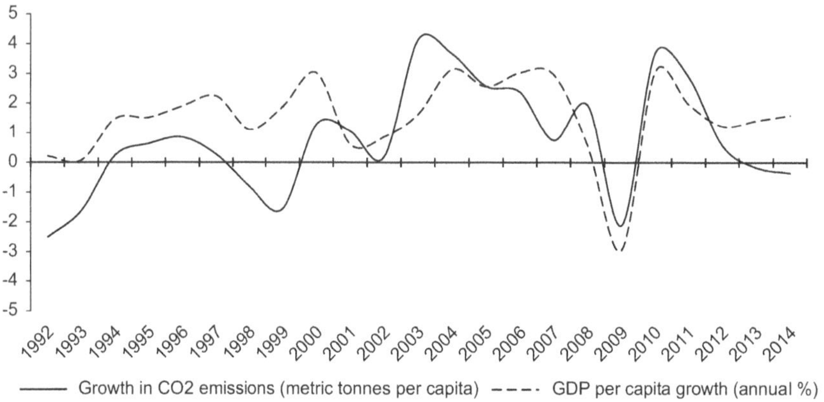

Fig. 6.4 Carbon dioxide emissions and PCGDP growth trends globally. (*Source*: Author's own elaboration on the basis of the World Bank's World Development Indicators (2016))

in access to social services, and the close association between growth and greenhouse gas emissions. Recognising the importance of sustainability to the development discourse, the 2030 Agenda for Sustainable Development as well as Africa's Agenda 2063 include several goals, targets, and indicators that capture the three dimensions of sustainability.

They also underline the importance of factoring economic, social, and environmental sustainability into national planning frameworks (AUC 2014; UN 2015b). Furthermore, the climate agreement reached in Paris during the 21 Conference of the Parties calls on countries to commit to actions that are consistent with maintaining temperature levels to no more than 2 degrees Celsius above preindustrial levels.

The development initiatives cited earlier call for a paradigm shift in the development discourse and refute the practice whereby countries initially focus on growth and address issues of inclusion and environmental degradation later. Attempts to integrate sustainability into the development discourse can be traced to the Rio conference or earth summit in 1992 (UN 1992). Subsequently in 2012, Rio+20 reaffirmed the commitment of countries to the implementation of a sustainable development agenda. Specifically, member states agreed to establish an inclusive and transparent intergovernmental process on SDGs that should 'incorporate in a balanced way all three dimensions of sustainable development and their interlinkages' and 'should be coherent with and integrated into the UN development agenda beyond 2015'.[3]

The implication of the 2030 Agenda for Sustainable Development is that Africa's structural transformation must be anchored by the principles of sustainable development. This raises the question of how efforts to advance economic, social, and environmental sustainability will impact on Africa's structural transformation agenda. What is the likely impact of the 2030 Agenda for Sustainable Development on structural transformation in Africa?

The theory of the relationship between sustainable development and transformation can be gleaned from the seminal work of Simon Kuznets (1955) who hypothesised that the relationship between inequality and transformation is U-shaped, with inequality increasing in the early stages of development as capital holders become relatively more wealthy due to new opportunities to invest, and subsequently decreasing as the benefits of transformation trickle down to the population allowing an average level of income to be achieved through notably better wages. Incidentally, the Kuznets theory is in line with the 'Grow First and Redistribute Later' school of thought. In reality, the process of structural transformation observed in some Asian countries such as South Korea, Taiwan, and Singapore was actually contrary to the Kuznets theory. The rapid economic growth which was achieved mainly through substantial increases in the manufacturing and export sectors during the 1960s–1990s period resulted in declines in absolute poverty without corresponding increases in inequality. According to Stiglitz (1996), this outcome was due to the fact that these countries immediately re-invested the benefits from rapid growth into land reform and universal education all of which supported an inclusive growth process.

A contrasting theory on the relationship between growth and social inclusion thought which is consistent with the sustainable development thinking argues that addressing inequalities promotes growth for at least three reasons. First, inequality in resource endowments affects the ability of poor but competent agents to borrow in order to carry out productive investments because they lack collateral (Aghion and Bolton 1992; Banerjee and Newman 1993; Piketty 1997). As a result, inequality results in misallocation and underutilisation of resources, slower economic growth, and possibly a weaker industrialisation and structural transformation process. Second, inequality affects growth through its impact on human capital and fertility. Poor parents with relatively low level of education often fail to invest in the education of their children, partly because of inadequate resource endowments. Low income also affects decisions about fertility (Ehrhart 2009). Parents with low income tend to have more children with the hope of increasing the aggregate household income. However, high fertility further curtails the ability of the parents to invest in quality education for their children (Perotti 1996) thus contributing to intergenerational poverty and inequality. Third, inequality affects the size and composition of aggregate demand and hence reduces the market for industrial production (Nurkse 1953).

Grossman and Krueger's (1995) Environmental Kuznets curve concept provides theoretical framework for the analysis of the relationship between growth and the environment. The authors hypothesise that industrial development initially leads to greater emissions of greenhouse gases; however, net emissions eventually decline as the increase in income associated with further development leads to technological innovations to curb such emissions. Once again, this theory is at odds with the current thinking on sustainable development which requires countries to integrate or decouple greenhouse gas emissions from their growth process even in the initial stages of development. In line with the sustainable development paradigm, there is a growing body of research that suggests that relative to other continents, the potential for economic greening is higher in Africa (Collier and Venables 2012; Simon 2013). These authors argue that Africa's vast untapped water resources and location gives it a competitive edge in the production of renewable energies such as hydro and solar power. Furthermore, as a latecomer, Africa can leapfrog fossil fuels to clean energy by immediately benefitting from advanced countries' greening experiences and associated technologies. In this context, synergies between environmental and economic development are likely even at the early stages of the development process. Indeed, some Africa countries are among the global front runners in terms of greening their economies (UNEP 2015).

In addition to issues of social inclusion and environmental degradation, institutions can play an important role in fostering growth and sustainable transformation. The 'institutional quality hypothesis' contends that sustainable growth is significantly influenced by the institutions within which agents interact. In effect, institutions can play a mutually reinforcing role in supporting structural transformation (Rodrik et al. 2002; Veeman and Politylo 2003; Osman et al. 2012).

The foregoing analysis of the literature suggests that in theory, structural transformation and growth can be achieved without necessarily sacrificing social inclusion or the environment. In effect, countries need not adopt a 'Grow First and Achieve Social Inclusion and Environmental Conservation Later' approach. On the contrary, promoting inclusion and addressing environmental degradation as part of a development strategy can lead to growth and sustainable transformation. Does the evidence support this theory in Africa? To gain insights to this question, this study empirically analyses the impact of institutional quality and selected indicators of economic, social, and environmental sustainability on structural transformation in Africa.

Methodology and Data

Using data for a group of 29 African countries[4] spanning the period 1995–2011, this study employs a panel data econometric model to investigate the empirical link between selected indicators of sustainable development and structural transformation.

Selection of Indicators

The choice of appropriate indicators to measure economic, social, and environmental sustainability is a complex one given the variety of interpretations of sustainable development. At the international level, measurement indicators have been developed by various agencies, notably the UN. Similarly, most European countries, including the UK, Germany, Belgium, and Switzerland have developed their own measurement indicators that are linked to national development frameworks (UN 2008).

For the purposes of this study, the choice of indicators is based on the methodology adopted by the Commission on Sustainable Development (UN 2001), to define the three dimensions of sustainable development. The economic dimension comprises indicators of growth, investment, green economy, and energy use. Indicators of social sustainability comprise poverty, access to improved sanitation, life quality, and maternal health. The environmental dimension consists of climate change, agriculture land, forest cover, and access to improved water sources. Institutional quality is measured by indicators of government stability, socio-economic conditions, corruption, and democratic accountability.

The number of indicators chosen was small enough to be manageable but sufficiently large in number to ensure that the three dimensions of sustainable development were adequately captured in the analysis. The choice of variables was also informed by data availability (Osman et al. 2012).

Economic Dimension

GDP growth is one of the most widely used indicators for measuring progress towards economic growth (Giddings et al. 2002). Related to

GDP growth is per capita income which takes into account the population size. Second, the investment to GDP ratio performance reflects the infusion of capital to accelerate economic development. However, not all growth or investments are sustainable especially if they deplete a nation's natural capita. For this reason, we include two additional indicators: renewable energy consumption and the efficiency and intensity of energy use. Renewable energy consumption can be a catalyst for economic growth while minimising environmental stress (Bugaje 2006). Energy efficiency and intensity levels are also important indicators of sustainability (Vera and Langlois 2007). For this study, the energy supply to GDP ratio is a proxy for energy efficiency; lower ratios imply greater efficiencies in the use of energy.

Social Dimension

The unemployment rate, life expectancy, access to improved sanitation, and the adolescent fertility rate are used as proxies for social sustainability in this study. The unemployment rate is a key determinant of the poverty level (Jarvis and Jenkins 1998). Life expectancy is a component of two widely acknowledged measures of social development: the physical quality-of-life index (PQLI) and the UN Development Programme's Human Development Index (UNDP-HDI) (Morris 1979; Ray 2008). Access to improved sanitation is also a globally accepted indicator of social development (Katukiza et al. 2010). Furthermore, the adolescent fertility rate or unintended pregnancies is a significant cause of school dropout, which in turn undermines productive capacities and contributes to poverty (Klima 1998; McQueston et al. 2012; Kibret et al. 2014; Akbarzadeh et al. 2016).

Environmental Dimension

Indicators of environmental sustainability comprise carbon dioxide emissions, forest cover, access to improved water sources, and access to arable land. Carbon dioxide emissions are the leading cause of global warming and hence a relevant indicator of environmental degradation. Forest cover on the other hand constitutes a carbon sink that mitigates global warming (Foody 2003). Access to improved water sources is influenced by extreme weather conditions including droughts. Hectares of arable land per capita is an indicator of the quality of land management practices (Zander and Kächeleb 1999).

Institutional Dimension

In addition to the three dimensions, institutional quality indicators capturing four broad issues—government stability, socio-economic pressures, corruption, and accountability—are also included in the analysis. The rationale for including the institutional dimension is that structural transformation requires effective institutions. Institutional effectiveness is a function of government stability, the pervasiveness of corruption, democratic accountability, and socio-economic conditions (Veeman and Politylo 2003; Acemoglu and Robinson 2012; Osman et al. 2012).

Measuring Structural Transformation

The study constructs a structural transformation index comprising seven variables selected on the basis of the key features of the structural transformation process which are a declining GDP share of agriculture accompanied by increasing productivity in the sector, a rapid increase in the pace of urbanisation, rising relative GDP shares of the manufacturing and service sectors, and declining fertility rates (Christiaensen et al. 2011; Timmer et al. 2012; Dabla-Norris et al. 2013; Armah et al. 2014). In addition to these key components, improvements in ICT infrastructure, better functioning of financial markets, and reductions in the infant mortality rate are key to a successful structural transformation agenda (LDC IV Monitor 2015).

Consistent with these processes, the seven indicators that constitute the transformation index are cereal yield per hectare, per capita gross agricultural production, the GDP share of manufacturing value-added, services value-added as a per cent of the GDP, domestic credit provided by the financial sector as a per cent of GDP, the infant mortality rate, and

Internet users per 100 inhabitants. The cereal yield and agriculture production index measure agricultural productivity. Shifts in sectoral shares are measured by the relative contributions of the manufacturing and services sectors to GDP, while the number of Internet users per 100 people is used as a proxy for ICT infrastructure development (Kenney 1995; Mansell and Wehn 1998; Chang and Baek 2010). The infant mortality rate is included in the index to capture the inclusiveness of the structural transformation process (Jain 1985; Fayissa 2001). The volume of credit provided to the banking sector reflects the important role of finance to economic development and structural transformation (Cetorelli and Gambera 2001; Calderon and Liu 2003). The list of variables is provided in Table 6.1.

Modelling Approaches

The study uses factor analysis to construct a structural transformation index which is regressed against proxies for the three dimensions of sustainable development as well as indicators of institutional effectiveness.

Factor Analysis

The methodological approach to constructing the multidimensional index is similar to that of the UNDP-HDI. We use factor analysis to group the seven components of the structural transformation variable into a single index. The composite index retains as much information as possible from the seven underlying variables. Each factor is estimated using the maximum likelihood method to identify a set of strongly associated indicators (Dempster et al. 1977).

Constructing a weighted composite indicator of structural transformation enables cross-country comparisons of the status of structural change and collapses several proxies of structural change from one dependent variable into one variable while retaining most of the underlying information (Saltelli 2007; Kelbore 2014).

Table 6.1 Classification and variable description

	Dimension	Area of importance	Proxy indicator and description	Source
Independent variable	Economic	Economic growth	PCGDP (annual %)	WDI
		Investment	Total investment (% of GDP)	WEO
		Green economy	Renewable energy consumption (% of total final energy consumption)	WDI
		Energy use	Energy intensity level of primary energy	WDI
	Social	Poverty	Unemployment rate (% of total labour force)	WDI
		Sanitation	Sanitation facilities (% of population with access)	WDI
		Life quality	Life expectancy at birth (years)	WDI
		Maternal health	Adolescent fertility rate (births per 1000 women ages 15–19)	WDI
	Environmental	Climate change	Emissions of carbon dioxide (metric tons per capita)	WDI
		Agriculture land	Arable land (hectares per person)	WDI
		Forests	Forest area (square km)	WDI
		Water	Improved water source (% of population with access)	WDI
	Institutional quality	Government stability	Government's ability to carry out its declared program and policy	ICRG
		Socio-economic conditions	Socio-economic pressures at work in society that could constrain government action or fuel social dissatisfaction	ICRG
		Corruption	Corruption within the political system	ICRG
		Democratic accountability	How responsive government is to its people in terms of widespread participation of nation's citizens	ICRG

(*continued*)

Table 6.1 (continued)

	Dimension	Area of importance	Proxy indicator and description	Source
Dependent variable	Structural transformation	Higher agricultural productivity by achieving a higher cereal yield	Cereal yield (kg per hectare) Agriculture gross per capita production index (2004–2006 = 100)	WDI FAOSTAT
		Higher share of manufacturing in value-addition	Manufacturing value-added (% of GDP)	WDI
		Increases in the share of ICT in services exports	Services value-added (% of GDP)	WDI
		Better telecommunications infrastructure	Internet users (per 100 people)	WDI
		Improving health services by decreasing infant mortality rate	Infant mortality rate (per 1000 live births)	WDI
		More developed financial market	Domestic credit provided by financial sector (% of GDP)	WDI

Source: Author's own elaboration

Note: WDI stands for the World Bank's World Development Indicators (2015); WEO stands for the IMF's World Economic Outlook Database (2015); FAOSTAT stands for the FAO's Statistics Division Database (2015); and ICRG stands for the Political Risk Services Group International Country Risk Guide Database (2015)

The use of factor analysis to construct a weighted multidimensional structural transformation indicator involves the following five steps: the first is to check for the correlation structure of the underlying variables; the next step identifies and retains the sub-set of the factors that explain the largest amount of variance in the underlying variables (based on the estimated eigenvalues); the third step is to rotate the eigenvalues of the retained factors using the Oblimin with Kaiser Normalisation method in order to determine the contribution of each of the underlying variables to each of the retained factors; the fourth step aggregates the contribution of each underlying variable to all the retained factors, these weights are then normalised to arrive at the contribution of each variable to the structural transformation index; and based on the latter weights, the final step is to compute a weighted average per country of each variable's contribution to the structural transformation index. The computed weighted average represents the composite structural transformation index.

Panel Data Modelling

The panel model approach facilitates control for unobserved heterogeneity in the adjustment dynamics between countries (Bond 2002). A critical consideration when employing such models is whether or not to use a lagged dependent variable. A growing body of researchers, including Baltagi (2013) and Hsiao (2014), argues that longitudinal panel data with lagged values cause critical estimation problems including biased coefficients. This study does not apply a lagged dependent variable. We use the natural logarithm of variables that show a positive skew to normalise their distributions. The model specification is as follows:

$$y_{it} = \beta_{0i} + \beta_1 \text{ECON}_{it} + \beta_2 \text{SCO}_{it} + \beta_3 \text{ENV}_{it} + \beta_4 \text{INST}_{it} + \varepsilon_{it},$$

where i is country dimension (i = 1 ... N), t represents the time dimension (t = 1 ... T); y_{it} is the composite structural transformation index; ECON_{it} is a vector of indicators representing the economic dimension of sustainability; SOC_{it} represents a vector of indicators of social sustainability; ENV_{it} is a vector of variables reflecting the environmental dimension of sustainable development; and INST_{it} is a vector of variables representing institutional quality.

Empirical Results and Discussions

Composite Structural Transformation Index

The first step in constructing the structural transformation index is to understand the underlying relationships among the variables used to construct the index. For this purpose, we estimate the correlation between the variables used to construct the composite index. The correlation matrix reveals that the most highly correlated variables are services value-added and the infant mortality rate (−0.559) followed by service value-added and manufacturing value-added (0.552) (see Table 6.2). Variables that exhibit low levels of correlation are less likely to share common factors.

Using factor analysis, we employ a set of generally accepted criteria to identify factors in which the seven structural transformation variables with high correlations form one underlying variable. (Kelbore 2014). They are factors with eigenvalues closer to or greater than 1, factors with individual contributions to the overall variance that exceed 10 per cent, and factors with cumulative contributions to the overall variance that are greater than 70 per cent.

Based on the aforementioned criteria, we retain three factors with the following characteristics: statistical significant at 0.001 per cent level;

Table 6.2 Correlation matrix

	CREY	APIN	SERVA	INTU	MANVA	CREDIT	MORTI
CREY	1.000						
APIN	0.085	1.000					
SERVA	0.137	0.159	1.000				
INTU	0.245	0.216	0.243	1.000			
MANVA	0.352	−0.026	0.552	0.175	1.000		
CREDIT	0.484	−0.068	0.376	0.390	0.521	1.000	
MORTI	−0.311	−0.223	−0.559	−0.468	−0.339	−0.353	1.000

Source: Author's own elaboration on the basis of the modelling results

Note: CREY stands for cereal yield; APIN stands for agriculture gross per capita production index; SERVA stands for services value-added; INTU stands for Internet users; MANVA stands for manufacturing value-added, CREDIT stands for domestic credit provided by financial sector; and MORTI stands for infant mortality rate

goodness-of-fit, Chi-Square (31.471), Kaiser-Meyer-Olkin Measure of Sampling Adequacy (0.680), and Bartlett's Test of Sphericity (953.938). Factor 1 has an eigenvalue of 2.887 and individual and cumulative contributions of 41.240, which implies that this factor group explains approximately 41 per cent of the overall variance of the seven structural transformation indicators. Factor 2 has an eigenvalue of 1.214, an individual contribution of 17.349 per cent and a cumulative contribution of 58.589 per cent. Factor 3 has an eigenvalue of 0.959 which is very close to 1 and an individual contribution of 13.695 per cent. Overall, these three factors cumulatively explain 72.284 per cent of the overall variance (see Table 6.3). Factors 4, 5, 6, and 7 are not used for constructing composite index because they fail to meet the criteria outlined earlier.

In the next step, we rotate the three factors using the Oblimin with Kaiser Normalisation method in order to estimate the relative contribution of each variable to each of the retained factors. We use the Oblimin method because unlike other methods, it assumes the underlying variables are correlated which is normally the case in the field of social science. In contrast, orthogonal rotation methods such as Varimax assume that factors in the analysis are uncorrelated; however, this assumption is only valid in the field of physics or natural science (Gorsuch 1983). The resulting factor loadings represent the strength of association between a factor and a variable. By squaring and scaling the factor loadings, we are able to evaluate the importance of a variable to a factor (Field 2000).

Table 6.3 Eigenvalues of structural transformation data set

	Eigenvalues	% of Variance explained	Cumulative %
Factor 1	2.887	41.240	41.240
Factor 2	1.214	17.349	58.589
Factor 3	.959	13.695	72.284
Factor 4	.753	10.764	83.048
Factor 5	.531	7.579	90.627
Factor 6	.371	5.305	95.932
Factor 7	.285	4.068	100.000

Source: Author's own elaboration on the basis of the modelling results

Note: Extraction method is maximum likelihood. The eigenvalue is a measure of how much of the variance of the observed variables a factor explains. The eigenvalue of 2.8 for factor 1 means that factor explains as much variance as 2.8 of the observed variables

Table 6.4 shows the rotated factor loadings for each indicator as well as the three retained factors. Factor 1 contains two variables: services value-added (0.996) and manufacturing value-added (0.517). The other five variables have relatively smaller factor loadings (i.e., less than 0.500). Factor 2 contains only two variables with factor loadings exceeding 0.500: domestic credit (0.763) and cereal yield (0.618). The two variables have a correlation coefficient of 0.484. Factor 3 contains relatively lower factor loadings since their eigenvalue do not exceed 1. Hence, although none of factor loadings are greater than 0.500, we can retain the three variables that are closer to 0.500: infant mortality rate (−0.499), Internet users (0.493), and agriculture production index (0.450). The other four variables cannot be retained in Factor 3 since their factor loadings are smaller than 0.200. Nonetheless, these variables are already allocated to Factors 1 and 2 with higher factor loadings.

To estimate the relative weight or contribution of each variable to a factor, we square the factor loadings and scale them to unity to facilitate comparison among the variables. The results reveal three intermediate indicators. The first intermediate indicator includes service value-added (with a weight of 0.703) and manufacturing value-added (with a weight of 0.189). The second intermediate indicator includes domestic credit (with a weight of 0.425) and cereal yield (with a weight of 0.278). The third group includes mortality

Table 6.4 Factor and weight analysis

	Factor loading			Squared factor loading (scaled to sum to unity)			Weight
	Factor 1	Factor 2	Factor 3	Factor 1	Factor 2	Factor 3	
SERVA	.996	-.127	.170	.703	.012	.038	.292
MANVA	.517	.356	-.160	.189	.092	.034	.118
CREDIT	.205	.763	-.118	.030	.425	.018	.180
CREY	-.048	.618	.028	.002	.278	.001	.109
MORTI	-.323	-.243	-.499	.074	.043	.326	.116
INTU	-.055	.445	.493	.002	.144	.318	.125
APIN	.028	-.088	.450	.001	.006	.265	.059

Source: Author's own elaboration on the basis of the modelling results

Note: Extraction method is maximum likelihood. Rotation method is Oblimin with Kaiser normalisation. Rotation converged in seven iterations. CREY stands for cereal yield; APIN stands for agriculture gross per capita production index; SERVA stands for services value-added; INTU stands for Internet users; MANVA stands for manufacturing value-added; CREDIT stands for domestic credit provided by financial sector; and MORTI stands for infant mortality rate

rate (with a weight of 0.326), Internet use (with a weight of 0.318), and agricultural production (with a weight of 0.265). These weights are not used for the calculation of the index but simply for illustration purposes.

The weights or relative contribution of each of the seven variables to the structural transformation index, presented in the last column of Table 6.4, are computed as follows:

$$W_j = \sum STI_{ij}^{\ 2} / \begin{pmatrix} \sum STI_{i1}^{\ 2} + \sum STI_{i2}^{\ 2} + \sum STI_{i3}^{\ 2} + \sum STI_{i4}^{\ 2} + \sum STI_{i5}^{\ 2} + \\ \sum STI_{i6}^{\ 2} + \sum STI_{i7}^{\ 2} \end{pmatrix},$$

where W is estimated weight, STI represents the variables that comprise the structural transformation index; i is the factor dimension (i = 1, 2, 3), and j is variable dimension (j = 1 represents services value-added, 2 is manufacturing value-added, 3 is domestic credit provided by financial sector, 4 is cereal yield, 5 is infant mortality rate, 6 is Internet users, and 7 is agriculture gross per capita production index).

Services value-added contributes the largest weight to the structural transformation index (29.2), followed by domestic credit (18.0 per cent), Internet use per 100 inhabitants (12.5 per cent), manufacturing value-added (11.8 per cent), infant mortality rate (11.6 per cent), cereal yield (10.9 per cent), and agriculture gross per capita production index (5.9 per cent). The composite structural transformation index (CSTI) is then computed as follows:

$$\text{CSTI}_{it} = \sum W_{cit} \cdot STI_{cit},$$

where c is indicator dimension (c = 1 … C), i is country dimension (i = 1 … N), t is time dimension (t = 1 … T); W_{cit} is estimated weight; STI_{cit} represents the variables that comprise the structural transformation index; and CSTI_{it} is composite structural transformation index.

Figure 6.5 displays evolution of structural transformation for 29 African countries for the period 1995–2011 based on our estimated composite index for each country. Overall, structural transformation in the selected countries has improved by 41.4 per cent on average with the exception of the Gambia and Tanzania.

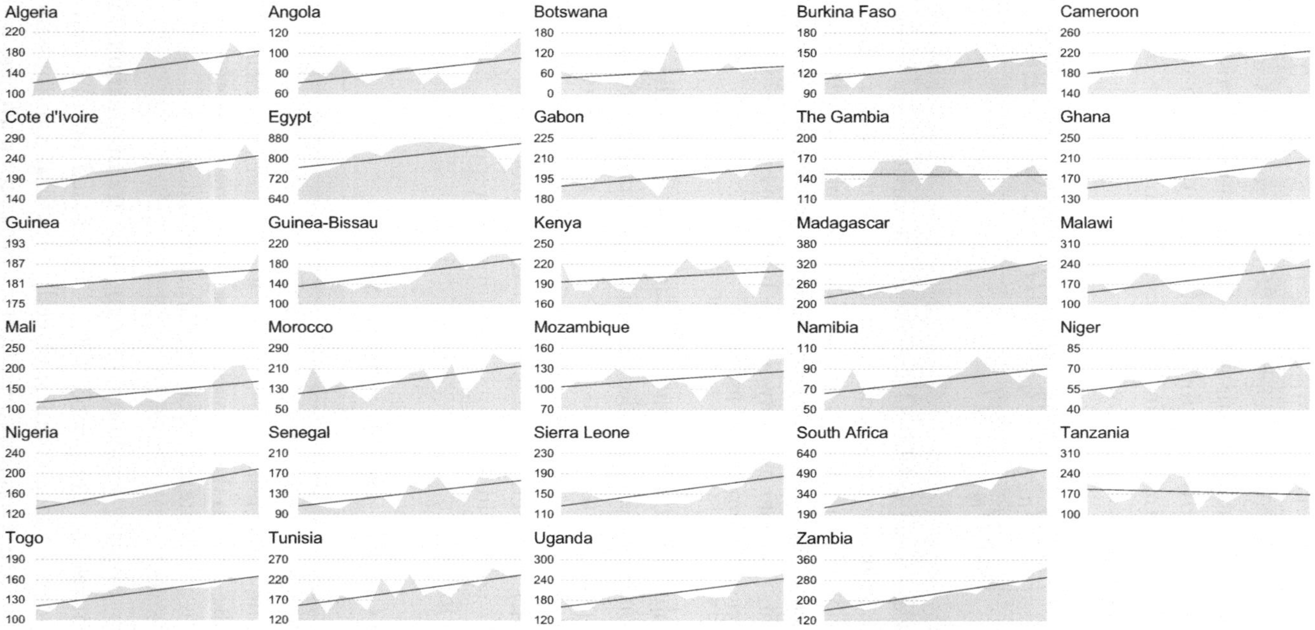

Fig. 6.5 Evolution of African structural transformation index. (*Source*: Author's own elaboration on the basis of the modelling results. *Note*: Grey line is the best-fitting regression line)

In the Gambia, cereal yield was down by 15 per cent for the period 1995–2011 and agriculture gross per capita production index declined from 85.27 to 72.97. It is important to note that the country has virtually no natural resources and is highly dependent on development assistance. During the period under study, country was significantly affected by external shocks, including massive cuts in external aid by the international community that had an adverse impact on investments and tourism activities.[5] These may have been contributing factors to *De-industrialisation* as its share of manufacturing value-added declined from 8.15 per cent to 5.48 per cent over the same period.

The decline in Tanzania's structural transformation index can be attributed to a significant fall in cereal yields (from 1702 to 1390 kg per hectare) over the period. Meanwhile, the country experienced only a marginal increase (1 percentage point from 7.1 per cent to 8.1 per cent) in the share of manufacturing value-added over the 16-year period.

Morocco experienced the greatest improvement in structural transformation (199.2 per cent) followed by South Africa (150.8 per cent). Both countries experienced substantial increases in cereal yield (261 per cent in Morocco and 83 per cent in South Africa), gross per capita agriculture production (93 per cent in Morocco and 25 per cent in South Africa) and domestic credit (112 per cent in Morocco and 26 per cent in South Africa).

Individual Effects by Fixed Effect Model

The composite structural transformation index is used as a dependent variable in a panel regression model. The model can be constructed based on three standard approaches, pooled ordinary least squares regression, fixed effect (or least squares dummy variables; LSDV), or random effect analysis.

The pooled regression pools all the data series to avoid the cross section and country-specific effects that may exist among 29 African countries. The random effect analysis assumes that individual effects are uncorrelated with observed predictors by applying common mean value for the intercept (Akbar et al. 2011; Osman et al. 2012). Based on the findings of the Hausman and Wald specification tests, we use the fixed effect analysis (Hsiao 2014). This is appropriate because it allows for heterogeneity among selected 29 African countries with 16 dependent variables. Differentiating intercepts for each country the model controls for country variations in achieving structural transformation (see Fig. 6.5). The results of the Hausman test are presented in Table 6.5.

From the panel data analysis, we observe that the three dimensions of sustainable development have significant impacts on African structural transformation. In Model I that focuses only on the economic dimension, investment and the energy intensity level are found to be statistically significant at 1 per cent level, which means that higher levels of investment have a positive impact on the process of structural transformation. Furthermore, lower energy intensity level or more efficient use of energy also has a positive and significant impact on the structural transformation process.

The findings however highlight potential trade-offs between clean energy, as proxied by the renewable energy indicator, and structural transformation. The coefficient on renewable energy consumption carries negative sign indicating the relative short-term cost-effectiveness of traditional energy in the production process. Nevertheless, this variable is weakly significant at 10 per cent level suggesting that it may not be a binding constraint to structural transformation. However, rising per capita incomes do not have a statistically significant impact on structural transformation.

In Model II, only the social dimension was tested, and three independent variables are highly significant except unemployment rate. To be more specific, the provision of good sanitation services and related infrastructure supports a country's structural transformation efforts. Similarly, longer life expectancy is positively associated with the structural transformation process while rising adolescent fertility rates have a negative and a statistically significant impact on of structural transformation. This is consistent with the findings of previous studies (Klima 1998; McQueston et al. 2012). But our analysis does not confirm the hypothesis that the unemployment level has a statistically significant impact on structural transformation.

Table 6.5 Results (coefficients) of panel data analysis

	Variable	Model I	Model II	Model III	Model IV	Model V
ECON	PCGDP	0.0034			0.0038 (2.003)**	0.0035 (1.863)*
	INV	0.0060 (4.369)***			0.0038 (2.851)***	0.0039 (2.962)***
	REC	−0.0043 (−1.766)*			0.0049 (1.983)**	0.0047 (1.878)*
	EIL	−0.3031 (−4.942)***			−0.1684 (−2.520)**	−0.1053
SOC	UEMPL		0.0566		0.0452	0.0873
	SANIT		0.0144 (4.104)***		0.0115 (2.760)***	0.0121 (2.899)***
	LIFE		0.0073 (2.333)**		0.0049	−0.0042
	AFR		−0.6066 (−6.268)***		−0.4074 (−3.993)***	−0.4303 (−4.220)***
ENV	CO_2			0.0881 (2.614)***	0.0302	0.0095
	ARAL			−0.2895 (−4.766)***	−0.1824 (−2.857)***	−0.2179 (−3.365)***
	FORE			0.0000	0.0000	0.0000
	WATS			0.0138 (8.327)***	0.0063 (2.848)***	0.0064 (2.643)***
INST	GOVS					0.0005
	SOCIO					0.0199 (2.027)**
	CORR					−0.0007
	DEMO					0.0229 (2.429)**
	C	5.8540 (36.236)***	6.8667 (12.160)***	3.7593 (14.894)***	5.8918 (9.454)***	5.5994 (8.593)***
Hausman test						
Chi-Sq. statistic		10.6523**	47.4023***	14.0502***	58.4661***	55.5932***
Wald test						
F-statistic		145.7811***	136.6526***	101.1344***	61.7847***	62.4995***
Chi-square		3936.091***	3689.621***	2730.628***	1668.188***	1687.487***

(*continued*)

Table 6.5 (continued)

Variable	Model I	Model II	Model III	Model IV	Model V
Appropriate model	Fixed effect	Fixed effect	Fixed effect	Fixed effect	Fixed effect
Adjusted R-squared	0.8965	0.9120	0.9086	0.9187	0.9199
Log likelihood	191.4405	231.3463	221.9623	255.1122	261.2915
F-statistic	134.2430	160.3597	153.8327	139.9591	129.5693
Prob(F-statistic)	0.0000	0.0000	0.0000	0.0000	0.0000
Akaike info criterion	−0.6428	−0.8047	−0.7666	−0.8686	−0.8774
Schwarz criterion	−0.3616	−0.5235	−0.4854	−0.5193	−0.4940
Hannan-Quinn criter.	−0.5324	−0.6943	−0.6562	−0.7314	−0.7269
Durbin-Watson stat	1.2874	1.5254	1.4547	1.5991	1.6294

Source: Author's own elaboration on the basis of the modelling results

Note: *** $p < 0.01$; ** $p < 0.05$; * $p < 0.1$; and robust *t*-statistics in brackets. ECON stands for economic dimension; SOC stands for social dimension; ENV stands for environmental dimension; INST stands for institutional dimension; PCGDP stands for per capita GDP; INV stands for investment; REC stands for renewable energy consumption; EIL stands for energy intensity level of primary energy; UEMPL stands for unemployment rate; SANIT stands for sanitation facilities; LIFE stands for life expectancy at birth; AFR stands for adolescent fertility rate;CO_2 stands for emissions of carbon dioxide; ARAL stands for arable land; FORE stands for forest area; WATS stands for improved water source; GOV stands for government stability; SOCIO stands for socio-economic conditions; CORR stands for corruption; DEMO stands for democratic accountability; and C stands for constant

Model III that focuses on indicators of environmental efficiency reveals that structural transformation is enhanced by increased carbon dioxide emissions and improved access to water. However, access to arable land has a significantly negative impact on transformation while forest cover has no significant impact as well.

Unlike the other models, Model IV, which takes into account the economic, social, and environmental dimensions of sustainable development, reveals the positive and significant contribution of per capita incomes to structural transformation. However, renewable energy now has a positive and significant impact on structural transformation while carbon dioxide emissions no longer have a significant impact when the economic and social dimensions are taken into account. This points to the potential role of renewable energy in facilitating transformation in Africa.

Including institutional variables in the model (V) further improves the explanatory power of the model as evidenced by increased value of both the adjusted R-squared and log likelihood estimates. However, the level of significance of per capita incomes and renewal energy consumption are lower implying a weaker impact on transformation when institutional factors are taken into account. Democracy and accountability and socio-economic conditions have a positive and statistically significant bearing on structural transformation. Government stability and corruption carry the expected signs but are not statistically significant.

Conclusion

The most important insight and findings gained from our analysis are that an inclusive and sustainable structural transformation agenda requires tackling the economic, social, and environmental dimensions of sustainable development in an integrated way. A silo approach that focuses on one dimension at the expense of the other has a less optimal impact on transformation. For instance, our results show that when the economic and social dimensions are ignored, structural transformation is associated with statistically significant emissions of carbon dioxide. An integrated approach also minimises trade-offs between transformation and environmental preservation. The findings also reveal synergies between transformation and adult fertility rates as well as transformation and access to social services such as sanitation.

An important policy implication of our analysis is that mainstreaming the three dimensions of sustainable development into national and regional planning frameworks is imperative for successful implementation of the 2030 Agenda for sustainable development. However, several African countries and governments tend to have weak capacities to mainstream such priorities and policies in their national planning frameworks and to analyse the trade-offs associated with their policy choices (UN 2015a). Hence, strengthening capacities for development planning and evidence-based policymaking is imperative. However, this will require improved access to reliable data which can be achieved by strengthening the capacities of National Statistical Offices for data collection and analysis.

Acknowledgements Reprinted from Development, *Can the SDGs Promote Structural Transformation in Africa? An Empirical Analysis*, Vol. 58, Issue 4, 2015, pp. 473–491, Bartholomew Armah and Seung-Jin Baek, (original copyright notice as given in the publication in which the material was originally published), with permission of Springer.

Notes

1. Analysis based on the OECD International Development Statistics Database (2016).
2. According to the UNSD (United Nations Statistics Division) (2015), 30 African countries reached completion point and received significant debt relief under the enhanced HIPC initiative. Other three African countries, including Eritrea, Somalia, and Sudan reached pre-decision point under the initiative. The MDRI, which was provided a 100 per cent debt cancellation to low-income African countries, has further mitigated the effects of external debt burden, amounting cumulatively US$31.7 billion in end-2013 net present value terms.
3. The UN General Assembly Resolution 66/288. "The future we want," July 27, 2012, paragraph 246, accessed November 18, 2015, from: http://www.un.org/en/ga/search/view_doc.asp?symbol=%20A/RES/66/288
4. They are Algeria, Angola, Botswana, Burkina Faso, Cameroon, Cote d'Ivoire, Egypt, Gabon, the Gambia, Ghana, Guinea, Guinea-Bissau,

Kenya, Madagascar, Malawi, Mali, Morocco, Mozambique, Namibia, Niger, Nigeria, Senegal, Sierra Leone, South Africa, United Republic of Tanzania, Togo, Tunisia, Uganda, and Zambia.
5. Analysis based on the World Bank's World Development Indicators (2015).

References

Acemoglu, D., and J. Robinson. 2012. *Why Nations Fail: The Origins of Power, Prosperity, and Poverty*. New York: Crown Publishing Group.

AfDB (African Development Bank). 2012. *Income Inequality in Africa*, Briefing Note 5. Abidjan: AfDB. Available from: http://www.afdb.org/fileadmin/uploads/afdb/Documents/Policy-Documents/FINAL%20Briefing%20Note%205%20Income%20Inequality%20in%20Africa.pdf. Accessed 10 Nov 2015.

Aghion, P., and P. Bolton. 1992. Distribution and Growth in Models of Imperfect Capital Markets. *European Economic Review* 36 (3): 603–611.

Akbar, A., M. Imdadullah, M.A. Ullah, and M. Aslam. 2011. Determinants of Economic Growth in Asian Countries: A Panel Data Perspective. *Pakistan Journal of Social Sciences* 31 (1): 145–157.

Akbarzadeh, M., Z. Yazdanpanahi, L. Zarshenas, and F. Sharif. 2016. The Women's Perceptions About Unwanted Pregnancy: A Qualitative Study in Iran. *Global Journal of Health Science* 8 (5): 189–196.

Armah, B., M. Keita, A. Gueye, V. Bosco, J. Ameso, and Z. Chinzara. 2014. Structural Transformation for Inclusive Development in Africa: The Role of Active Government Policies. *Development* 57 (3–4): 438–451.

AUC (African Union Commission). 2014. *Agenda 2063: The Africa We Want*. Addis Ababa: AUC. Available from: http://archive.au.int/assets/images/agenda2063.pdf. Accessed 18 Nov 2015.

Baltagi, B.H. 2013. *Econometric Analysis of Panel Data*. 5th ed. New York: John Wiley & Sons.

Banerjee, A.V., and A.F. Newman. 1993. Occupational Choice and the Process of Development. *Journal of Political Economy* 101 (2): 274–298.

Bond, S.R. 2002. Dynamic Panel Data Models: A Guide to Micro Data Methods and Practice. *Portuguese Economic Journal* 1 (2): 141–162.

Bugaje, I.M. 2006. Renewable Energy for Sustainable Development in Africa: A Review. *Renewable and Sustainable Energy Reviews* 10 (6): 603–612.

Calderon, C., and L. Liu. 2003. The Direction of Causality between Financial Development and Economic Growth. *Journal of Development Economics* 72 (1): 321–334.

Cetorelli, N., and M. Gambera. 2001. Banking Market Structure, Financial Dependence and Growth: International Evidence from Industry Data. *Journal of Finance* 56 (2): 617–648.

Chang, Y.S., and S.J. Baek. 2010. Limit to Improvement: Myth or Reality? Empirical Analysis of Historical Improvement on Three Technologies Influential in the Evolution of Civilization. *Technological Forecasting and Social Change* 77 (5): 712–729.

Christiaensen, L., L. Demery, and J. Kuhl. 2011. The (Evolving) Role of Agriculture in Poverty Reduction: An Empirical Perspective. *Journal of Development Economics* 98 (2): 239–254.

Collier, P., and A.J. Venables. 2012. Greening Africa? Technologies, Endowments and the Latecomer Effect. *Energy Economics* 34 (S1): 75–84.

Dabla-Norris, E., A. Thomas, R. Garcia-Verdu, and Y. Chen. 2013. *Benchmarking Structural Transformation Across the World*, IMF Working Paper, (WP/13/176). Available from: https://www.imf.org/~/media/Websites/IMF/imported-full-text-pdf/external/pubs/ft/wp/2013/_wp13176.ashx. Accessed 10 Nov 2015.

Dempster, A.P., N.M. Laird, and D.B. Rubin. 1977. Maximum Likelihood from Incomplete Data via the EM Algorithm. *Journal of the Royal Statistical Society Series B (Methodological)* 39 (1): 1–38.

ECA (Economic Commission for Africa), and AUC. 2012. *Economic Report on Africa 2012: Unleashing Africa's Potential as a Pole of Economic Growth*. Addis Ababa: ECA. Available from: https://www.uneca.org/sites/default/files/uploaded-documents/CoM/cfm2012/issuespaper-unleashingafricapotential-as-a-newpole.pdf. Accessed 4 Aug 2015.

ECA, AUC, AfDB, and UNDP (United Nations Development Programme). 2011. *MDG Report 2011 – Assessing Progress in Africa Toward the Millennium Development Goals*. Addis Ababa: ECA. Available from: https://www.uneca.org/sites/default/files/PublicationFiles/mdgreport2011_book_eng.pdf. Accessed 10 Sept 2015.

ECA, AUC, AfDB, and UNDP. 2014. *MDG Report 2014 – Assessing Progress in Africa Toward the Millennium Development Goals: Analysis of the Common African Position on the Post-2015 Development Agenda*. Addis Ababa: ECA. Available from: https://www.uneca.org/sites/default/files/PublicationFiles/mdgreport2014_book_eng.pdf. Accessed 10 Sept 2015.

ECA, AUC, AfDB, and UNDP. 2015. *MDG Report 2015 – Assessing Progress in Africa Toward the Millennium Development Goals: Lessons Learned in Implementing the MDGs*. Addis Ababa: ECA. Available from: https://www.

uneca.org/sites/default/files/PublicationFiles/mdg-2015_eng_16sep_rev2.pdf. Accessed 19 Nov 2015.

Ehrhart, C. 2009. *The Effects of Inequality on Growth: A Survey of the Theoretical and Empirical Literature*, ECINEQ WP 2009–107. Society for the Study of Economic Inequality. Available from: http://www.ecineq.org/milano/WP/ECINEQ2009-107.pdf. Accessed 22 Oct 2015.

FAO (Food and Agriculture Organization). 2015. *FAO Statistics Division Database* [Online]. Available from: http://faostat3.fao.org. Accessed 10 Nov 2015.

Fayissa, B. 2001. The Determinants of Infant and Child Mortality in Developing Countries: The Case of Sub-Sahara Africa. *Review of Black Political Economy* 29 (2): 83–100.

Field, A. 2000. *Discovering Statistics Using SPSS for Windows: Advanced Techniques for the Beginner*. London: Sage.

Foody, G.M. 2003. Remote Sensing of Tropical Forest Environments: Towards the Monitoring of Environmental Resources for Sustainable Development. *International Journal of Remote Sensing* 24 (20): 4035–4046.

Giddings, B., B. Hopwood, and G. O'Brien. 2002. Environment, Economy and Society: Fitting Them Together into Sustainable Development. *Sustainable Development* 10 (4): 187–196.

Gorsuch, R.L. 1983. *Factor Analysis*. 2nd ed. Hillsdale: Erlbaum.

Grossman, G.M., and A.B. Krueger. 1995. Economic Growth and the Environment. *Quarterly Journal of Economics* 110 (2): 353–377.

Hsiao, C. 2014. *Analysis of Panel Data*. 3rd ed. London: Cambridge University Press.

ILO (International Labour Organization). 2014. *Global Employment Trends 2014: The Risk of a Jobless Recovery*. Geneva: ILO. Available from: http://www.ilo.org/wcmsp5/groups/public/---dgreports/---dcomm/---publ/documents/publication/wcms_233953.pdf. Accessed 21 Oct 2015.

IMF. 2015. *World Economic Outlook Database* [Online]. Available from: http://www.imf.org/external/ns/cs.aspx?id=28. Accessed 10 Nov 2015.

Jain, A.K. 1985. The Impact of Development and Population Policies on Fertility in India. *Studies in Family Planning* 16 (4): 181–198.

Jarvis, S., and S. Jenkins. 1998. How Much Income Mobility Is There in Britain? *Economic Journal* 108 (447): 428–443.

Katukiza, A.Y., M. Ronteltap, A. Oleja, C.B. Niwagaba, F. Kansiime, and P.N.L. Lens. 2010. Selection of Sustainable Sanitation Technologies for Urban Slums: A Case of Bwaise III in Kampala, Uganda. *Science of the Total Environment* 409 (1): 52–62.

Kelbore, Z.G. 2014. *Multidimensional Structural Transformation Index: A New Measure of Development*, MPRA Paper No. 62920 [Online]. Available from: https://mpra.ub.uni-muenchen.de/62920/1/MPRA_paper_62920.pdf. Accessed 22 Sept 2015.

Kenney, G.I. 1995. The Missing Link – Information. *Information Technology for Development* 6 (1): 33–38.

Kibret, A., H. Bayu, and M. Merga. 2014. Prevalence of Unintended Pregnancy and Associated Factors Among Pregnant Women Attending Antenatal Clinics in Debre-Markos Town, North West Ethiopia 2012. *Journal of Women's Health Care* 4 (3): 1–6.

Klima, C.S. 1998. Unintended Pregnancy: Consequences and Solutions for a Worldwide Problem. *Journal of Nurse-Midwifery* 43 (6): 483–491.

Kuznets, S. 1955. Economic Growth and Income Inequality. *American Economic Review* 45 (1): 1–28.

LDC IV Monitor. 2015. *Istanbul Programme of Action for the LDCs (2011–2020): Monitoring Deliverables, Tracking Progress – Analytical Perspectives*. London: Commonwealth Secretariat.

Mansell, R., and U. Wehn. 1998. *Knowledge Societies: Information Technology for Sustainable Development*. Oxford: Oxford University Press.

McQueston, K., A. Glassman, and R. Silverman. 2012. *Adolescent Fertility in Low and Middle-Income Countries: Effects and Solutions*, Working Paper 292. Centre for Global Development.

Morris, M.D. 1979. *Measuring the Condition of the World's Poor: The Physical Quality of Life Index*. New York: Pergamon.

Nurkse, R. 1953. *Problems of Capital Formation in Underdeveloped Countries*. Oxford: Oxford University Press.

OECD. 2016. *International Development Statistics Database* [Online]. Available from: http://www.oecd.org/development/stats/idsonline.htm. Accessed 6 Mar 2016.

Osman, R.H., C. Alexiou, and P. Tsaliki. 2012. The Role of Institutions in Economic Development: Evidence from 27 Sub-Saharan African Countries. *International Journal of Social Economics* 39 (1/2): 142–160.

Perotti, R. 1996. Growth, Income Distribution and Democracy: What the Data Say. *Journal of Economic Growth* 1 (2): 149–187.

Piketty, T. 1997. The Dynamics of the Wealth Distribution and the Interest Rate with Credit Rationing. *Review of Economic Studies* 64 (2): 173–189.

Political Risk Services Group. 2015. *International Country Risk Guide Database* [Online]. Available from: http://www.prsgroup.com/ICRG.aspx. Accessed 10 Nov 2015.

Ray, A.K. 2008. Measurement of Social Development: An International Comparison. *Social Indicators Research* 86 (1): 1–46.

Rodrik, D., A. Subramanian, and F. Trebbi. 2002. *Institutions Rule: The Primacy of Institutions over Geography and Integration in Economic Development*, NBER Working Paper No. 9305. Cambridge: National Bureau of Economic Research.

Saltelli, A. 2007. Composite Indicators between Analysis and Advocacy. *Social Indicators Research* 81 (1): 65–77.

Simon, D. 2013. Climate and Environmental Change and the Potential for Greening African Cities. *Local Economy* 28 (2): 203–217.

Stiglitz, J.E. 1996. Some Lessons from The East Asian Miracle. *World Bank Research Observer* 11 (2): 151–177.

Timmer, C.P., M.M.O. Badiane, D. Rodrik, H. Binswanger-Mkhize, and F. Wouterse 2012. *Patterns of Growth and Structural Transformation in Africa: Trends and Lessons for Future Development Strategies*, IFPRI Thematic Research Note 2. Washington, DC: International Food Policy Research Institute. Available from: http://www.ifpri.org/cdmref/p15738coll2/id/126946/filename/127157.pdf. Accessed 28 June 2015.

UN (United Nations). 1992. *Earth Summit, Agenda 21: The United Nations Programme of Action from Rio*. New York: UN.

UN. 2001. *Indicators of Sustainable Development: Framework and Methodologies*, Division of Sustainable Development Background Paper No. 3. New York: UN.

———. 2008. *Measuring Sustainable Development*. New York: UN (Report of the Joint UNECE/OECD/Eurostat Working Group on Statistics for Sustainable Development).

———. 2015a. *Addis Ababa Action Agenda of the Third International Conference on Financing for Development*. UN: Addis Ababa. Available from: http://www.un.org/esa/ffd/wp-content/uploads/2015/08/AAAA_Outcome.pdf. Accessed 2 Sept 2015.

———. 2015b. *Transforming Our World: The 2030 Agenda for Sustainable Development*, A/RES/70/1. New York: UN. Available from: https://sustainabledevelopment.un.org/content/documents/21252030%20Agenda%20for%20Sustainable%20Development%20web.pdf. Accessed 1 June 2015.

UNEP (United Nations Environment Programme). 2015. *Building Inclusive Green Economies in Africa: Experience and Lessons Learned 2010–2015*. Nairobi: UNEP.

UNSD (United Nations Statistics Division). 2015. *Millennium Development Goals Database* [Online]. Available from: http://mdgs.un.org/unsd/mdg/default.aspx. Accessed 10 Nov 2015.

Veeman, T.S., and J. Politylo. 2003. The Role of Institutions and Policy in Enhancing Sustainable Development and Conserving Natural Capital. *Environment, Development and Sustainability* 5 (3): 317–332.

Vera, I., and L. Langlois. 2007. Energy Indicators for Sustainable Development. *Energy* 32 (6): 875–882.

World Bank. 2015. *World Development Indicators* [Online]. Available from: http://data.worldbank.org/datacatalog/world-development-indicators. Accessed 10 Nov 2015.

———. 2016. *World Development Indicators* [Online]. Available from: http://data.worldbank.org/datacatalog/world-development-indicators. Accessed 5 Apr 2016.

Zander, P., and H. Kächeleb. 1999. Modelling Multiple Objectives of Land Use for Sustainable Development. *Agricultural Systems* 59 (3): 311–325.

Part III

Theory of the Developmental S-Curve in Process of Development

7

Limit to Improvement: Myth or Reality?

Empirical Analysis of Historical Improvement on Three Technologies Influential in the Evolution of Civilisation

Introduction

The evolution of civilisation appears to have primarily resulted from continuous improvement made possible by technological advances. A group of social evolutionists and economists have identified, among others, energy (White 1949; Nordhaus 1997), transport (Kipling 1917; Lewis 1936), as well as information and communication (Morgan 1877; Lenski 1974, 1984; Bell 1980) to be the three most influential technologies.

On the other hand, a number of eminent scholars beginning with Thomas Robert Malthus (1798) have cited several forces, natural, physiological, technological, as well as environmental, which can place a limit on on-going improvement (Modis 2002; Jonathan 2005; Turner 2008). Perhaps the best-known work is *The Limits to Growth* by Donella H. Meadows et al. (1972), published in 1972, which argues that the world may run short of resources, placing a limit on growth. The most recent updated revision of *The Limits to Growth* was published in 2004 (Meadows et al. 2004). In 2008, Graham Turner's 'A Comparison of *The Limits to Growth* with Thirty Years of Reality' found that 'changes in industrial production, food production, and pollution are all in line with

S. J. Baek, *The Political Economy of Neo-modernisation*,
https://doi.org/10.1007/978-3-319-91394-0_7

the book's predictions of economic and social collapse in the twenty-first century'.[1]

In direct opposition to the concept of limit to improvement, it has been proposed that improvement will continue and will accelerate without limit (Good 1965; Vinge 1993; Hanson 2008; Jorgenson and Stiroh 2000). Futurist Ray Kurzweil (2005) popularised the idea of technological singularity by publishing *The Singularity Is Near*. In it, he argued that through the law of accelerating returns, technology is progressing towards the singularity to create super-intelligence.

The purpose of this study is to explore this paradox between improvement versus limit. We plan to empirically analyse historical data available on the three technologies that have been mentioned. In order to conduct such an analysis, we have developed a new framework of a disconnected multiple technology S-curve as well as the measure defined as X-factor.

This chapter is organised as follows. First, we will review the concept of multiple technology S-curves in order to develop connected versus disconnected S-curves. The performance difference of emerging over mature technology will then be measured by X-factor. Second, published sources of historical data on the three technologies will be briefly explained. Third, a statistical analysis will be conducted for each of the three selected technologies. Fourth, we will generate four propositions from our analysis. Finally, we will discuss a number of policy implications and suggest areas for future research.

Multiple Technology S-Curves

In 1986, Foster attempted to operationalise Schumpeter's concept of *Creative Destruction* (1939) by proposing that the technology S-curve may explain how a new technology may replace existing technology. For a single technology S-curve for a given technology, Foster (1986), Dosi (1982), Sahal (1985), Utterback (1996), and Becker and Speltz (1983) have identified three stages of evolution from introduction, growth, and maturity.

In the introduction stage, the progress of technology is often slow due to the fact that technology faces several technical bottlenecks yet to be resolved. Also, not many researchers are attracted, as technology is not well-known. The growth stage takes place after the technology has proven its utility. The emergence of dominant design (Utterback and Abernathy 1975; Suarez and Utterback 1995) and subsequent acceptance of the standard increase the volume of sales which, in turn, increases support for research. Thus, the pace of improvement quickens during this stage. Finally, the performance level of technology enters its maturity stage when progress occurs very slowly or even stops reaching a ceiling (Foster 1986; Brown 1992; Chandy and Tellis 2000). For example, the speed of propeller aircraft and transmission speed of cooper wire reach their speed limit determined by physical and natural law. Or progress may slow due to limits of scale or system complexity that technology has reached. When it slows, some innovators introduce new technology, which initially has lower performance than that of old one (Foster 1986; Christensen 2003). This implies that the first technology S-curve is likely to be followed by second, third, and other technology S-curves as shown in Fig. 7.1.

Firms are advised to make a timely switch from old to new and to follow the line to continue their technology improvement. More recent evidence in such industries as electronic components, computers, hard disks, and communications indicates that the new generation of technologies is likely to be pioneered by new firms which did not exist before or by firms from outside the industries involved (Foster 1986; Christensen 2003). The succession of individual technologies over time may be viewed as fractal aspect of S-curve whereupon several cascading S-curves can give rise to a single large-scale S-curve (Modis 1994). In such a large-scale S-curve, there are several overlapping regions representing the transition from one S-curve to the next one. Analysis of these regions in detail can be the key to learn how technology advances with cascading S-curves.

In order to use the multiple technology S-curve as our framework of analysis, it would be more useful to differentiate two types of multiple technology S-curves. One is the traditional case where the performance

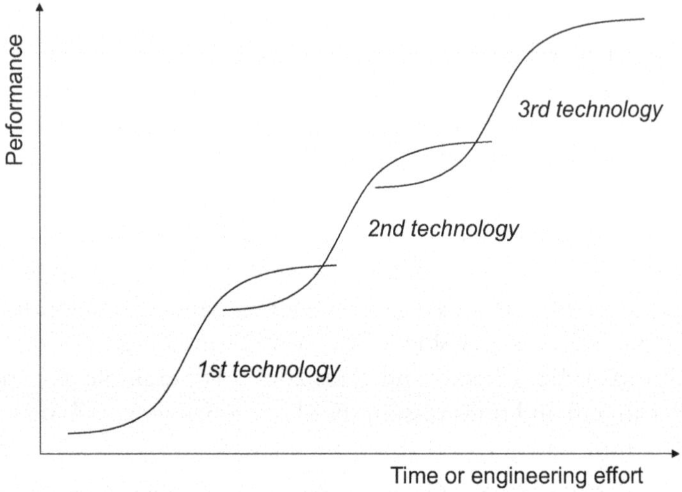

Fig. 7.1 Multiple technology S-curves. (*Source*: Author's own elaboration on the basis of Christensen (1992))

of new technology begins below that of old technology. In time, however, the performance of new technology makes rapid advancement to overtake and disrupt the old technology (Christensen 2003; Christensen and Raynor 2004; Christensen et al. 2003). Another type is where the performance of new technology begins above that of old technology. We shall name the first type as connected multiple technology S-curves, and the second type as disconnected multiple technology S-curves. These two different concepts are shown in Fig. 7.2. In the study of long-term technology evolution, we believe that the use of both connected and disconnected multiple technology S-curves will provide much more important and significant insights, as we shall demonstrate later in this study.

Several previous studies have compared the performance difference between emerging technology over mature technology. For example, Anderson and Tushman (1990) have observed that the 1903 Lubbers machine can produce 700 square feet of flat-glass per hour, or 4.67 times more than an artisan blowing cylinders by hand. More recently, Sood and Tellis (2005) reported that in four out of ten technologies they have exam-

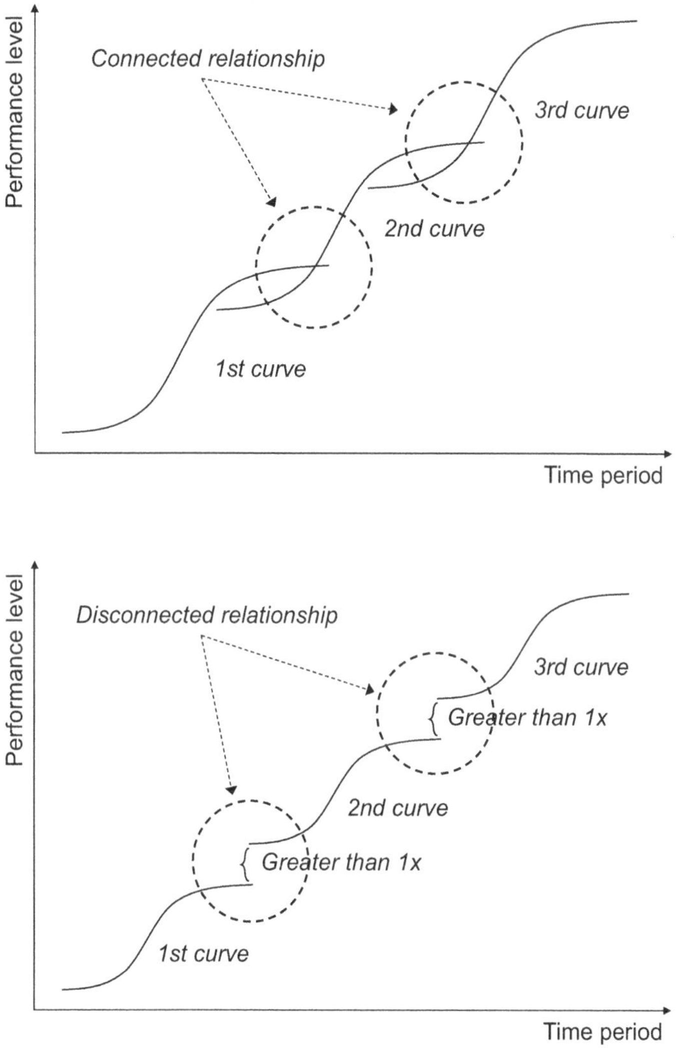

Fig. 7.2 Two types of multiple technology S-curves. (*Source*: Author's own elaboration)

ined, new technologies performed better than the old from the time they were introduced. For example, Fig. 7.3 shows no crossing between the pair of fibre optic (new) over Cu/Al wires (old) and wireless technologies. Clearly, fibre optic has performed much better than old technologies from its start.

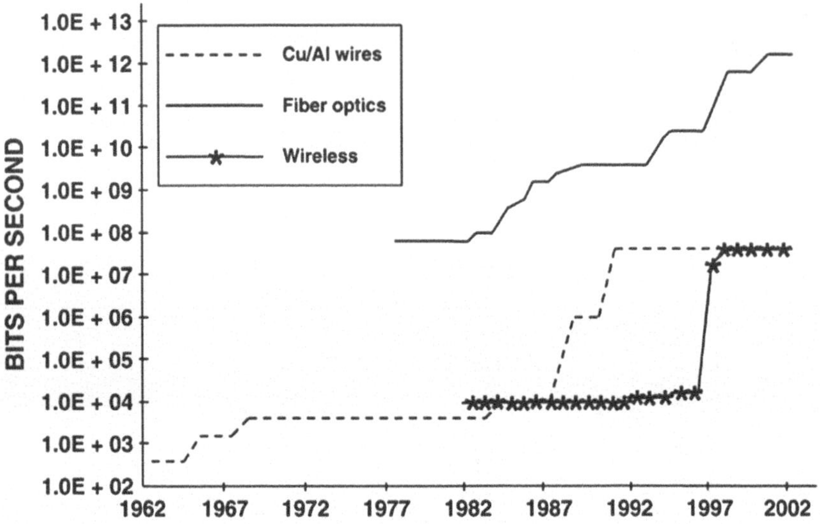

Fig. 7.3 Disconnected technology S-curve in data transfer. (*Source*: Author's own elaboration on the basis of Sood and Tellis (2005))

According to Foster, McKinsey's guideline on the performance difference between new and old technology is in the order of 20 or 30 to 1 in electronic industries and 5 to 1 in other industries.[2]

Our analysis requires a more formal definition of the performance difference between emerging new technology and mature old technology as well as the connected versus disconnected technology S-curve.

First, we shall name the performance difference as X-factor. X-factor is then:

$$\text{X-factor} = \frac{Es}{Ms},$$

where Es represents performance level of emerging technology at the time of its 'start' while Ms represents performance level of mature technology at the time of the 'start' of emerging technology.

And then,

when X-factor ≥ 1, we have disconnected S-curves
and when X-factor < 1, we have connected S-curves.

It is particularly important that the performance difference of emerging technology over mature technology be measured at the same time period that emerging technology begins. If published data on mature technology are not available at the time of the 'start', the performance measure of mature technology needs to be extrapolated from the regression models that will be explained next.

For our regression analysis of the S-curve, we have first chosen to use the logistic curve (Sood and Tellis 2005; James and Sood 2006), which is shown as Formula (7.1).[3]

$$y = \alpha + \frac{(\eta - \alpha)}{1 + \exp^{-\beta(x-\mu)}}, \tag{7.1}$$

where α is lower asymptote, η is target height, β is growth rate,[4] μ is time of maximum growth.

We recognised the possible errors in estimating both upper and lower asymptote from the logistic analysis (Young 1993; DeBecker and Modis 1994). Therefore, when the result of the logistic curve fit appears poor or unreliable, we have then used the exponential growth curve as defined in Formula (7.2)[5]:

$$y = \exp(\beta_0 + \beta_1 x), \tag{7.2}$$

where β_0 is constant, β_1 is annual growth rate.

Sources of Historical Data

As we have decided to analyse the three technologies (energy, transport, and information and communication), which appeared to have significantly influenced the evolution of civilisation, we will briefly explain the reason for selecting these three elements as well as published sources of historical data we have used.

The first element to be analysed is energy. In his book, *The Evolution of Culture: The Development of Civilisation to the Fall of Rome*, anthropologist Leslie A. White claimed that 'Culture evolves as the amount of energy harnessed per capita per year is increased, or as the efficiency of the instrumental means of putting the energy to work is increased'.[6] This became the so-called White's Law.[7] Since one of the most important uses of energy is lighting, we have decided to focus on the history of lighting. The primary sources of data are from Nordhaus's article (1997), supplemented by *The Museum of Electric Lamp Technology* (Hooker 2009) and others, including conference papers and academic articles (Brodrick 2008; Narukawa et al. 2007; Denbaars 2008; Horner 2008; Young 2008).

The second element influencing the evolution of civilisation selected is transport. Rudyard Kipling remarked that 'Transportation is Civilisation'.[8] William Lewis once said, 'No great civilisation has been built without some well-defined system of transportation.'[9] Both have emphasised that the significance of transportation to civilisation is critical. For historical data on trains, cars, aircraft, and spacecraft, we have used data from *Railway speed records, Les chiffres clés des records sur rail dans le monde and SVT 137 153ff. Leipzig* for trains, *The history of the coveted world record* for cars, and *All the world's rotorcraft, and Apollo 10* for aircraft and spacecraft (Procter 1997; Les chiffres clés des records sur rail dans le monde 2009; Starostin 2009; Wilbrink 2009; The History of the Coveted World Record 2009; Wade 2009).

The last element selected is information and communication. Sociologist Gerhard Lenski (1974, 1984) claims that the more information in use, the more advanced the society. Daniel Bell said, 'In the coming century, the emergence of a new social framework based on telecommunications may be decisive for the way in which economic and social exchanges are conducted, the way knowledge is created and retrieved, and the character of work and occupations in which men engage.'[10] Here, we collected data on computing speed from Nordhaus's article, the undersea cable system from Koh and Magee's article and Internet backbone systems from Kurzweil's book (Nordhaus 2001; Kurzweil 2005; Koh and Magee 2006).

Analysis of Historical Data

Energy

The history of improvement in lighting is as old as the history of civilisation. The origin of lighting sources goes back to when Australopithecus discovered fire in the earliest time. The history of lighting[11] then included the candle, lamp, incandescent lamp, fluorescent lamp, and light-emitting diode (LED) lamp. Figure 7.4 describes the progress of lighting performance, using luminous efficacy[12] shown as the performance measure on the vertical axes. This graph shows several disconnected points such as open fire to Neolithic lamp, Neolithic lamp to Babylonian lamp, Babylonian lamp to candle, candle to oil lamp, gas lamp or kerosene lamp to tungsten filament lamp, tungsten filament lamp to compact fluorescent lamp, and compact fluorescent lamp to white LED lamp. However, data available for a statistical analysis are limited to three tech-

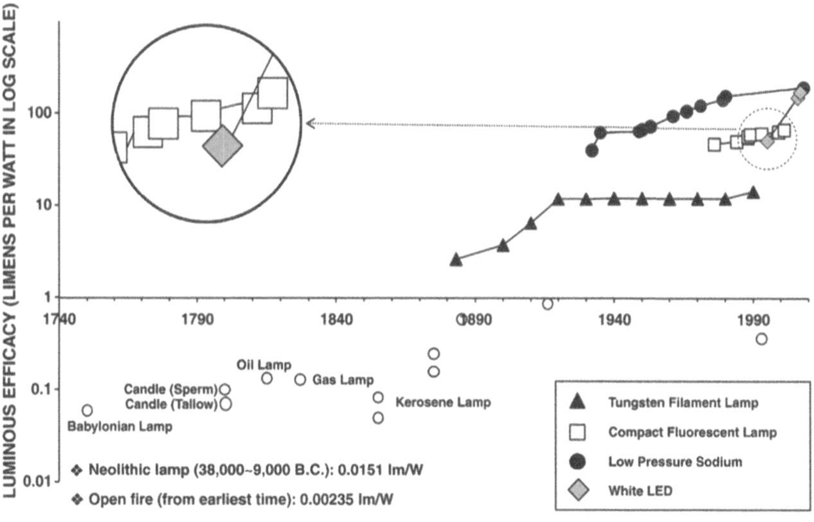

Fig. 7.4 Progress of lighting efficacy. (*Source*: Author's own elaboration on the basis of Nordhaus (1997), Brodrick (2008), Narukawa et al. (2007), Denbaars (2008), Horner (2008), Young (2008), and Hooker (2009). *Note*: Detailed backup data are available in Appendix 7)

nologies, the tungsten filament lamp and compact fluorescent lamp for indoor lighting and the low-pressure sodium lamp for outdoor lighting.

Figure 7.5 depicts each lighting technology's progress along the logistic curve. For indoor lighting, both tungsten filament lamp and compact fluorescent lamp technologies follow respective logistic curves with an R^2 of 99.6 per cent for the tungsten filament lamp and 96.3 per cent for the compact fluorescent lamp. When the compact fluorescent lamp appeared in 1976, its luminous efficacy was 47.2, which exceeded the tungsten filament lamp's efficacy of 12 at the same time period by an X-factor of 3.94X. This demonstrates clear evidence of disconnected S-curves between the tungsten filament lamp and the compact fluorescent lamp.

Data available from the next technology of the white LED lamp are limited, so a statistical analysis was not performed. By late 1990, the performance from the white LED lamp began to exceed that of the compact fluorescent lamp and has been increasing ever since. However, the experimental nature of the technology does not allow us to estimate the precise magnitude. When we compare the technological performances between the white LED lamp in 1995 and the compact fluorescent lamp at the same time period, an X-factor is calculated as 0.80X. This shows that connected technology S-curves may have existed. Data analysed for low-pressure sodium technology for outdoor lighting has also shown a good fit with a logistic curve with R^2 of 98.8 per cent. The results of the regression and summary of the X-factor analysis are shown in Fig. 7.5 and Tables 7.1 and 7.2.

In summary, a number of different technologies have accounted for a dramatic improvement in lighting. An overall improvement generated from the beginning of the tungsten filament lamp, compact fluorescent lamp, and white LED lamp amounts to a 6400 per cent increase or 65 times better performance during the period of 125 years. Our statistical analysis of the tungsten filament lamp, compact fluorescent lamp, and white LED lamp indicates that the continuous progress of both connected and disconnected technologies played the critical role of improvement. To put it another way, this overall evolution may be viewed as the result of a single large-scale S-curve (Marchetti 1979). Now, we will move to the second element that has significantly contributed to civilisation: transport.

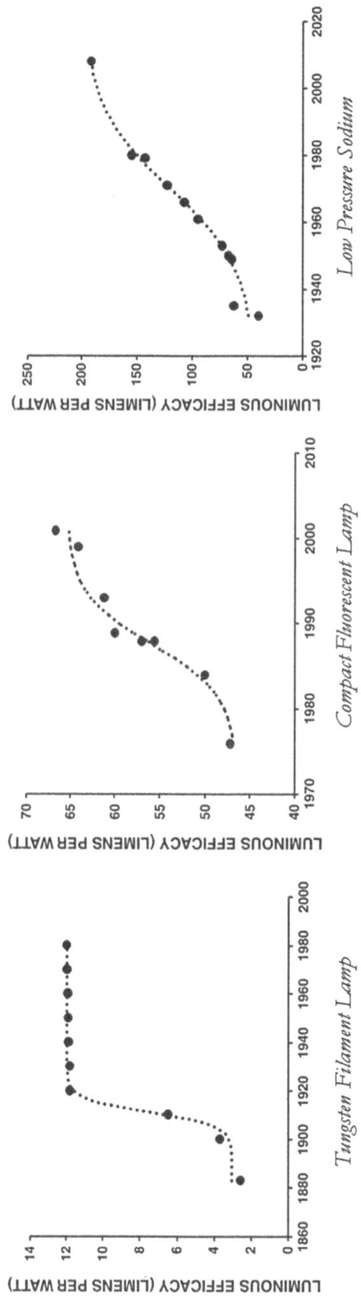

Fig. 7.5 Logistic curve analysis of three lighting technologies. (*Source*: Author's own elaboration on the basis of Nordhaus (1997) and Hooker (2009))

Table 7.1 Summary of results from the regression analysis on lighting

Elements	# of years	# of samples	Logistic function Growth rate-gradient (t-value)	Upper limit (t-value)	R^2
Tungsten filament lamp	98	10	0.355 (3.748)*	11.9683 (98.830)*	0.996
Compact fluorescent lamp	26	8	0.335 (2.104)**	65.3491 (38.661)*	0.993
Low-pressure sodium lamp	77	11	0.081 (5.711)*	199.1 (21.451)*	0.988

Source: Author's own elaboration on the basis of the modelling results
Note: * and ** denote the p-value significant at the 1 per cent and 5 per cent levels, respectively

Table 7.2 Summary of X-factor analysis on lighting

Comparison of two technologies	Year of comparison	X-factor	Type
Compact fluorescent lamp over tungsten filament lamp	1976	3.94	Disconnected
White LED lamp over compact fluorescent lamp	1995	0.80	Connected

Source: Author's own elaboration on the basis of the modelling results

Transport

The evolution of transport is also as old as the beginning of civilisation from human walking and running in the earliest times. The war chariots used by the Mesopotamians around 3000 BC to the American stagecoach in the early 1800s have all relied on the speed of horses to support the evolution of transport. Moving on to the train, automobile, aircraft, and eventually spacecraft, overall improvement can best be explained by each invention's reliance on totally different types of technology. Meis and Leta (1997) have dramatically described this overarching trend of transportation as shown in Fig. 7.6. By incorporating empirical studies by others, we have tried to conduct quantitative analysis of this overarching trend of transport. Another similar study by Arnulf Grubler (1990) documented the applicability of S-curves to length of transport infrastructures.

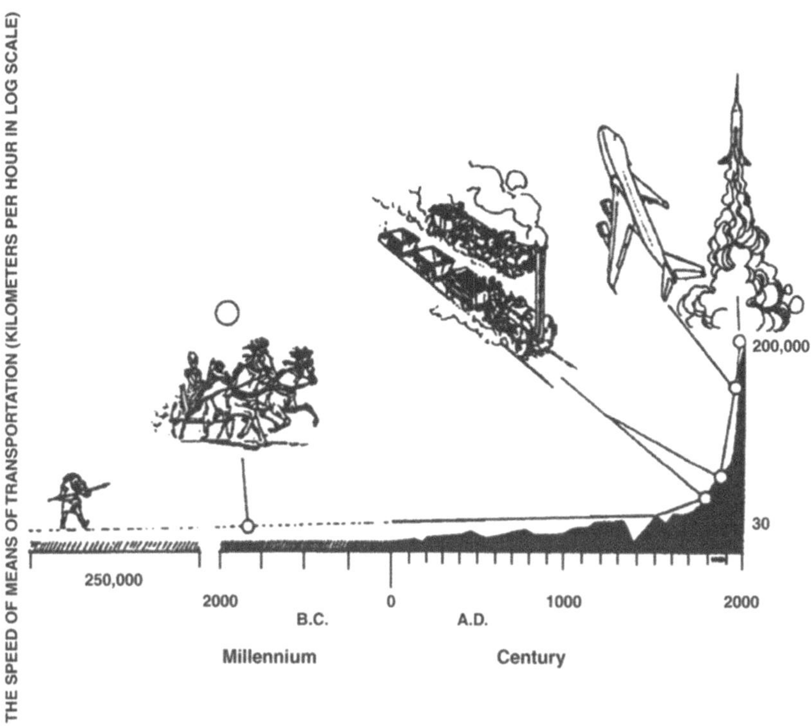

Fig. 7.6 Evolution of means of transportation. (*Source*: Meis and Leta (1997))

Figure 7.7 depicts the progress of transport according to its maximum speed in the basis of the most dominate means such as human walk through spacecraft. First, Bohannon (1997) analysed the physical maximum speed of the human walk to be approximately 9.11 kilometre per hour (km/h). Next, the maximum speed of a stagecoach is approximately 30 km/h.[13] Then, train, car, aircraft, and spacecraft have followed with a faster speed.

Figure 7.8 presents a logistic curve analysis of the improvement pattern of technologies of transport. Transport technologies analysed include the train, car, and aircraft.

As for the train, the diesel train surpassed in 1930s the steam train in speed. Within 20 years, the diesel train was superseded by the electric multiple unit train (EMU). These different technologies are responsible for the entire evolution of train technology, which again appears to follow a logistic curve.

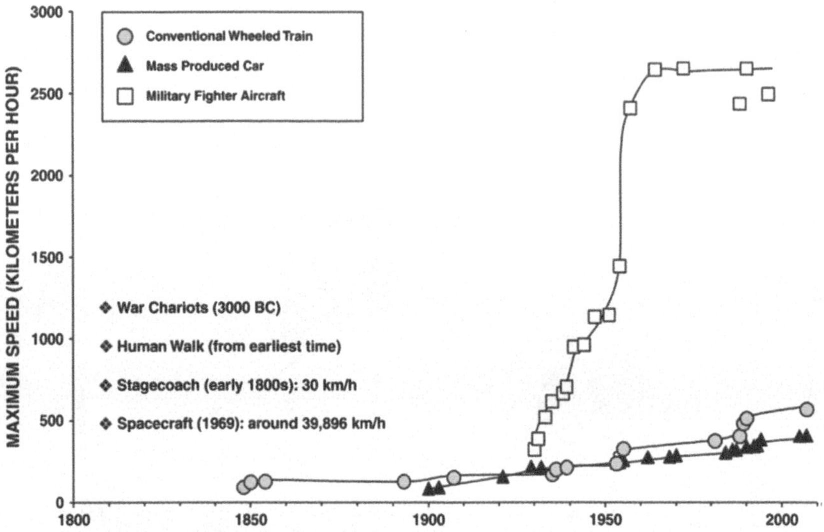

Fig. 7.7 Evolution of transport (maximum speed progress). (*Source*: Author's own elaboration on the basis of Bohannon (1997), Procter (1997), September Collings Foundation Newsletter (2008), Les chiffres clés des records sur rail dans le monde (2009), Starostin (2009), Wilbrink (2009), The History of the Coveted World Record (2009), and Wade (2009))

As for the passenger car, the Mercedes 35 hp, which was first introduced in 1900, had its maximum speed at 85 km/h. Since then, the maximum speed of a car has increased to 412.16 km/h. The overall improvement pattern of cars follows a logistic curve (R^2 = 0.937). Next, military fighter aircraft technology has dramatically improved its speed over the train or car. However, its performance seemed to have reached a certain limit during the 1970s. Thus, aircraft improvement appears to follow a logistic curve as well (R^2 = 0.957). Again, two different types of engines, piston and jet, have been combined into a single aircraft technology S-curve.

As shown in Tables 7.3 and 7.4, available data allow us to estimate three logistic models with respect to the train (R^2 = 0.968, growth rate of 0.034), car (R^2 = 0.937, growth rate of 0.029), and aircraft (R^2 = 0.957, growth rate of 0.215). In addition, the piston engine and jet engine under aircraft technology also follow a logistic curve with respective R^2 of 99.4 per cent and 98.8 per cent.

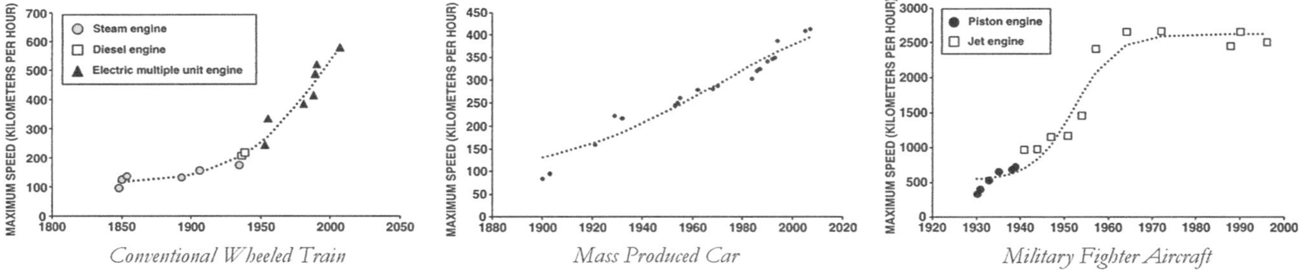

Fig. 7.8 Logistic curve analysis of three transport technologies. (*Source*: Author's own elaboration on the basis of Procter (1997), Les chiffres clés des records sur rail dans le monde (2009), Starostin (2009), Wilbrink (2009), and The History of the Coveted World Record (2009). *Note*: Detailed backup data are available in Appendix 8, 9, and 10)

Table 7.3 Summary of results from the regression analysis on transport

Elements	# of years	# of samples	Logistic function: Growth rate – gradient (t-value)	Upper limit (t-value)	R^2
Conventional wheeled Train	160	15	0.034 (3.167)*	869.9 (2.500)**	0.968
Mass produced car	108	20	0.029 (4.138)*	500.5 (4.740)*	0.937
Military fighter aircraft	67	17	0.215 (2.780)**	2621.4 (26.258)*	0.957
Piston engine	12	6	0.602 (2.347)	701.2 (31.017)	0.994
Jet engine	56	11	0.995 (3.783)*	2582.1 (57.327)	0.988

Source: Author's own elaboration on the basis of the modelling results
Note: * and ** denote the p-value significant at the 1 per cent and 5 per cent levels, respectively

Table 7.4 Summary of X-factor analysis on transport

Comparison of two technologies	Year of comparison	X-factor	Type
Conventional wheeled train over stagecoach	1848	3.22	Disconnected
Mass produced car over conventional wheeled train	1900	0.60	Connected
Military fighter aircraft (piston) over mass produced car	1930	1.79	Disconnected
Military fighter aircraft (jet) over military fighter aircraft (piston)	1941	1.37	Disconnected
Spacecraft over military fighter aircraft (jet)	1969	15.45	Disconnected

Source: Author's own elaboration on the basis of the modelling results

The aircraft's initial speed (327 km/h) in 1930, when compared with the train (189 km/h) and car (182 km/h) in the same time period, indicates that the aircraft represents the disconnected technology S-curve over those of the train and car by an X-factor of 1.79X. Further analysis of aircraft powered by a jet over piston engine indicates the existence of another disconnected S-curve in operation by an X-factor of 1.37X, as supported by a previous study (Lee and Nakicenovic 1988). Another comparison between spacecraft over a jet engine represents the disconnected technology S-curve by an X-factor of 15.45X. In contrast, a performance comparison between the train and car suggests the existence of connected multiple S-curves with an X-factor of 0.6X. This may be due to the fact that both are subjected to speed limits imposed by authorities. We can conclude that several disconnected technologies combined to make another dramatic improvement in transport, as shown in Table 7.4.

From the nearly 149-year period covering the use of stagecoach to jet aircraft, an overall improvement generated amounts to another dramatic increase of nearly 8750 per cent or 89 times faster speed of transport during this time period. And this overall improvement has been realised because of several disconnected technologies as well as the continuous progress made on both connected and disconnected technologies. It is interesting that the magnitude of this increase is comparable to that observed for the case of lighting technology. Now, we will cover the final element, which is the most influential technology in modern civilisation.

Information and Communication

Early years in the history of civilisation, the numbering system was the first tool developed in the history of calculation. By using numbers, humans invented the abacus, the earliest known calculation tool in 2400 BC. Then, many other tools followed with much greater speed of calculation. In the history of communication, human delivered message by using speech in earliest time. Then, symbols were developed to accelerate the speed of communication. As a modern tool, the telegraph was invented in 1844, showing its communication speed of 50–60 words per minute (Marchionini 1999).

The progress of modern information and communication technology may be measured by the speed of computer calculations and by the data transfer rate in telecommunication. The rate of improvement in modern information and communication has undergone a revolutionary explosion, as indicated by both Moore's law (computing power doubling every 18 months) and Gilder's law (communication power doubling every six months). More importantly, this revolution appears as though it will continue for some time to come (Moore 1965; Gilder 2002; Warren 2002; Aizcorbe 2005; Hutcheson 2005).

Figure 7.9 reproduces the analysis of an important pattern of computer calculations conducted by Nordhaus (2001) where speed of computing is measured by MSOPS.[14] He has categorised computer systems by different technologies in different time periods such as 1900, 1940, and 1950. For performance of computer speed, he calculated geometric means of different computer systems to generate the following performance difference of numbers of 8.81, 12.97, 1979.17, 27.89, 4.39, 26.88, and 248.00. For example, Nordhaus has calculated the performance difference of the microprocessor (during the 1990s) over that of the early microprocessor (during the 1980s) to be 26.88X. However, the long time gap of 10 years or more existing between the two technologies makes it not possible to use these as the value of X-factor. Although these numbers cannot be used to measure X-factors as we defined them, they represent meaningful findings, because they indicate the critical importance of new technologies over old ones.

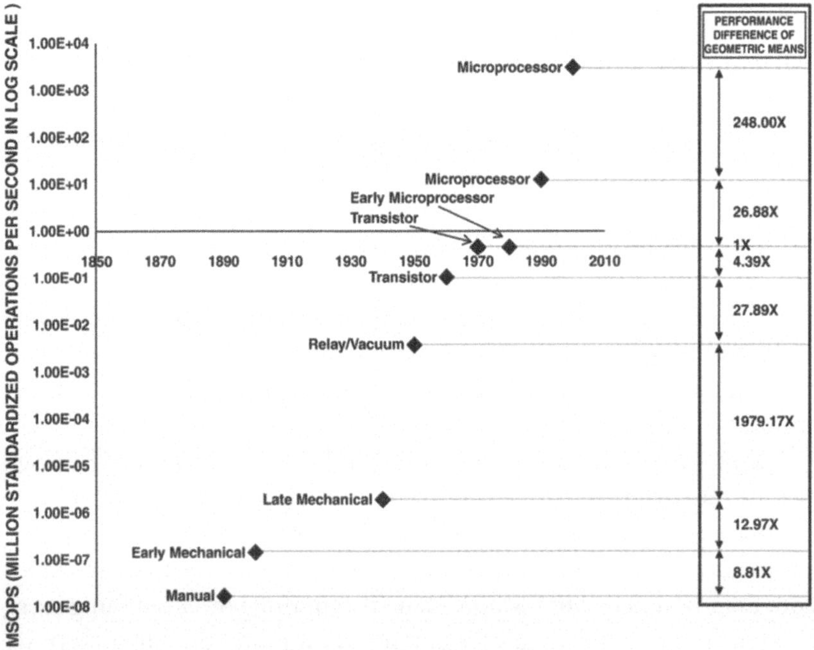

Fig. 7.9 History of computing speed. (*Source*: Author's own elaboration on the basis of Nordhaus (2001). *Note*: Detailed backup data are available in Appendix 11)

Figure 7.10 shows the full data from Nordhaus's study. For instance, he has recorded from a 1909 report from Burroughs, which compared the speed of trained clerks adding up long columns of numbers by hand with a Burroughs calculator, the comparison showing that the calculator had an advantage of about a factor of six.[15] On the other hand, the mechanical and relay/vacuum tube, and transistor and microprocessor appear to follow connected patterns. The mix of many different types of improving technologies in computer systems appears to have generated a very rapid growth rate of computing speed with no sign of slowing down. Thus, the use of exponential model has yielded a better fit to the data provided by Nordhaus. The resulting exponential model yielded the estimated annual growth rate of 22.1 per cent with R^2 of 0.813.

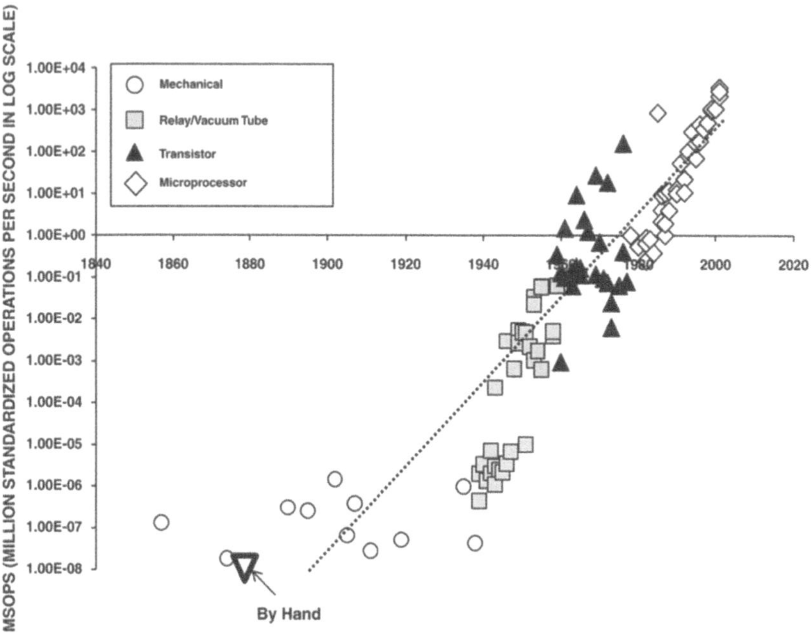

Fig. 7.10 Exponential growth curve analysis of computer speed. (*Source*: Author's own elaboration on the basis of Nordhaus (2001). *Note*: The dotted line represents the predicted values from growth curve regression. Moreover, detailed backup data are available in Appendix 12)

As for improvement in communication, we have used two meaningful data on undersea cable and Internet backbone systems. An undersea cable (a submarine communications cable) is a cable that is laid beneath the sea to carry telecommunications between countries. This is one of the most important factors in telecommunication progress (Jorgenson and Wessner 2007). Figure 7.11 depicts the improvement in data transfer rates of the undersea cable system. For performance on the vertical axes, bandwidth measurement was used.[16]

Historical improvement in undersea cable technology according to Koh and Magee (2006) has been developed by three different technologies: single cable, coaxial cable, and optical cable. The technological improvement in undersea cables started in the 1850s with a single cable system. However, this system appeared to have reached its limit around 1880s, although it continued to be used. In order to respond to the vast increase in demand

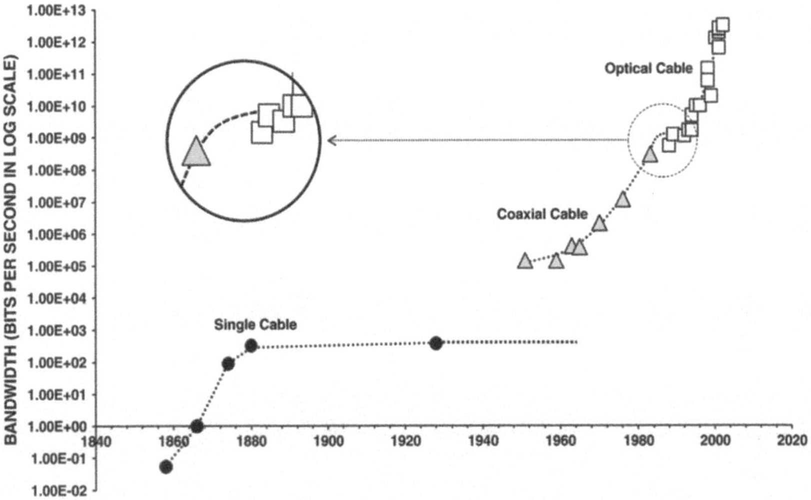

Fig. 7.11 Logistic curve analysis of the undersea cable system. (*Source*: Author's own elaboration on the basis of Koh and Magee (2006). *Note*: This graph is mainly derived from the article, 'A Functional Approach for Studying Technological Progress: Application to Information Technology'. We only focused on main trends such as single cable, coaxial cable, and optical cable. Detailed backup data are available in Appendix 13. Moreover, the dotted line represents the predicted values from logistic regression)

for data communication, a radically new type of coaxial cable system was developed in the 1950s. We have calculated the X-factor of 361X for the coaxial over the single cable system. In 1988, the optical cable system was introduced as the next generation to replace the coaxial cable system. However, the X-factor of 0.29 we have calculated indicates that the optical cable system is a connected emerging technology. Therefore, we can observe that improvement in undersea cable systems may be made up of both connected and disconnected technologies.

Internet backbone systems, as shown in Fig. 7.12, refer to the high-speed Internet networks that carry data across major cities, countries, and continents. In the early stage of Internet backbone, data transfer was made through the telephone line or the so-called dial-up Internet access. We simply refer to it as modem in this study. From 1969 to 1984, the modem speed had improved to 56,000 bits per second (bps). And then, in 1985, a new

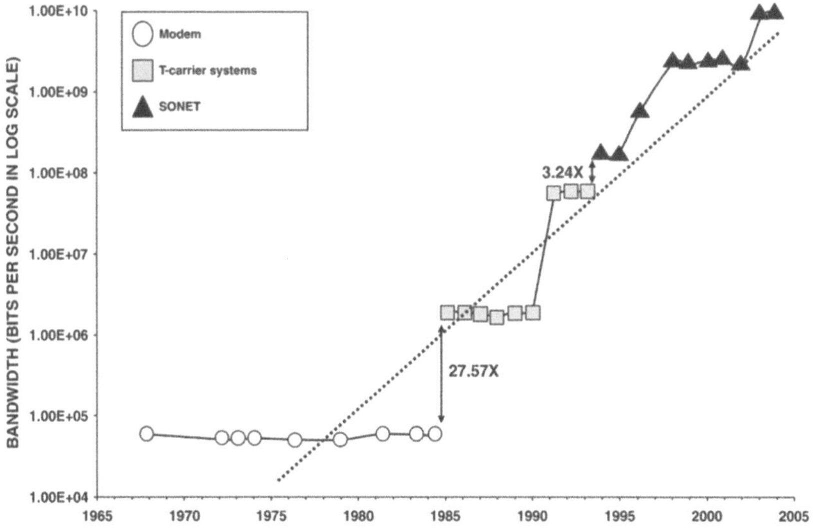

Fig. 7.12 Exponential growth curve analysis of Internet backbone systems. (*Source*: Author's own elaboration on the basis of Kurzweil (2005) and *The Singularity Is Near* (2009). *Note*: Detailed backup data are available in Appendix 14. Moreover, the dotted line represents the predicted values from growth curve regression. And data were derived from Kurzweil (2005) and a web site the singularity is near, available from http://singularity.com)

T-carrier system (a series of wideband digital data transmission formats) was introduced that could transmit data 27.57 X times faster than the previous system. Then, in 1994, the first version of the Synchronous Optical Networking (SONET) was introduced, with an X-factor of 3.24X. Therefore, we can conclude the dominant role of disconnected technologies in the progress of Internet backbone systems.

As in the case of computing speed, rapid growth rates for both undersea cable and Internet backbone systems continue without showing any sign of slowing down. Our exponential regression model generated good fit with the annual growth rate of 18.9 per cent for undersea cable with R^2 of 0.882, and the annual growth rate of 41.9 per cent for Internet backbone systems with R^2 of 0.889, as summarised in Table 7.5.

In fact, an overall improvement generated from information and communication technologies has been incredibly fast with more than twenty trillion per cent or two hundred billion times faster speed during 122 years for calculation speed, with more than five thousand

Table 7.5 Summary of results from the regression analysis on information and communication

			Logistic function		
Elements	# of years	# of samples	Growth rate – gradient (t-value)	Upper limit (t-value)	R^2
Computing speed (total)	146	86	0.221 (21.385)*	N/A	0.813
Undersea cable system (total)	145	30	0.189 (14.437)*	N/A	0.882
Single cable system	71	5	N/A	398.66 (52.858)**	0.996
Coaxial cable system	33	7	N/A	5.65E+10 (4.270)**	0.993
Optical cable system	15	18	N/A	1.12E+15 (3.928)*	0.882
Internet backbone system	37	28	0.419 (14.413)*	N/A	0.889

Source: Author's own elaboration on the basis of the modelling results
Note: * and ** denote the p-value significant at the 1 per cent and 5 per cent levels, respectively

Table 7.6 Summary of X-factor analysis on information and communication

Comparison of two technologies	Year of comparison	X-factor	Type
Coaxial cable over single cable	1951	361.21	Disconnected
Optical cable over coaxial cable	1988	0.29	Connected
T-carrier systems over modem	1985	27.57	Disconnected
Sonet over T-carrier systems	1994	3.24	Disconnected

Source: Author's own elaboration on the basis of the modelling results

trillion per cent or fifty trillion times faster speed during 145 years for the undersea cable system, and with nearly twenty million per cent or two hundred thousand times faster speed during 37 years for Internet backbone systems.

And Table 7.6 is a summary of the X-factors we calculated for information and communication technologies. By reviewing the factors in Table 7.6, we can observe that the general trend in information and communication is a mixture of both connected and disconnected technologies. The continuous progress of both connected and disconnected technologies in information and communication sectors showing a rapid exponential growth appears to account for substantial part of the whole improvement.

Summary of Analyses for the Three Elements

All the results of our regression analysis and X-factor analysis are summarised in Tables 7.7 and 7.8. Historical data on lighting and transport yields good results from a logistic curve analysis with R^2 of greater than 0.85 while information and communication produce better results from a growth curve analysis with R^2 of greater than 0.80.

As for the X-factor analysis, we have calculated the overall average X-factor of 38X with a large standard deviation of 107X. This average is made up of the average of 52X from disconnected S-curves and 0.56X from the average of connected S-curves. In some cases, the role of new disconnected technology appears to be very important. New disconnected technologies with a large X-factor could destroy the basis of existing technologies with the gale force of destruction that Schumpeter referred to (Schumpeter 1939).

Table 7.7 Summary of results from the regression

Elements	# of years	# of samples	Annual growth rate (t-value)	Logistic function: Growth rate – gradient (t-value)	Logistic function: Upper limit (t-value)	R^2
Tungsten filament lamp	98	10	N/A	0.355 (3.748)*	11.9683 (98.830)*	0.996
Compact fluorescent lamp	26	8	N/A	0.335 (2.104)**	65.3491 (38.661)*	0.993
Low-pressure sodium lamp	77	11	N/A	0.081 (5.711)*	199.1 (21.451)*	0.988
Conventional wheeled train	160	15	N/A	0.034 (3.167)*	869.9 (2.500)**	0.968
Mass produced car	108	20	N/A	0.029 (4.138)*	500.5 (4.740)*	0.937
Military fighter aircraft	67	17	N/A	0.215 (2.780)**	2621.4 (26.258)*	0.957
Piston engine	12	6	N/A	0.602 (2.347)	701.2 (31.017)	0.994
Jet engine	56	11	N/A	0.995 (3.783)*	2582.1 (57.327)	0.988
Computing speed	146	86	0.221 (21.385)*	N/A	N/A	0.813
Undersea cable system	145	30	0.189 (14.437)*	N/A	N/A	0.882
Single cable system	71	5	N/A	N/A	398.66 (52.858)**	0.996
Coaxial cable system	33	7	N/A	N/A	5.65E+10 (4.270)**	0.993
Optical cable system	15	18	N/A	N/A	1.12E+15 (3.928)*	0.882
Internet backbone system	37	28	0.419 (14.413)*	N/A	N/A	0.889

Source: Author's own elaboration on the basis of the modelling results
Note: * and ** denote the p-value significant at the 1 per cent and 5 per cent levels, respectively

Table 7.8 Summary of X-factor analysis

Comparison of two technologies	Year of comparison	X-factor	Type
Compact fluorescent lamp over tungsten filament lamp	1976	3.94	Disconnected
White LED lamp over compact fluorescent lamp	1995	0.80	Connected
Conventional wheeled train over stagecoach	1848	3.22	Disconnected
Mass produced car over conventional wheeled train	1900	0.60	Connected
Military fighter aircraft (piston) over mass produced car	1930	1.79	Disconnected
Military fighter aircraft (jet) over military fighter aircraft (piston)	1941	1.37	Disconnected
Spacecraft over military fighter aircraft (jet)	1969	15.45	Disconnected
Coaxial cable over single cable	1951	361.21	Disconnected
Optical cable over coaxial cable	1988	0.29	Connected
T-carrier systems over modem	1985	27.57	Disconnected
SONET over T-carrier systems	1994	3.24	Disconnected
Avg. X-factor (standard deviation)	38.13 (107.48)		
Avg. X-factor in disconnected(standard deviation)	52.22 (125.18)		
Avg. X-factor in connected(standard deviation)	0.56 (0.26)		

Source: Author's own elaboration on the basis of the modelling results

However, the overall improvement of a given technology is generated by the combination of both X-factor and continuous improvement from that technology. Table 7.9 summarises these overall improvement rates for five technologies of lighting, transport, calculation, undersea cable, and Internet backbone system. Both lighting and transcript have experienced large improvement rates of 6400 per cent during a 125 year period and 8.750 per cent during a 149 year period, respectively. Impacts from both X-factors and incremental improvement from the continuing technologies are responsible for these large improvement rates.

Even though both lighting and transport systems have generated a radical improvement over a long-term period, the magnitude of improvements taking place in modern information and communication systems is both so huge and super fast that it may be unprecedented

Table 7.9 Summary of overall improvement rates

Technologies	Year	Performance	Time period	Per cent improvement
Lighting				
Tungsten filament lamp	1838	2.6 lm/W	125	6400 per cent
White LED lamp	2007	169 lm/W		
Transport				
Concord Stagecoach	1848	30 km/h	149	8750 per cent
Lockheed F-22 raptor	1996	2655 km/h		
Information and communication				
Calculation				
Manual calculation	1880	1.68E−08 bps	122	20,357,142,857,043 per cent
Gateway Athlon	2001	3.42E+03 bps		
Undersea cable system				
Telegraph	1858	5.56E−02 bps	145	5,732,088,214,329,330 per cent
Apollo	2002	3.18E+12 bps		
Internet backbone system				
Modem	1968	5.00E+04 bps	37	19,903,900 per cent
OC-193	2004	9.95E+09 bps		

Source: Author's own elaboration on the basis of the modelling results

in the history of civilisation. These unprecedented improvements of twenty trillion per cent, more than five thousand trillion per cent and nearly twenty million per cent summarised in Table 7.9 appear to have come from a very large number of competing technologies of both connected and disconnected types.

These X-factors summarised in Table 7.8 are subjected to a number of limitations. In some cases, data during the earliest stage of a new technology are often unavailable until the new technology becomes firmly established. This will often result in a greater value to be calculated for X-factor. In other cases, the nature of metrics available is unique to each technology. These metrics when cross tabulated to calculate an averaged X-factor across industries can create a misleading result. Furthermore, what constitutes a new and emerging versus existing and continuing technology is sometime difficult to identify when the transition is taking place. Thus, it would be extremely difficult to pinpoint exact timing when existing technology reaches its limit and emerging technology begins. All of these limitations make our calculated X-factor values to be preliminary and tentative

Four Propositions

As we have discovered some important common patterns from our analysis of these three elements, we are ready to summarise these in the following five propositions:

1) *The first proposition is that in the case of individual technology, the limit of improvement will be ultimately reached over time. For example, improvements of individual lighting technologies such as the tungsten filament lamp and the compact fluorescent lamp have reached their ultimate physical limit in time. Also, single technology S-curves are usually made up of series of several connected S-curves. For example, train technology was made up of three different engines of steam, diesel, and EMU technologies.*
2) *Continuous improvement in the three elements of civilisation in general appears to have been realised from the operation of multiple technology S-curves of both connected and disconnected types over a very long time period.*
3) *The third proposition is that the longer the time period of improvement involved, the more critical becomes the role of disconnected technology S-curves over connected S-curves. For example, the evolution of lighting and transport technologies from the beginning of civilisation appears to have been sustained by several radically different types of disconnected technologies.*
4) *The fourth proposition is that when the rate of improvement is super fast, especially during a shorter time period, there will be a correspondingly large number of competing technologies as in the modern information and communication sectors. In other words, the super fast rate of improvement*

Table 7.10 Multiple S-curve grid

	Rate of improvement	
Time period	Super fast	Moderate
Short	4th proposition: mixed S-curves	1st proposition: connected S-curves
Long	2nd and 3rd propositions: disconnected S-curves	2nd and 3rd propositions: disconnected S-curves

Source: Author's own elaboration

attracts an unusually large number of competing technologies of both connected and disconnected types, and the winning technologies are being selected on the basis of generating the most significant performance difference over the other technologies. Again, the impact from these disconnected technologies will be most critical to overall improvement.

Combining the time period with the rate of improvement, we create the 2 × 2 grid of multiple S-curves and place these four propositions in Table 7.10:

Policy Implications and Future Research

What are some of the important policy implications from this study? Out of many implications, we have chosen to focus on the six following implications.

1) *Over a long-term period, disconnected multiple technologies appear to sustain continuous improvement overcoming limits. Disconnected emerging technologies are likely to come from outside the firm, the industry, or the nation. Thus, policymakers need actively to search, monitor, evaluate, and invest for emerging outside technologies.*
2) *The time to begin the search for disconnected emerging technologies is when the rate of the process of mature technology begins to decrease.*
3) *Performance differences from emerging technology over the mature technology or X-factor are large with the average of 38X. Thus, the use of the conventional cost-benefit measure will be ineffective. Instead, a very high*

payoff, high risk measure will be needed especially in information and communication areas.

4) *As the pace of improvement in modern technologies enters into a fast rate over a relatively short time period, the number of competing technologies will get very large. Thus, it will be difficult to predict where the winning technology will come from. Thus, the use of a mixed strategy to promote both disconnected and connected S-curves will be more appropriate. In addition to looking outside, policymakers need to focus within the firm, the industry, and the nation as well.*
5) *The time to search for emerging technologies in the super-fast improvement era should begin much sooner than in the case of the long-term improvement period. In fact, the search for emerging technologies should be on-going when improvement is super fast.*
6) *With the average X-factor of 0.56 in connected S-curves, a conventional evaluation measure may be used for emerging connected S-curves. However, as noted earlier, a very high payoff measure should be used for disconnected S-curves.*

In summary, the most important insight gained from our tentative analysis is that, in general, there appear to be no limits to improvement in the evolution of civilisation. Whenever one technology matures, emerging technology will come to replace the mature technology. These emerging technologies can be either connected or disconnected from the existing mature technologies. Our preliminary analysis shows that much of the past improvement comes from new technologies that on first serious application appear to be substantially superior from earlier technologies. In addition, enormous continuous improvement, which has accompanied both connected and disconnected new technologies, appears to have played the critical role in sustaining the evolution of civilisation.

There are several areas that deserve further research. We believe that the concept of X-factor we developed in this study will be useful in future studies of emerging technology in general and radical or disruptive technologies in particular. Additional empirical studies will be needed to better understand the magnitude and range of X-factors in different technologies and reasons for their variation. Furthermore, more research may be conducted to search for a deeper understanding of how the X-factor may be related to the mechanisms of S-curve transitions.

As for the source of future disconnected technology, further studies will be needed to provide a more concrete guideline going beyond the general guideline of looking outside the firm, the industry, and/or the country. For example, further studies on patterns of technology convergences (Fidler 1997; Walters 2002; Shepard 2002; Alger 2005; Kurzweil 2005; Jenkins 2006; Ludes 2008; Schmidt 2008) may yield a more effective and concrete searching guideline for emerging technologies. Although we realised the critical role of incentive and competition in determining the rate of technology progress (Smith 1776; Schumpeter 1939; Romer 1990; Nickell 1996; Keddy 2000; Aghion et al. 2005; Hashmi 2005), we were not able to explicitly incorporate this factor in this study.

Finally, the concept of improvement and its limit is one of the most important issues in modern management of organisations. Just in the last generation, many organisations have attempted to implement a number of improvement approaches. A partial list of these approaches include total quality improvement, lean thinking, six sigma, knowledge management, enterprise resource planning, and so forth. What are the limits of these improvement approaches? What are differences and similarities when examining firm-level improvement in contrast to this study? These topics should remain a rich arena for many empirical studies in the future.

Acknowledgements Reprinted from Technological Forecasting and Social Change, Vol. 77, Issue 5, Yu Sang Chang and Seung-Jin Baek, *Limit to Improvement: Myth or Reality? Empirical Analysis of Historical Improvement on Three Technologies Influential in the Evolution of Civilisation*, pp. 712–729, Copyright (2010), with permission from Elsevier.

Notes

1. Quoted in Hecht (2008), Prophesy of economic collapse 'coming true', available from http://www.newscientist.com/article/dn16058-prophesy-of-economiccollapse-coming-true.html.
2. Quoted in Foster (1986), Innovation: The Attacker's Advantage, pp. 108–109.

3. For logistic functions, we have used non-linear regression in SAS to fit data to logistic curve. In fact, for non-linear regression, those four parameters above will be estimated through the regression analysis. However, before parameters will be estimated, we need to estimate them in advance and then apply them into logistic model. Thus, we calculated the starting value of each parameter based on our own logic. For example, we inputted real value for parameters, α and η. Moreover, for β, we used compound annual growth rate formula to get input value. Lastly, for μ, we inputted the value of the year, which shows the biggest improvement of performance over previous year.
4. Growth rate (β) refers to a gradient, which sets the length of time the curve is stretched out over, the time of maximum growth.
5. For growth function, we have used regression analysis package from SPSS.
6. Quoted in White (1949), The Science of Culture, pp. 368–369.
7. White's Law describes that C equals to 'E times T', where E is a measure of energy consumed per capita per year, where T is the measure of efficiency of technical factors utilising the energy and where C represents the degree of cultural development.
8. Quoted in Kipling (1917), A Diversity of Creatures, p. 240.
9. Quoted in Lewis (1936), The Significance of Transportation to Civilization, p. 1.
10. Quoted in Bell (1980), The Social Framework of the Information Society, p. 500.
11. Table of lighting history is presented in Appendix 6.
12. This concept is the ratio of luminous flux emitted to electric power consumed and is generally accepted measurement for lighting efficiency. Moreover, Lumens per watt (lm/W) is a unit of luminous efficacy.
13. The 30 km/h is derived from the source, the Collings Foundation's Newsletter (2008). However, we could not measure the certain maximum speed because there are many variables that affect the top speed of stagecoach, such variables as the number of horses or of passenger. Instead, we know that no stagecoach was able to move faster than 30 km/h (1997). Thus, we can fairly assume that the maximum speed of stagecoach is around 30 km/h.
14. MSOPS refers to a million standardised operations per second that is the ability to add 20 million 32-bit integer numbers in one second. Thus, it can be interpreted as how fast the computer operates as the ability improves.

15. Quoted in Nordhaus (2001), The Progress of Computing, p. 37.
16. Bandwidth is a measure of consumed data expressed in bits per second (=data transfer rate).

References

Aghion, P., N. Bloom, R. Blundell, R. Griffith, and P. Howitt. 2005. Competition and Innovation: An Inverted-U Relationship. *Quarterly Journal of Economics* 120 (2): 701–728.

Aizcorbe, A. 2005. Moore's Law, Competition, and Intel's Productivity in the Mid-1990s. *American Economic Review* 95 (2): 305–308.

Alger, D. 2005. *Build the Best Data Center Facility for Your Business: A Comprehensive Guide to Designing and Operating Reliable Server Environments*. Indianapolis: Cisco Press.

Anderson, P., and M.L. Tushman. 1990. Technological Discontinuities and Dominant Designs: A Cyclical Model ofTechnological Change. *Administrative Science Quarterly* 35 (4): 604–633.

Becker, R.H., and L.M. Speltz. 1983. Putting the S-Curve Concept to Work. *Research Management* 26 (5): 31–33.

Bell, D. 1980. The Social Framework of the Information Society. In *The Microelectronics Revolution*, ed. T. Forrester, 500–549. Oxford: Blackwell.

Bohannon, R.W. 1997. Comfortable and Maximum Walking Speed of Adults Aged 20–79 Years: Reference Values and Determinants. *Age Ageing* 26 (1): 15–19.

Brodrick, J.R. 2008. *Emerging Lighting Technologies*, National ENERGY STAR Lighting Partner Meeting. Available from: http://www.energystar.gov/ia/partners/downloads/meetings/EmergingLightingTechologies_Brodrick.pdf. Accessed 8 Dec 2008.

Brown, R. 1992. Managing the 'S' Curves of Innovation. *Journal of Consumer Marketing* 9 (1): 61–73.

Chandy, R.K., and G.J. Tellis. 2000. The Incumbent's Curse? Incumbency, Size, and Radical Product Innovation. *Journal of Marketing* 64 (3): 1–17.

Christensen, C.M. 1992. Exploring the Limits of the Technology S-Curve. Part I: Component Technologies. *Production and Operations Management* 1 (4): 334–357.

———. 2003. *The Innovator's Dilemma: The Revolutionary Book that Will Change the Way You Do Business*. New York: Harper Collins.

Christensen, C.M., and M.E. Raynor. 2004. *The Innovator's Solution: Creating and Sustaining Successful Growth*. Cambridge: Harvard Business School Press.

Christensen, C.M., S.D. Anthony, and E.A. Roth. 2003. *Seeing What's Next: Using the Theories of Innovation to Predict Industry Change*. Cambridge: Harvard Business School Press.

Collings Foundation. 2008. *Newsletter* [Online]. Available from: http://forum.armyairforces.com/m159817-print.aspx. Accessed 3 De 2008.

DeBecker, A., and T. Modis. 1994. Determination of the Uncertainties in S-Curve Logistic Fits. *Technological Forecasting and Social Change* 46 (2): 153–173.

Denbaars, S. 2008. *What Is LED Lighting: Technology Overview and Introduction*. DOE SSL Market Introduction Workshop. Available from: http://www.ledoptics.com/images/pdf/what%20is%20led%20lighting.pdf. Accessed 9 Dec 2008.

Dosi, G. 1982. Technological Paradigms and Technological Trajectories. *Research Policy* 11 (3): 147–162.

Fidler, R.F. 1997. *Mediamorphosis: Understanding New Media*. Thousand Oaks: Pine Forge Press.

Foster, R.N. 1986. *Innovation: The Attacker's Advantage*. New York: Summit.

Gilder, G. 2002. *Telecosm: The World After Bandwidth Abundance*. New York: Free Press.

Good, I.J. 1965. Speculations Concerning the First Ultra-Intelligent Machine. *Advances in Computers* 6: 31–88.

Grubler, A. 1990. *The Rise and Fall of Infrastructures: Dynamics of Evolution and Technological Change in Transport*. New York: Springer-Verlag.

Hanson, R. 2008. Economics of the Singularity. *IEEE Spectrum* 45 (6): 37–42.

Hashmi, A.R. 2005. Competition and Innovation: The Inverted-U Relationship Revisited. In *The 20th Annual Congress of the European Economic Association*. Available from: http://www.chass.utoronto.ca/~ahashmi/pdf_files/20050812p1.pdf. Accessed 6 Apr 2009.

Hecht, J. 2008. Prophesy of Economic Collapse Coming True. *New Scientist*. Available from: http://www.newscientist.com/article/dn16058-prophesy-of-economic-collapsecoming-true.html. Accessed 26 Dec 2008.

Hooker, J.D. 2009. *The Museum of Electric Lamp Technology* [Online]. Available from: http://www.lamptech.co.uk. Accessed 6 Apr 2009.

Horner, P. 2008. What Have We Accomplished in Lighting Efficiency? *Consortium for Energy Efficiency*. Available from: http://www.cee1.org/cee/

mtg/06-08mtg/Wed%20June%2011/General%20Session/01aHornerNEMA.pdf. Accessed 22 Dec 2008.

Hutcheson, D.G. 2005. Moore's Law: The History and Economics of An Observation that Changed the World. *Electrochemical Society Interface* 14 (1): 17–21.

James, G.M., and A. Sood. 2006. Performing Hypothesis Tests on the Shape of Functional Data. *Computational Statistics and Data Analysis* 50 (7): 1774–1792.

Jenkins, H. 2006. *Convergence Culture: Where Old and New Media Collide*. New York: New York University Press.

Jonathan, H. 2005. A Possible Declining Trend for Worldwide Innovation. *Technological Forecasting and Social Change* 72 (8): 980–986.

Jorgenson, D.W., and K.J. Stiroh. 2000. Raising the Speed Limit: U.S. Economic Growth in the Information Age. *Brookings Papers on Economic Activity* (1): 125–235.

Jorgenson, D.W., and C.W. Wessner. 2007. *Enhancing Productivity Growth in the Information Age*. Washington, DC: National Academies Press.

Keddy, P. 2000. *Competition*. 2nd ed. Boston: Kluwer Academic Publishers.

Kipling, R. 1917. *A Diversity of Creatures*. New York: Doubleday, Page and Company.

Koh, H.B., and C.L. Magee. 2006. A Functional Approach for Studying Technological Progress: Application to Information Technology. *Technological Forecasting and Social Change* 73 (9): 1061–1083.

Kurzweil, R. 2005. *The Singularity Is Near: When Humans Transcend Biology*. New York: Penguin.

———. 2009. *The Singularity Is Near* [Online]. Available from: http://singularity.com. Accessed 22 Feb 2009.

Lee, T., and N. Nakicenovic. 1988. Technology Life Cycles and Business Decisions. *International Journal of Technology Management* 3 (4): 411–426.

Lenski, G.E. 1974. *Human Societies: An Introduction to Macrosociology*. New York: McGraw-Hill.

———. 1984. *Power and Privilege: A Theory of Social Stratification*. Chapel Hill: University of North Carolina Press.

Les chiffres clés des records sur rail dans le monde. 2009. *Retrouvez sur cette page les vidéos des différents records du monde de vitesse sur rail de la SNCF ainsi que les chiffres clés des records sur rail dans le monde* [Online]. Available from: http://tgv16orange.free.fr/videos.html. Accessed 3 Mar 2009.

Lewis, W. 1936. The Significance of Transportation to Civilization. *Annals of the American Academy of Political and Social Science* 187 (1): 1–6.

Ludes, P. 2008. *Convergence and Fragmentation: Media Technology and the Information Society*. Bristol: Intellect Books.

Malthus, T.R. 1798. *An Essay on the Principle of Population*. London: J. Johnson.

Marchetti, C. 1979. Energy Systems – The Broader Context. *Technological Forecasting and Social Change* 14 (3): 191–203.

Marchionini, G. 1999. *Expanding Library Services in the Digital Age: The Search for [Almost] Equilibrium*, DLF Forum on Digital Library Practices. Available from: http://www.diglib.org/forums/summer1999/presentations/marchionini-1999-07.pdf. Accessed 5 Apr 2009.

Meadows, D.H., D.L. Meadows, J. Randers, and W.W. Behrens III. 1972. *The Limits to Growth*. New York: Universe Books.

Meadows, D.H., J. Randers, and D.L. Meadows. 2004. *Limits to Growth: The 30-Year Update*. White River Junction, Vermont: Chelsea Green.

Meis, L., and J. Leta. 1997. Modern Science and the Explosion of New Knowledge. *Biophysical Chemistry* 68 (1–3): 243–253.

Modis, T. 1994. Fractal Aspects of Natural Growth. *Technological Forecasting and Social Change* 47 (1): 63–73.

———. 2002. Forecasting the Growth of Complexity and Change. *Technological Forecasting and Social Change* 69 (4): 377–404.

Moore, G.E. 1965. Cramming More Components onto Integrated Circuits. *Electronics Magazine* 38 (8): 114–117.

Morgan, L.H. 1877. *Ancient Society*. London: Macmillan.

Narukawa, Y., M. Sano, M. Ichikawa, S. Minato, T. Sakamoto, T. Yamada, and T. Mukai. 2007. Improvement of Luminous Efficiency in White Light Emitting Diodes by Reducing a Forward-Bias Voltage. *Japanese Journal of Applied Physics* 46 (2): 36–40.

Nickell, S.J. 1996. Competition and Corporate Performance. *Journal of Political Economy* 104 (4): 724–746.

Nordhaus, W. 1997. Do Real-Output and Real-Wage Measures Capture Reality? The History of Lighting Suggests Not. In *The Economics of New Goods*, ed. T.F. Bresnahan and R.J. Gordon, 29–66. Chicago: University of Chicago Press.

———. 2001. *The Progress of Computing*, Discussion Paper No. 1324. New Haven: Yale University. Available from: http://cowles.econ.yale.edu/P/cd/d13a/d1324.pdf. Accessed 3 Mar 2009.

Procter, G.L. 1997. *Railway Speed Records* [Online]. Available from: http://homepages.ihug.co.nz/~procter/SPEED.HTM. Accessed 3 Mar 2009.

Romer, P.M. 1990. Endogenous Technological Change. *Journal of Political Economy* 98 (5): 71–102.
Sahal, D. 1985. Technological Guideposts and Innovation Avenues. *Research Policy* 14 (1): 61–82.
Schmidt, S. 2008. *The Coming Convergence: Surprising Ways Diverse Technologies Interact to Shape Our World and Change the Future.* New York: Prometheus.
Schumpeter, J.A. 1939. *Business Cycles: A Theoretical, Historical, and Statistical Analysis of the Capitalist Process.* New York: McGraw-Hill.
Shepard, S. 2002. *Telecom Convergence: How to Bridge the Gap Between Technologies and Services.* New York: McGraw-Hill.
Smith, A. 1776. *An Inquiry into the Nature and Causes of the Wealth of Nations.* London: W. Strahan.
Sood, A., and G.J. Tellis. 2005. Technological Evolution and Radical Innovation. *Journal of Marketing* 69 (3): 152–168.
Starostin, M. 2009. *All the World's Rotorcraft* [Online]. Available from: http://www.aviastar.org. Accessed 3 Mar 2009.
Suarez, F.F., and J.M. Utterback. 1995. Dominant Designs and the Survival of Firms. *Strategic Management Journal* 16 (6): 415–430.
The History of the Coveted World Record. 2009. *The History of the Coveted World Record.* Available from: http://www.worldrecordcarauction.com/car-history.php. Accessed 22 Feb 2009.
Turner, G.M. 2008. A Comparison of The Limits to Growth with 30 Years of Reality. *Global Environmental Change* 18 (3): 397–411.
Utterback, J.M. 1996. *Mastering the Dynamics of Innovation.* 2nd ed. Cambridge: Harvard Business School Press.
Utterback, J.M., and W.J. Abernathy. 1975. A Dynamic Model of Process and Product Innovation. *International Journal of Management Sciences* 3 (6): 639–656.
Vinge, V. 1993. Technological Singularity. *Whole Earth Review* 81: 89–95.
Wade, M. 2009. *Apollo 10* [Online]. Available from: http://www.astronautix.com/flights/apollo10.htm. Accessed 22 Feb 2009.
Walters, S.J.K. 2002. *The New Telephony: Technology Convergence, Industry Collision.* New York: Prentice Hall.
Warren, P. 2002. The Future of Computing: New Architectures and New Technologies – Part 1: Biology versus Silicon. *Engineering Science and Education Journal* 13 (2): 61–65.
White, L.A. 1949. *The Science of Culture: A Study of Man and Civilization.* New York: Farrar, Straus and Giroux.

Wilbrink, J. 2009. *SVT 137 153 ff. Leipzig* [Online]. Available from: http://www.dbtrains.com/en/trainsets/epochII/SVT_Leipzig. Accessed 3 Mar 2009.

Young, P. 1993. Technological Growth Curves: A Competition of Forecasting Models. *Technological Forecasting and Social Change* 44 (4): 375–389.

Young, R. 2008. *Specifying Food Service Lighting for Energy Efficiency*. Food Service Lighting Seminar. Available from: http://www.fishnick.com/education/presentations/Food_Service_Lighting_Jul-08.pdf. Accessed 9 Dec 2008.

8

Is the Universal Consensus on 'Technology Drives Development' Analytic or Synthetic?

Introduction

As knowledge-based economies are today more advanced, and the importance of economic growth led by innovation based on knowledge and information is further emphasised, countries around the world are competing with innovative systems and building clusters to foster technological innovation more than ever before. Such emphasis was made in the 1990s by new economic growth theorists who called technological innovation the key to endogenous economic growth—for example, Romer (1990), Coe and Helpman (1995), Aghion and Howitt (1998), Lumenga-Neso et al. (2005). This theme was studied from various perspectives: the study of the importance of the causality between knowledge and economic growth, the impact on total productivity from created knowledge and embedded technological innovation, and the relationship between the diffusion of knowledge through trade and economic growth. These verities have led to a global consensus that technological innovation is undoubtedly a key internal driver of economic growth.

In addition to its contribution to growth, technological development may also either directly or indirectly affect the reduction of inequality in

S. J. Baek, *The Political Economy of Neo-modernisation*,
https://doi.org/10.1007/978-3-319-91394-0_8

various ways. Woodhouse and Sarewitz (2007) argue that technology contributes to mitigating inequalities by strengthening people's knowledge through education, research, professions, and skills. They further note that capital investment and R&D for technological progress enhances the economic and welfare system. This school of thought may have a strong kind of *Utilitarian* assumption.

However, this utilitarianism is highly likely to be denied if the essential nature of technological development is considered. Effectively, technology has two features: skill-biased technological change and routine-biased technological change. The former can be associated with increasing the premium on skilled technology, while the latter is likely to have the capacity to replace the repetitive work domain of middle-class workers, which may lead to intensifying income inequality or the polarisation of society (Acemoglu and Autor 2011; Jaumotte et al. 2013; Goos et al. 2014).

As most countries are building innovation systems or technology-intensive clusters in response to their transition to a knowledge-based economy, their policy intervention may lead to an increase in income inequality by accumulating wealth in certain areas and creating high-wage jobs corresponding to technology-driven high productivity. As a result, income inequality in society can widen even further. In this regard, Daron Acemoglu (2002) argues that the main reason for the increase in the wage gap between college graduates and non-college graduates, even if the supply of college graduates increases (even to an excessive level), is that the productivity of college graduates has risen sharply. In other words, recent technological advances have largely been associated with the further increase in the productivity of highly skilled (educated) workers.

Regardless of the positive or negative correlation between technological progress and inequality, what makes it more complicated is that there is a much more complex relationship between inequality and economic growth. Baek (2017) hypothesised that reducing income inequality through effective redistribution could have a positive effect on economic growth on the basis that income support for the poor and tax cuts have a positive influence on growth resulting in an increase in effective demand.

As the (re)distribution of income increases, the income of the poor increases who tend to consume more than those who are wealthier. Total demand will therefore increase as redistribution takes effect.

It was however argued by neo-classical economists that the income redistribution policy had become an obstacle to economic growth, as many developed countries in the West experienced a recession due to the steep slump in oil prices. Most representatively, Okun (1975) insists that there is a trade-off relationship between income redistribution and economic efficiency. In this context, the stronger the income redistribution, the lower the economic efficiency and, finally, the more likely economic growth is subdued.

Overall, it is plausible that technological innovation has had a positive impact on economic growth, but there could be much debate as to whether technological progress mitigates inequality challenges or exacerbates them. Even if technological advances tackle inequality problems, it cannot be said that such a reduced level of inequality necessarily contributes to economic growth (Galor and Tsiddon 1997; Forbes 2000). On the other hand, even if technological innovation affects the inequality problem negatively, it cannot be concluded that such a serious inequality level necessarily hampers growth dynamics (Li and Zou 1998; Bell and Freeman 2001). I would say that the relationship between the two is much more complicated. In conclusion, the universal belief that technological innovation has a positive effect on economic growth shall remain merely a proposition, and one that can be denied.

In dealing with such controversies, the rest of the chapter is made up of five sections. First, the study begins with a discussion on the ambivalence towards technological innovation with the case of Silicon Valley. Second, various types of technology S-curves are reviewed in order to explore a more detailed evolution of technological advances. The following section attempts to preliminarily identify some regularities in the dynamics among technology, economic growth, and inequality. Having explored such a complex nexus, an integration of the three dimensions on the basis of S-curve theory is attempted by proposing the *Developmental S-Curve*. The chapter ends with a summary and implications for the twenty-first-century context.

Ambivalence to Technological Innovation

In Silicon Valley, one may feel like all the innovations in the world are gathered all at once in one place. Beginning with a simple Internet search engine, Google is now at the heart of many people's lives through email and browser services. Google has also had a massive impact on political, economic, and social development around the world, through cloud computing, big data, and many other innovations. In recent years, Google has tried to further share the benefits of the Internet with more than half of the planet's population that remain disconnected, through *Project Loon*.[1]

Another innovation giant, Apple, is also changing many aspects of our lives through disruptive technological innovation. Apple invented the iPod and released it to the world, creating a new distribution structure in the world's music market, and subsequently introduced the iPhone, which completely changed the concept of the mobile phone. Everyday life for many without this type of phone today is hard to imagine.

Besides Google and Apple, of course, Silicon Valley houses hundreds of other innovative companies, including Facebook, Intel, Visa, Twitter, Oracle, Netflix, Yahoo, eBay, Cisco, Hewlett Packard, Adobe Systems, Symantec, Uber, and many others.

Measuring the degree of innovation based on the number of patents registered, as of 2015 Silicon Valley accounted for 13.5 per cent of all innovations made in the entire US. This share was 9.1 per cent in 2000 and has been steadily increasing ever since (see Fig. 8.1). Silicon Valley is only 1500 square miles in size yet includes 39 of the Fortune 1000 companies. Similarly, Silicon Valley also accounts for one-third of all venture capital investment in the US. According to *The Economist* (2015), the number of Silicon-Valley-based technology firms with market values of over US$1 billion was 99 in 2015 with a total worth of some US$2.8 trillion, an increase of 75 per cent in just 30 months. Furthermore, these 99 companies account for around 6 per cent of all corporate profits in the US.

As a result, per capita personal income in Silicon Valley in 2015 was US$86,976, nearly double the national average of US$48,112, according

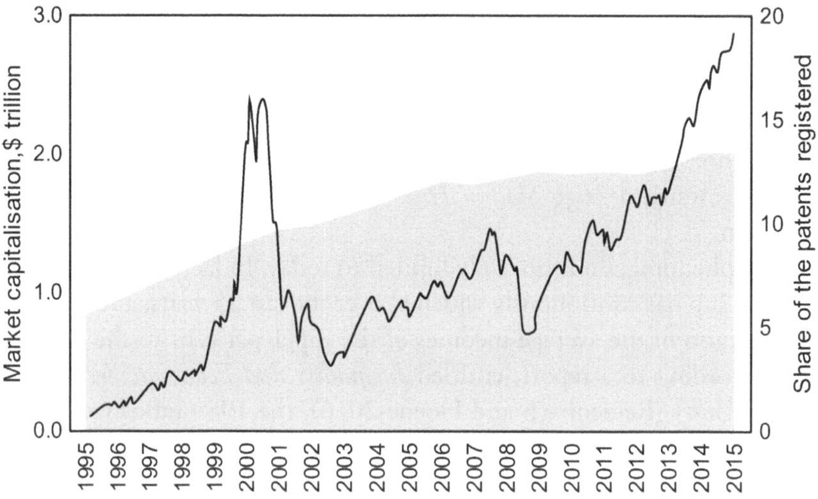

Fig. 8.1 Economic and technological performance in Silicon Valley. (*Source*: Author's own elaboration on the basis of the US Patent and Trademark Office (2017) and *The Economist* (2015). *Note*: The market capitalisation was skyrocketed during the period 1997–2001 due to burst of dot-com bubble)

to the US Department of Commerce (2017).[2] Moreover, median household income was US$55,775 in the country, while in Silicon Valley, this reached US$92,094.

Silicon Valley, despite being the world's most technologically advanced and innovative area, is in fact one of the most unequal areas in the US. It can also be considered as the grave of venture companies where only one per cent of such firms can survive. Furthermore, it can be widely observed that the benefits of technological advancement have been concentrated in a small group of the population, in that many have lost their jobs due to some technology automation, while others who make smart use of technology have experienced a significant increase in their incomes.

According to the Silicon Valley Institute for Regional Studies' report (2017) entitled *2017 Silicon Valley Index*, the wage gap among the five main occupational groups in Silicon Valley is huge: Management, Business, Science and Arts (MBSA); Natural Resources, Construction and Maintenance (NRCM); Sales and Office (SO); Production, Transportation, and Material Moving (PTMM); and Services (S). The median wage in the MBSA group is US$107,776, while the median wages in the other four groups are much lower—US$29,370 for the S group, US$35,524 for the PTMM group, US$42,575 for the SO group, and US$58,872 for the NRCM group. From this huge gap in wages, it can be inferred that technology contributes to job polarisation—shifting society towards a *High Skill = High Wage* and *Low Skill = Low Wage* situation.

This phenomenon is not only limited to today. In fact, Silicon Valley's income gap has continuously widened over the last 25 years, and is based on the ratio of the average incomes of the top 1 per cent to the bottom 99. According to a report, entitled *Inequality and Economic Security in Silicon Valley* (Reidenbach and Hoene 2016), the 1989 ratios were 16.7 per cent, 19.6 per cent, and 10.8 per cent in San Mateo county, San Francisco county, and Santa Clara county, respectively. However, these ratios have been increasing and reached 46.2 per cent, 43.2 per cent, and 30.8 per cent in 2013, respectively. Along with such changes in the ratios, Silicon Valley's middle-class households have been substantially replaced with low- and upper-income households over the same period. The share of middle class was 56 per cent in 1989 and had reduced to 45.7 per cent by 2014, while the shares of low- and high-income group have risen by 4.5 per cent (from 30.3 per cent to 34.8 per cent) and 5.8 per cent (from 13.7 per cent to 19.5 per cent), respectively. On these empirical grounds, Silicon Valley's high-income earners grew in number and earned more, while the number of low-income earners increased and they earned less. As a result, it can be interpreted that Silicon Valley's technology firms appear to have aggravated inequality rather than reduce it.

To generalise the case of Silicon Valley, it is true that, over the centuries, technological advancement and innovation have played a major role in mankind's prosperity. In the past, if horses and trams were our only means of transport, today, we are able to fly to the other side of the world to make friends and to build personal networks with people of all races. Furthermore, we can easily use thousands of social networking tools to analyse potential market developments of other regions on a remote basis to maximise profitability. Smartphones can be used to make international phone calls or video conference calls with local experts and business stakeholders, and the results of such meetings can be quickly documented and stored in the chief executive officer's folder within seconds. It should be noted that today this may seem unremarkable, but even 20 years ago, it would be unimaginable.

On the other hand, it could also be true that technological advances have in many ways replaced our tasks and roles, which in some instances has even resulted in homelessness. Considering Ford's *Mass Production* system using conveyor belts, this has indeed led to the popularisation of automobiles by lowering the cost of their production, so in this sense the general public, particularly the middle- or low-income groups have benefitted from this technological advance. At the same time, innovative mass production has enabled capitalists to operate more efficiently in their economic production activities. However, the workers who were inevitably replaced by the conveyor belt innovation lost their jobs as a result. Thus, the wealthier have been able to accumulate wealth more rapidly, while those who have been disadvantaged, that is, those who have not benefitted from technological development have become poorer.

To generalise further the aforementioned phenomena, technological innovation is two-faced. Technological progress has enriched mankind, which has led many to foster economic development by sharing its benefits, but this may have contributed to making our society unequal. Besides, many fear the recent rapid technological advancements, and attempt to resist them. For instance, watching *AlphaGo* based on artificial intelligence winning a game of Go (an abstract board game) against Sedol Lee, the 18-time world champion, although there was great interest and fascination, there was also some fear that artificial intelligence may have already crossed a line (see Box 8.1).

Box 8.1 Is technology evolving too fast?

Among the influential hypotheses proposed by futurist Ray Kurzweil (2005), the *Law of Accelerating Returns* could in particular convey serious implications for our society. This concept is that as technology advances the speed of such improvement is accelerated, which becomes infinite. In line with this, he predicts that the *Age of Singularity*, the moment when technology transcends human beings, will come around in the year 2030.

He argues that 'Evolution creates human beings, human beings create technology, human beings combine evolving technologies to create next-generation technologies, and eventually the distinction between human beings and technology disappears.' This can be largely based on empirical evidence from *Moore's Law* (Fig. 8.2). This rule, introduced in 1965 by Gordon Moore, the founder of Intel, is explained by a simple statement in that the performance of an integrated circuit doubles every 18 months. For example, in January 1970, the number of transistors in one chip was 1000, which 18 months later increased to 2000 in June 1971, and then 4000, 8000, 16,000, 32,000, 64,000, 128,000, and so forth, which has been viewed as exponential improvement.

Given the fact that our smartphones now have a built-in chip with a larger computing capacity than what the whole world had 50 years ago, it is clear how amazing and exponential the pace of improvement of integrated circuits has been. In addition, through quantum computing, namely the *Dream of Computer*, we will be experiencing a world in which the time it takes to decompose a 300-digit integer by an existing supercomputer may be reduced to about 30 minutes. Considering these possibilities, Kurzweil's claim in *The Singularity Is Near* may come true, and a world in which humans are dominated by computers might become real very soon.

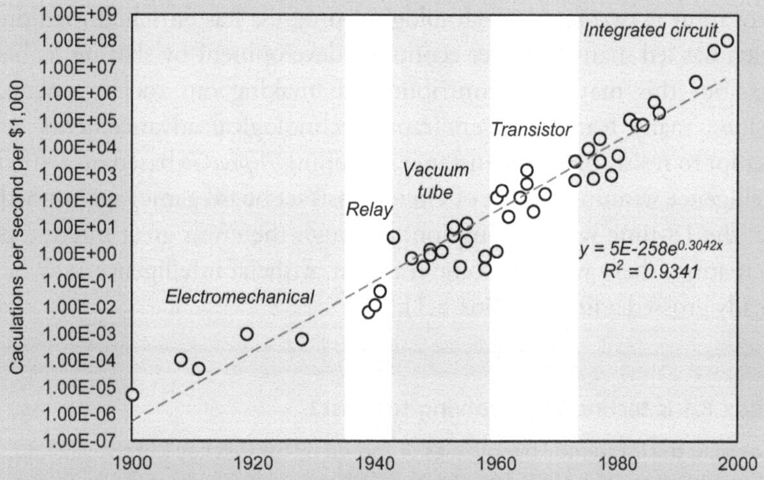

Fig. 8.2 Moore's law. (*Source*: Author's own elaboration on the basis of the Kurzweil (2009))

Varieties of S-Curves

From *Creative Destruction* theory by Schumpeter (1939) to *Disruptive Innovation* theory presented later by Christensen (1997), the underlying assumption of technological development shall begin with S-curve theory (Foster 1986; Schilling 2010). In principle, the S-curve appears to have a single line bent into a simple S-shape, but it could have a profound impact on areas related to not just technology but also humanities, society, economy, management, natural science, and even human civilisation (Chang and Baek 2010, 2011). In my view, the S-curve can be most widely applicable to the field of social sciences, and there are multiple aspects of S-curve, outlined in the following paragraphs.

For Technological Advances

It is interesting to first look at the technological improvement patterns of *Measurement-While-Drilling* (MWD) where a typical S-curve can be found in the developmental pattern. This technological progress is essentially associated with (un)conventional gas development and extraction activities. First of all, the S-curve of MWD technology can be composed of a function of R&D effort versus data transmission speed as shown in Fig. 8.3. In this figure, however, it is necessary to replace various highly technical terms with simple language (A, B, C, and D), so that it does not distort the meaning.

Initially, with the advent of technology A, the concept of MWD began to evolve. To extract gas more efficiently, the industry advanced technologies B and C by exerting continuous R&D efforts. Nevertheless, the degree of improvement for the data transmission speed of A, B, and C was not satisfactory (cost-effective) given the amount of investment devoted to R&D. Up until this period, it can be considered as the initial stage of the S-curve.

Competition among gas companies became increasingly fierce as consumers kept demanding energy at lower prices. Consequently, such

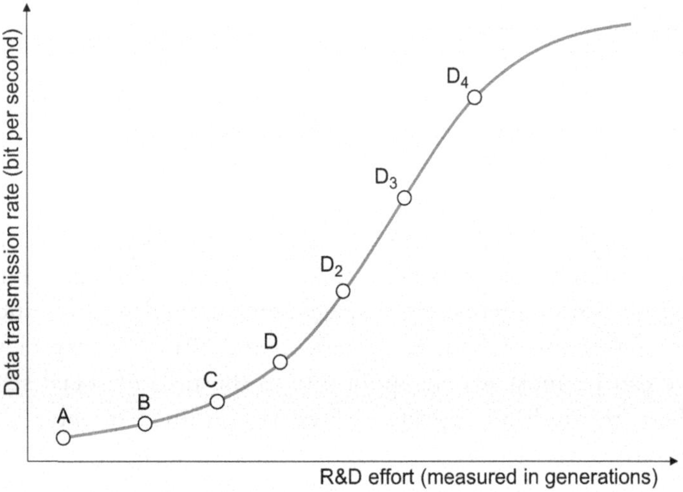

Fig. 8.3 Improvement of Measurement-While-Drilling technology. (*Source*: Author's own elaboration on the basis of Henderson (2005). *Note*: A is *Negative Mud Pulse*; B is *Positive Mud Pulse*; C is *Positive Mud Pulse 2nd Generation*; D is *Continuous M.P.-1G*; D2 is *Continuous M.P.-2G; D3 is Continuous M.P.BPSK 3G*; and D4 is *Continuous M.P.FSK3G*)

companies continued to make investments to advance, which meant having to endure not being cost-effective for a certain period. As a result, technology D was finally introduced and soon became the industrial standard technology. From then on, MWD technology was advanced in a way that gradually upgrades the level of the industrial standard technology. As technologies D2 and D3 were developed, the speed of the technological improvement of MWD accelerated. During this radical period, the industry was able to enjoy profitable business dynamics in accordance with the quality and speed of MWD technology, which encouraged re-investment for further improvements. In particular, many venture firms joined in this improvement process to develop more advanced technologies, which was often associated with substantial returns on investment by licensing new technologies to major players.

Then, with the introduction of D4, which can be considered the highest level of MWD technology, its progress gradually started to encounter limitations. Thus, regardless of how many inventors opted for technological advancement in this field, the performance of MWD did not improve much.

Overall, MWD technology has experienced rapid development from the beginning of A, through the dominant design D, to the moment of reaching D4 in which the technological progress enters the maturity stage. In recent years, the MWD technology widely employed in the unconventional gas extraction business has not been too far off D4 in terms of its advancement, and industrial market competition appears to be driven by economies of scale or management strategies such as supply chain management and marketing rather than technological progress itself. It should particularly be noted that the development of MWD technology would have greater potential to reconstruct the global energy market through increasing the supply of unconventional gas (e.g., shale gas).

Box 8.2 Does shale gas technology have the potential to change the global energy landscape?

Unconventional gas (e.g., shale gas) development was first pursued in the US. From that point on, the US has secured a technological advantage with particular focus on shale gas development. According to the US Energy Information Administration (2017), shale gas production in the US in 2035 is projected to rise sharply to about 1.6 billion m^3, which is estimated to account for 24 per cent of the country's projected natural gas production. The EIA's forecast for LNG imports in 2025 has also been adjusted to 36 billion m^3, 81 per cent down from its 2005 forecast of 1.8 billion m^3. This can be interpreted as a sign of the US' dream of stabilising its energy supply through shale gas extraction.

As the US' shale gas production has increased substantially in recent years, OPEC countries, who account for 40 per cent of world crude oil production, are increasingly facing a dilemma. They would lose market share by cutting production to protect market prices while they would suffer from a fall in prices if production increased to protect their market share.

As well as the US, some other major energy-importing countries, due to the radical improvement of shale gas technology, have attempted to realise their energy self-supply in recent years (or have even intended to go beyond their own supply and then enter the global gas market as an exporting player in the near future), and thus it is possible that the global oil and gas market could be rebalanced. In effect, unconventional gas reserves in China and the US, the world's first- and second-largest energy consumption countries, are estimated to be considerably high, and their reliance on energy imports is expected to decline in the near future. Therefore, the economic influence of major gas exporting countries such as those in the Middle East and Russia is likely to weaken. Besides, Western European countries who have largely been dependent on energy imports from Russia are seeking to diversify their gas importing sources, further affecting current market dynamics.

As seen in the Paris Agreement on climate change, interest in shale gas has begun to grow rapidly as the world considers environmental challenges such as global warming as the highest priority of their national planning frameworks. In fact, improvements in shale gas technology, which had been largely centred on high-tech-based venture firms (mostly small- and medium-sized companies), have recently been made by a number of major energy corporations. It is thus anticipated that the marketability of shale gas will be further strengthened if extraction and production costs lower in line with improved operational effectiveness and productivity. Such recent trends would bring not only a direct hit to global energy market dynamics, particularly for oil and gas industries, but also the possibility of slowing down the realisation of renewable energy (e.g., solar and wind power).

For Market Adoption

Looking at the process of the market penetration of the *Smartphone*, which has now become a necessity for so many people, could reveal another invaluable implication in exploring the S-curve. Recalling when the smartphone was first introduced to the market, many of those who were already familiar with the use of computers may have become quite frustrated when trying to send emails or search the web via a mobile phone. The smartphone's seemingly unnecessary advanced functions (e.g., a digital camera and digital multimedia broadcasting) were not appreciated much, but rather potential consumers often put the overly expensive market price down to such unnecessary functions.

Notwithstanding some of these negative experiences at the beginning, as soon as the smartphone was launched, young consumers in particular joined the smartphone boom. These so-called Early Adopters spread useful information across the world about purchasing tips and how to use the devices and listed the advantages of these products. Such rapid diffusion of information started to greatly change the landscape of the mobile phone market (Rogers 2003). Thanks to their efforts, mainstream consumers increasingly began to understand the complexities of smartphone features, and even those who had initially negative perceptions now joined the buying masses. The social environment saw people without a smartphone perceived as lagging behind (i.e., network externalities) which also played a big part in making the products stand out.

As such, the market dominating process of the smartphone indeed follows the S-curve pattern: early adopter leads the beginning stage of the curve, diffusion of information (innovation) drives the radical increasing stage, and the smartphone is used by almost all consumers, corresponding to the last stage of the S-curve.

In fact, many similar revolutionary processes have been practiced for a while. In the case of the radio, this has dominated the US market in the S-curve pattern since its release in 1920. Within just ten years of the launch of the radio, half of the potential consumers used one, while within the next 20 years, the radio became an indispensable necessity for all American consumers. But then the popularity of the radio suddenly declined as black-and-white television was invented. Not long after, the colour television was introduced, sealing the fate of the radio to some degree. Indeed, all electronic products evolve along a typical S-curve footprint during their birth and development (see Fig. 8.4).

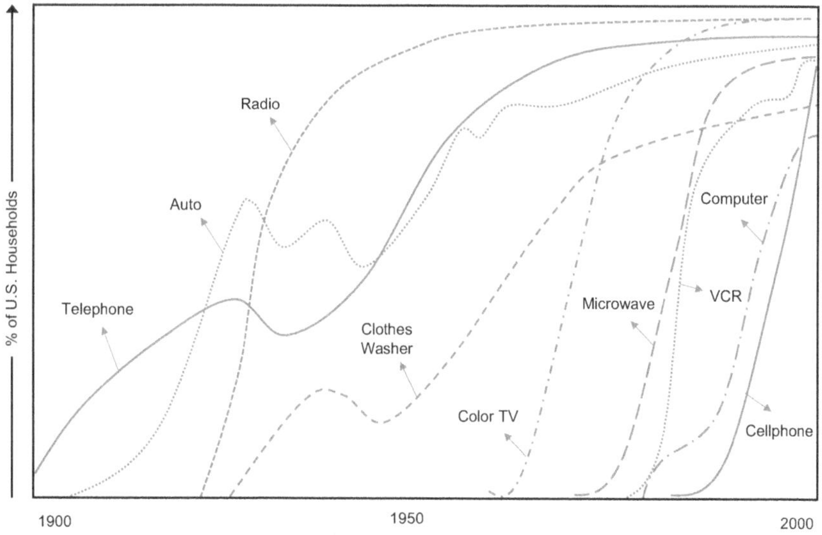

Fig. 8.4 Market adoption S-curve: The US' electronic market in the twentieth century. (*Source*: Author's own elaboration on the basis of McGrath (2013))

For Dominant Design

There have been many cases where the brand name became synonymous with the product (e.g., *Xerox* for copiers, *Kleenex* for tissues, etc.). *Kleenex*, for instance, underwent severe competition against other competitive tissue products at the outset but, being far ahead in its technology and marketing capabilities, it began to penetrate the market and eventually set the market standard. This can be regarded as *Dominant Design* (Utterback and Abernathy 1975).

Meanwhile, other products or services are still competing to set their specific market standard. In this regard, the motor vehicle industry is a good example. Although there are a number of major auto brands who compete with each other (e.g., General Motors, Toyota, Mercedes-Benz, Hyundai Motor, etc.), the automotive industry does not seem to have a dominant design yet. Particularly, Volkswagen has been pursuing for decades the dominant design of the automotive industry. Volkswagen's slogan *Das Auto* in German means *The Car* in English.[3]

Today, Volkswagen and other world-renowned automakers are devoting their efforts to invent environment-friendly vehicles. Among such eco-vehicles, electric cars have received major attention from mainstream customers largely due to recent innovative technologies allowing for the rapid charging of electric batteries.

However, it is a little-known fact that the electric car was actually invented much earlier than the gasoline-powered car (the current market's dominant design). This is mainly because the electric car was easier to start, compared to a gasoline-powered one, produced less noise and vibration, and did not emit much exhaust fumes. Therefore, from the late 1880s to the early 1890s, the electric car was strutting along streets, representing the exclusive property of the rich. However, as the technology of the internal combustion engine was rapidly advanced, the superior status of electric cars started to be overtaken by gasoline-powered ones.

Sooner or later, *Ford*, the largest automobile company in the US at that time, had succeeded in building a mass production system, which led to the popularisation of gasoline-powered cars. Since then, large oil companies have made efforts to prevent the entry of electric cars into the market through all kinds of lobbying to American politicians. Finally, an internal combustion engine prevailed as the dominant design over that of the electric car (see Fig. 8.5).

In recent years, however, the world has been enthusiastic about the concept of being *Environmentally Friendly*, while gasoline-powered cars are starting to be regarded as a source of old technology that pollutes the planet. As new green technologies that satisfy the environment-friendly requirements have begun to emerge, a dominant design war in the automotive industry is approaching a second round. In this context, the hegemony of the automobile market could soon be reconstructed drastically depending on what kind of eco-friendly technology the world's major auto companies produce. In other words, the concept of Dominant Design can explain newly emerging auto industrial dynamics. It may further be applied to measure other industries' competitiveness, even possibly that of nations.

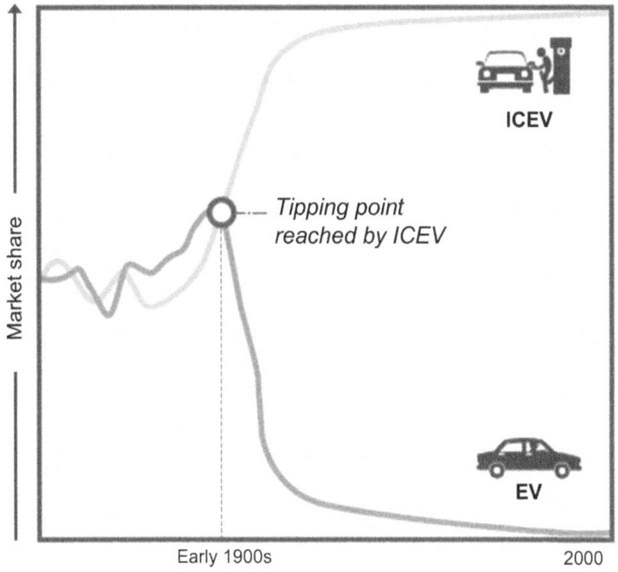

Fig. 8.5 Dominant design war: Gasoline car versus electric car. (*Source*: Author's own elaboration. *Note*: ICEV stands for internal combustion engine vehicle; EV stands for electric vehicle)

Through the Tipping Point

The concept of the S-curve discussed with regard to the three preceding perspectives has one common feature, namely that all of these imply that market dominance has been strengthened from a certain moment. In understanding such a radical phenomenon more analytically (see Fig. 8.6), in the early stages of the S-curve, the market performance appears to be only 5 points even if 25 points of marketing resources are invested. After that, the market still responds with an additional 10 points (i.e., from 5 to 15 in *Y*-axis), even if an additional marketing of 15 points (i.e., from 25 to 40 on the *X*-axis) is made. If firms' marketing efforts were to discontinue due to unsatisfactory performance, no additional market response would be generated. However, even if such ineffective marketing efforts were to continue in the market, the 'T' point of market performance would be reached at some point.

At the moment of crossing the 'T' point, additional marketing of 10 points (i.e., from 40 to 50 on the *X*-axis) tends to lead to an additional

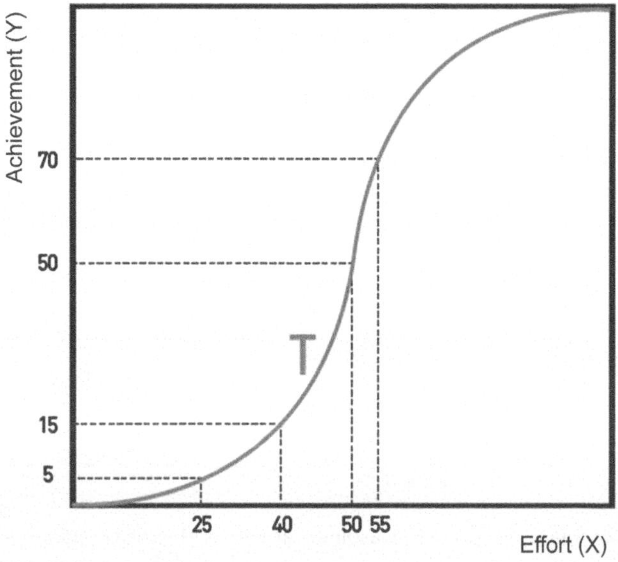

Fig. 8.6 Analytical presentation for 'Tipping Point'. (*Source*: Author's own elaboration)

market response of 35 points (i.e., from 15 to 50 in the *Y*-axis). From then on, only additional marketing of 5 points (i.e., from 50 to 55 on the *X*-axis) would be associated with an additional market response of 20 points (i.e., from 50 to 70 on the *Y*-axis). So, the 'T' point here in which the market reaction appears to change explosively is regarded as the *Tipping Point.*

The tipping point was originally used in American society in the 1970s. Sociologists observed that when the number of African-Americans (blacks) who had migrated to a particular area exceeded 20 per cent of the total population of the area, the white people in that area suddenly all left for another area. Thereafter, this 20 per cent point was increasingly used to convey the sense that a great change had occurred very rapidly. Going a step further, some sociologists explain the concept of the tipping point as that of when an unexpected phenomenon occurs; for instance, when a new epidemic emerges or an abrupt increase in crime occurs abruptly. Even in his book, Malcolm Gladwell (2000) described it as 'magic moment when an idea, trend, or social behaviour crosses a threshold, tips, and spreads like wildfire'. This concept can be used to interpret the social acceptance rate for new technologies. Indeed, the pace at which the tipping point is reached is likely to accelerate as convergence technology progresses more rapidly than expected.

Overall, technology has always served as a driver of social change and has indeed evolved and reacted with such social changes (Freeman and Perez 1988; Chang and Baek 2010). We have been observing that the first Industrial Revolution transformed the labour force along with the evolution of steam engines, while the second revolution created mass production systems, and the third Industrial Revolution has been leading societal change with the digital revolution through computer and information technology. The upcoming fourth revolution is expected to be associated with intelligent information technology, which would accelerate the convergence between technology and society by interconnecting people, things, and information.

If so, when would technologies with high ripple effects spread throughout society? This relates to the speed of the social acceptance of new technologies, that is, the tipping point. In this regard, the timing of the tipping point must be predicted in order to determine how to prepare for the integration of society and technology while minimising uncertainty (i.e., any risks that might arise). In other words, the analytical estimation of the tipping point shall be a critical step in preparing us for the coming years.

Preliminary Assessment of Possible Regularities

The implications of technological development that we have briefly seen through the case of Silicon Valley are, of course, difficult to generalise, so it is necessary to look at real country data. In selecting a proxy indicator for technology, domestic R&D expenditure is not an unreasonable factor to use. Moreover, economic growth can be at the level of PCGDP, while inequality using the income inequality index. With these three incorporated, Fig. 8.7 can present data from eight selected advanced countries in terms of technological innovation capacity, and from a period from 1960 onwards when modernising efforts started to accelerate. In mapping the three proxy indicators, there can be three commonalities identified as follows.

First, the level of PCGDP grew slowly at the beginning but, at some point, the level increased steeply, and then slowed again, which can be interpreted as presenting the S-curve pattern. Second, the income inequality index tended to decrease when the PCGDP level steeply increased while it tended to increase again when the level of PCGDP reached a stagnant period. Finally, the correlation between the level of PCGDP and gross domestic expenditure at R&D level is almost 98 per cent across all of the analysed countries. If gross domestic expenditure on R&D is agreed to be a proper proxy indicator to measure technological advancement, then the PCGDP level has to be closely associated with technological development, which has been largely derived from R&D efforts.

To understand these three commonalities in a more analytical way, the growth process of PCGDP is shown in an S-curve pattern. At the beginning, growth is slow, but this trend is accelerated suddenly right after

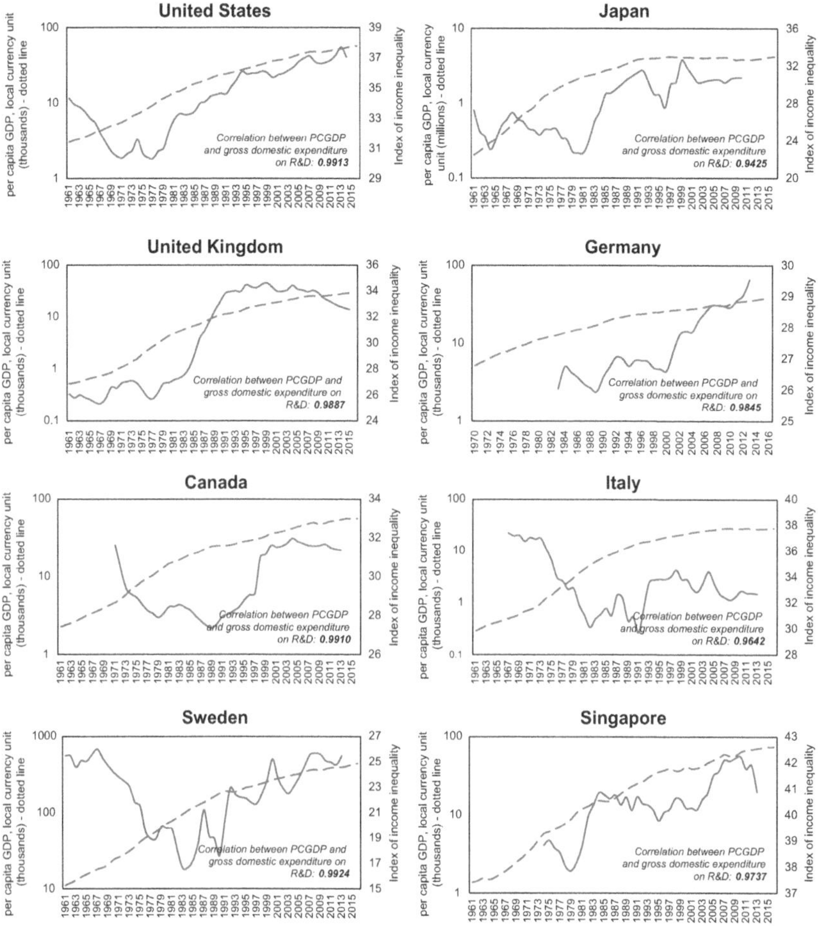

Fig. 8.7 Historical evolution for growth versus inequality in selected advanced economies. (*Source*: Author's own elaboration on the basis of the World Bank's World Development Indicators (2017), OECD (2017a), and Standardised World Income Inequality Database (SWIID) 5.1 (Solt 2016). *Note*: PCGDP data were retrieved from World Bank's World Development Indicators; gross domestic expenditure on R&D data for the purpose of correlation analysis were retrieved from OECD's Stat; income inequality data were retrieved from SWIID 5.1; and the dotted line denotes PCGDP at constant value (left-side scale); the line denotes index of income inequality (right-side scale))

passing through the tipping point, and later it stagnates. Of course, generalising such a pattern comes with risks, although the empirical data backs it up. Furthermore, given this observed tendency for narrowing income inequality during the period when the level of PCGDP passes through the tipping point, we may possibly hypothesise as follows: as the economy expands rapidly, the degree to which growth trickles down to the poor could be greater (utilitarian assumption).

Moreover, income inequality tends to rise sharply when economic growth stagnates because the economy, which can be shared across the population, shows sluggish growth, which means that the poor may have also been getting less or that the capital might have been concentrated within the rich during this period of stagnation. Last of all, there may be a variety of factors that trigger the tipping point; yet, the most significant of these is associated with technological development (Romer 1990; Coe and Helpman 1995; Aghion and Howitt 1998; Lumenga-Neso et al. 2005).

If these interesting relationships were to be acknowledged, a hypothesis developed one step further than the previous one can be presented. That is, when R&D activities become very active, leading to technological development and innovation, so as to substantially contribute to the growth of PCGDP, income inequality tends to decline. Conversely, when R&D is already undergoing congestion and no longer considered as radical innovation, the growth trend of PCGDP could enter the latter stage of the S-curve, which would lead to widening income inequality, partly due to skill-biased and routine-biased technological characteristics.

The given hypotheses derived from this very preliminary analysis of real data may possibly be making some generalisation errors. There is also controversy about the selection of proxy indicators and of observed countries as well as the period analysed. Nonetheless, the S-curve approach to exploring some complexities of the *Technology-Development-Inequality* nexus, which has not been done often in previous studies, would allow us to view such complexities from a different angle. In particular, there should be quite a different technological impact for our societies, between 'when technology progress is already in a stagnation phase' and 'when it is in a rapid growth phase'. So, without the stage-by-stage or cross-sectoral patterns of technological progress being taken into consideration, it may not be meaningful to discuss the impact of technological development on income inequality in a piecemeal fashion.

It is true that research on the effects of each of the stages of technological progress on inequality has received relatively little attention. It shall, however, be recognised that research efforts in this area have recently started to increase. For instance, Hyytinen and Toivanen (2011) have attempted to explore the relationship between income inequality and mobile technology advancements focusing on the beginning stage of technology diffusion. Their results show the existence of a positive association between the two, which is consistent with the hypothesis from the preliminary observation mentioned earlier. This interesting piece of empirical evidence on the positive relationship is mainly attributed to the fact that technology in the early phase is more likely to be consumed by the rich.

Proposal for the Theory of the Developmental S-Curve

Some different socio-economic impacts from each S-curve stage could have real implications for modernisation theories, which have been analysed theoretically throughout the preceding chapters. To explore this more in depth, let us discuss three broad groups of countries (i.e., least developed, middle-income, and advanced countries) based on the theoretical integration of the S-curve into such development theories.

Stage I: Dominated by Institutions

At the beginning phase of development, countries have a relatively poor factor endowment for economic growth (Lin 2012). For this reason, most of them are in the early stages of technological advancement and are yet to achieve considerable developmental performance. During this period, most of the poor begin to accumulate their own skills and knowhow while greatly focusing on farming activities, so productivity growth is not overly generated. Meanwhile, the rich or corporations also start to invest in technology R&D and infrastructure to drive future growth, but the associated results are still minimal. Thus, a small number of elite groups tend to exploit the poor workforce to succeed in accumulating capital. Moreover, companies with close ties to government occasionally grow through lobbying, embezzlement, and corruption. In addition, since the economic and social systems are controlled under a strong government, the concentration of wealth tends to appear only for a few elites, which is consistent with empirical evidence from Hyytinen and Toivanen (2011) that the beginning stage of the diffusion of technology tends to be associated with widening income inequality.

It is thus not easy for countries at this stage of development to accelerate economic growth as both the poor and the rich are unable to increase technology-driven productivity. Rather, other non-economic factors tend to exert a greater influence on the growth trajectory, such as institutions and economic and social structures. In other words, institutions may have a relatively greater impact than that of technological development on growth and inequality at this stage (Acemoglu and Robinson 2012).

Looking more closely at the role of institutions from a more historical perspective, cultural heritage, migrations of people and the legacy of colonisation have all shaped institutions in various countries differently. The influential role of European colonisation starting from the late fifteenth century is crucial in explaining the different institutional characteristics in different parts of the world. European, and particularly British, German, and Spanish, exploration and colonisation of Africa, Asia, Australia, Latin America, and North America impacted massively on the shaping of history and of the state of modern institutions in these places. Ideological wars between the promoters and opponents of slave trade and racial disaggregation, for example, led to the emergence of different interests and thus different types of institutions. Diseases, population density and the discovery of natural resources played important roles in pushing colonising powers to specific areas (Rotberg 2003).

In much of Africa, Central America, the Caribbean, and Asia, European powers set up extractive institutions that promoted their own interests as opposed to protecting property rights and establishing checks and balances on government (Acemoglu and Robinson 2012). The Europeans governed the 'new found lands' by institutionalising highly centralised systems to support the extraction of the abundant natural resources available. In particular, in the colonised nations of Africa and Latin America where natural resources were abundant and climatic conditions were favourable for growing crops but not for European settlement, the colonisers established extractive institutions to benefit their citizens back home (Acemoglu and Robinson 2012). In Latin America, where a large number of slaves were deployed, conditions were ripe for establishing plantations, which generated extreme inequality in terms of wealth, capital, and political power (Engerman and Sokoloff 2002).

As noted earlier, these types of institutions are associated with limited growth and exclusion leading to large levels of inequality in these regions. On the other hand, in countries where the colonisers (i.e., mostly European settlers and their descendants) established institutions that promoted property rights for the majority, in the likes of Australia, Canada, New Zealand, and the US, we see inclusive and sustained growth, underscored by innovation and technological advancement. In areas where colonisers settled in large numbers, the nature and structure of the institutions established were different. There was more focus on developing these areas for permanent settlement, rather than extractive institutions (Acemoglu and Robinson 2012). Protection of private property rights, particularly rights to land and trade under market institutions, were underscored. In addition, the relatively small size of farms in the northern parts of North America meant that more farmers were involved in agriculture, with low economies of scale in production (Engerman and Sokoloff 2002). The inclusive institutions established in these regions fostered a relatively fair (re)distribution of wealth, promoted innovation, and sustained economic growth.

Based on these broad regional categorisations, Africa, Latin America, and socialist groups that were shaped by strongly centralised and extractive institutions at their early stage of development are likely to experience relatively limited growth and retarded technological innovations (controlled innovations and talent in specific sectors), resulting in higher levels of inequality (Armah and Baek 2015).

In short, countries at this stage would experience the impact of development when the S-curve is overwhelmed by the impact of economic policies through institutions and governance systems. Thus, welfare outcomes such as economic growth and income inequality are largely affected by institutions and systems under a strong government as they have not yet benefitted from influential technological innovation through the pattern of S-curves.

Stage II: Accelerated through the Tipping Point

Countries that have gone through the aforementioned phase can enjoy the fruits of their technological innovation that had been invested in over several years. As the flow of technology advances beyond the tipping point under the framework of the S-curve, the scale of the economy could expand rapidly. As a result, many countries are successful in entering middle-income status.

Today, the poor are able to access information that has not been accessed before, greatly contributing to various economic activities (González et al. 2009). In addition, technological advancement has also made the likes of farming more efficient. In these countries, agricultural productivity, which can be measured by the agriculture value-added per worker, has exponentially increased over the past two decades (see Fig. 8.8). It can therefore be regarded as the period when a radical technological breakthrough has been realised. When the agricultural capacity of a country reaches its limit, further improvement to agricultural productivity tends to shift the labour force to other industries (e.g., manufacturing and services industries) rather than making an extra effort to further expand the agricultural economy. With radical improvements in agricultural productivity made over the past several decades, production indices have improved in all areas of production such as crops, food, and livestock, while the number of persons employed in agriculture has decreased over this period.

Furthermore, the benefit of technological advancements would be greater for the informal sectors many of whose employees are from vulnerable groups, such as family-based firms and small-scale producers where employment is dominated by young people and women in rural areas. This particular impact of technological progress is considered to potentially reduce inequality (Armah and Baek 2018).

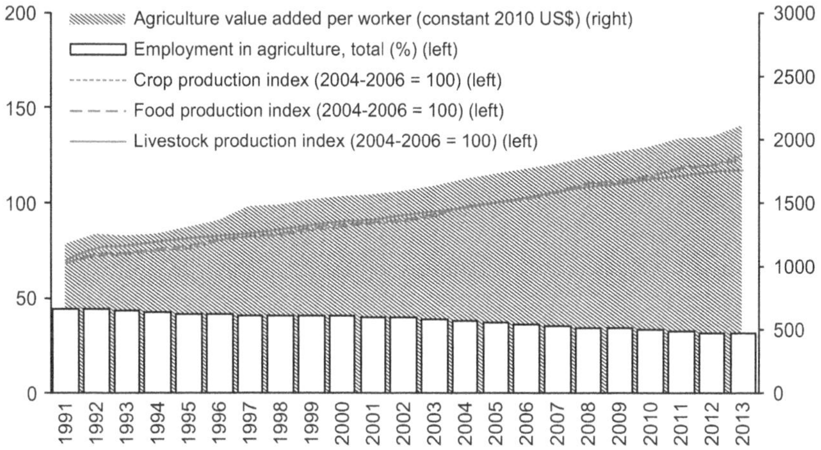

Fig. 8.8 Technology-driven agricultural development. (*Source*: Author's own elaboration on the basis of the ILO's Global Employment Trends (2014) and the World Bank's World Development Indicators (2017). *Note*: The Global Employment Trends 2014 is used for employment in agriculture data whereas the World Development Indicators is used for the others)

Particularly for countries at this stage, new elite groups tend to emerge. People of this group that are well-educated regardless of their background (rich versus poor) are able to create new wealth through innovative manufacturing processes, information made available by globalisation, and high-tech services activities. Their innovative efforts strongly prompt development of not only the information technology industry but also many other industries and sectors, which further create jobs. This transition from traditional to modern technology societies could accelerate socio-economic development, especially through the productive labour force moving from agricultural sectors.

Although there are jobs that can be replaced by technological automation (e.g., agricultural labourers, washers, and launderers), such job-destroying effects are counter-balanced or often overwhelmed by job-creation effects of innovative technologies (Stewart et al. 2015)—non-routine, cognitive tasks, particularly in knowledge-intensive sectors (e.g., accountants, business analysts, management executives, and information technology managers). In other words, a certain number of the poor, most of whom are relatively less educated, may be replaced by such technological advancement. Furthermore, due to the demands of multinational companies that seek to use cheaper labour resources benefitting from the acceleration of the globalisation of labour, the poor whose labour force is less competitive can also lose their jobs as a result. Nonetheless, the synergy effect may appear to be maximised through active cooperation and competition among industries across the economy so that the size of the economy could grow rapidly, following the S-curve pattern.

Many entrepreneurs are involved in the innovation race to take the lead among intense competition, and thus the tipping point of the S-curve is observed in various technologies. In particular, such entrepreneurial spirit contributes to realising a large number of tipping points with the intensive investment efforts that have been made since the early stages of development. Such efforts are frequently associated with a higher return on investment across the industry. As shown in Table 8.1, a survey of 96 industries and 7330 companies indicated that the average *Return on Equity* (adjusted for R&D) is 9.5 per cent, while such a return for most industries directly or indirectly related to technological innovation is above average.

Along with this improvement on returns on investment, economic expansion, largely driven by the synergies effects across industries, can result in the creation of new jobs to thicken the middle-class group of a population (largely supported by *Hidden Champions*).[4] Furthermore, the advent of new elite groups gives the poor more incentive to receive a better education in general and technical-skill training for information technology in particular. In many countries, from the 1990s through to the early 2000s, during which information technology greatly fuelled economic growth, the public's strong demand for higher education exploded as the number of public and private postsecondary institutions exponentially grew (see Fig. 8.9).

As technological innovation is also rapidly spread with the diffusion effect of information (Rogers 2003), citizens begin to demand transparency and fairness with regard to the practices of corporations and governments since information that was not of great interest in the past is now shared across the majority of households. As a result, lobbying and

Table 8.1 Business profitability by sector: Return on equity

Industry Name	# of firms	ROE (%)	Industry Name	# of firms	ROE (%)
Retail (Building Supply)	6	59.16	Drugs (Biotechnology)	426	10.35
Air Transport	18	40.87	Farming/Agriculture	37	10.14
Retail (Automotive)	25	35.64	Software (Entertainment)	13	10.02
Transportation	17	30.07	Environmental & Waste Services	89	9.73
Paper/Forest Products	23	29.59	Semiconductor	80	9.62
Cable TV	14	25.27	Drugs (Pharmaceutical)	164	9.52
Retail (Grocery and Food)	14	23.74	**Total Market (avg.)**	**7330**	**9.50**
Building Materials	41	23.68	Bank (Money Centre)	10	9.46
Computer Services	117	23.12	Reinsurance	3	9.41
Chemical (Diversified)	8	22.69	Chemical (Specialty)	100	9.38
Office Equipment & Services	24	20.51	Healthcare Information and Techno	125	9.18
Household Products	129	19.93	Real Estate (Development)	18	9.17
Entertainment	79	19.55	Utility (Water)	22	9.17
Computers/Peripherals	55	19.54	Banks (Regional)	645	9.03
Telecom. Services	67	19.23	Brokerage & Investment Banking	45	8.47
Information Services	64	19.03	Rubber& Tires	4	8.42
Auto & Truck	15	18.95	Chemical (Basic)	45	8.34
Beverage (Alcoholic)	25	18.56	Construction Supplies	51	8.24
Aerospace/Defense	96	18.22	Software (Internet)	297	8.16
Retail (General)	19	16.91	Insurance (Life)	22	8.06
Transportation (Railroads)	7	16.27	Hospitals/Healthcare Facilities	38	4.97
Healthcare Support Services	121	16.25	Power	68	4.34
Auto Parts	63	16.10	Publishing & Newspapers	37	3.60
Trucking	30	15.68	Insurance (General)	19	3.39
Recreation	66	15.52	Education	36	3.22
Retail (Special Lines)	108	15.44	Engineering/Construction	48	2.11
Packaging & Container	26	13.94	Oil/Gas (Integrated)	7	0.48
Retail (Distributors)	88	13.94	Financial Services (Non-bank & Insurance)	258	-0.10
Retail (Online)	57	13.76	Electronics (Consumer & Office)	24	-0.77
Homebuilding	33	13.12	Advertising	41	-1.06
Hotel/Gaming	69	12.60	Telecom (Wireless)	17	-3.90
Food Wholesalers	16	12.37	Green & Renewable Energy	25	-5.93
Electrical Equipment	119	11.76	Shipbuilding & Marine	11	-6.56
Machinery	127	11.58	OilfieldServices/Equip.	148	-8.24
Real Estate (Operations & Services	54	11.24	Precious Metals	109	-9.21
Investments & Asset Management	156	10.79	Tobacco	22	-14.86
Semiconductor Equip	45	10.67	Steel	38	-17.01
Food Processing	87	10.58	Coal & Related Energy	38	-32.97
Software (System & Application)	236	10.46	Oil/Gas (Production and Exploration)	330	-38.95
Telecom. Equipment	107	10.40	Metals & Mining	97	-43.00

Source: Author's own elaboration on the basis of Damodaran (2017)

corruption that were prevalent in the early stage of development are disappearing to some extent. Figure 8.10 shows the rapid spread of social networking in all age groups starting in 2005. The strong demand for better governance through such social networking can be a major

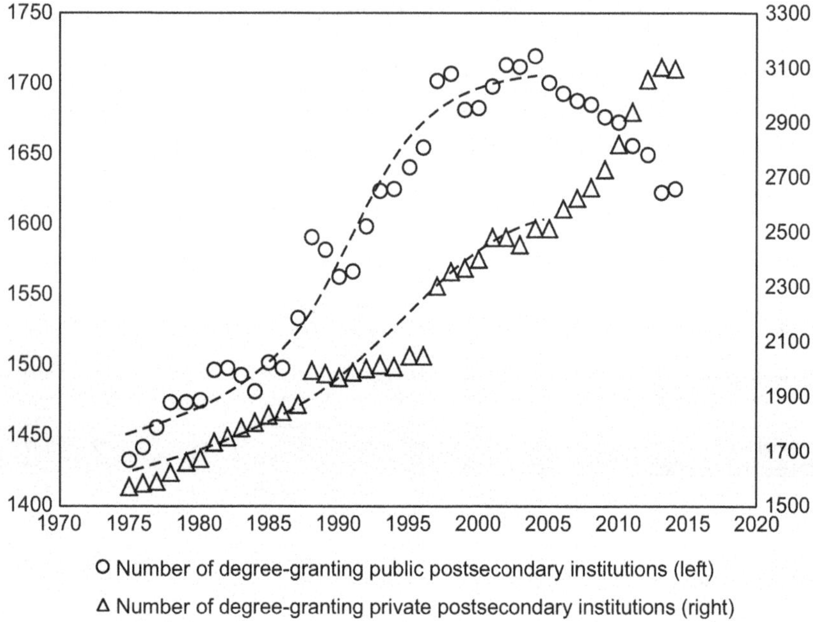

Fig. 8.9 Number of the US degree-granting postsecondary institutions: Public versus private. (*Source*: Author's own elaboration on the basis of the US National Centre for Education Statistics (2017))

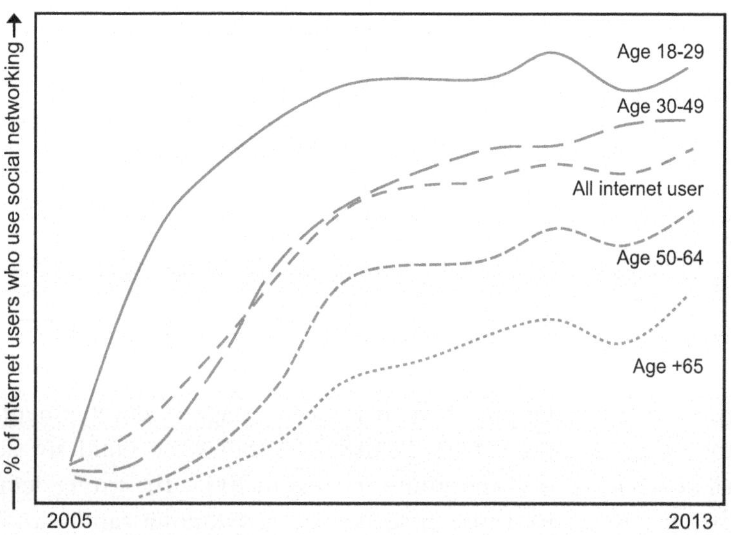

Fig. 8.10 Diffusion of social networking innovation in the US. (*Source*: Author's own elaboration on the basis of the Pew Research Centre (2013))

contributor to the widespread distribution of wealth, which is created rapidly during this stage of development. These countries thus tend ideally to achieve desired economic development and inequality reduction.

In sum, during this period when technological progress passes the tipping point in line with the S-curve framework, the impact of technological innovation along with the diffusion of information would overwhelm that of other socio-economic factors on economic growth and inequality. This is therefore the period when the proposition 'the bigger the number of tipping points are, the greater inclusive growth in which growth benefit are broadly shared across various income segments of population will be realised' can be valid.

Nevertheless, since socio-economic development tends to be so rapid when passing the tipping point of the S-curve, societies that have already felt the fruits of labour appear to pursue a sort of sustainable innovation so as to continue to enjoy such fruits. Such continuous innovation is accompanied by sustainable economic growth, while, in the meantime, the (re)distribution of wealth reaches a certain level. However, institutions and policies may not be effective enough to control the radical speed of S-curves, so an equitable and fair distribution system may not systematically be constructed in this period. In fact, when an economy is in a state of rapid growth, the government may be less motivated to set regulations with less demand from the public. This is consistent with the influential argument made by Vivek Wadhwa (2014), an American technology entrepreneur and academic, who states: 'The regulatory gaps between institutions/laws and technology exist because they cannot keep up with rapid technological advances.' This discrepancy among institutions, policies, and innovations will appear as a challenge in the next developmental stage.

Stage III: Restrained by Conflict

After a country has gone through the first two developmental stages, in the advanced stage of development, many people, regardless of being rich or poor, already have a higher education, so the intellectual level of society is becoming saturated. In other words, society's education level is sometimes inflated. Furthermore, the S-curve-driven economic expansion also begins to enter a period of stagnation. As a result, issues of unemployment become serious, and social issues begin to increase, such as widening inequality.

Figure 8.11 shows the mapping of the unemployment rate to the number of those enrolled in postsecondary institutions. In the enrolment trend, across the 1950–2010 period, two broad S-curve shapes can be identified based on the period of the 1980s as a point in time. A common phenomenon for two S-curve trends indicates that the unemployment rate drops at the point of passing the tipping point. On the contrary, the unemployment rate tends to increase rapidly when the enrolment number enters a period of stagnation. Thus, during this period, the unemployment situation of people with a higher education could exacerbate social unrest, further leading to distrust in politics and government, and eventually the deteriorating efficiency of economic policy.

As these social conflicts deepen, companies that have survived fierce competition in the previous stages appear to have a monopoly in their respective markets. This is also a frequently observed phenomenon in the S-curve of innovation as emphasised by Christensen (2003). That is, com-

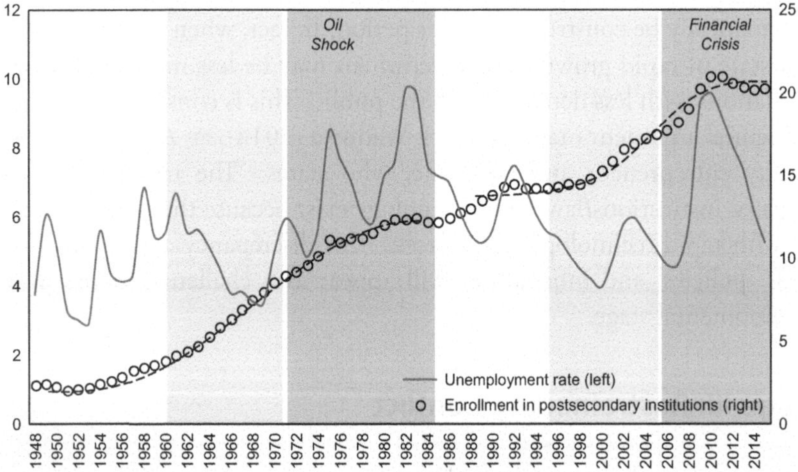

Fig. 8.11 Relationship between enrolment in degree-granting postsecondary institutions and unemployment rate in the US. (*Source*: Author's own elaboration on the basis of the US National Centre for Education Statistics (2017) and the US Bureau of Labour Statistics (2017))

panies that survived by securing a competitive advantage can accumulate more and more wealth, finally leading to the phenomenon of *Winner-Takes-All*. In other words, the economics of the Internet favour natural monopolies, and the absence of a competitive business environment can result in more concentrated markets, benefitting incumbent firms.

Figure 8.12 presents global-level empirical data for the direct correlation between the share of large companies and the total employment share from these companies. This can imply that large companies that have succeeded among intense competition tend to continue to expand their share of the workforce. For example, a recent emerging term *Amazon'd* speaks volumes in this regard. This refers to a situation where incumbent firms are losing their market share as Amazon approaches their market. Amazon started out as an online bookstore in 1994, and now competes for cloud services with Microsoft, e-commerce with eBay, and information technology with Google and Apple, and even artificial intelligence and medicines with many others. As such, Amazon is increasing its shares in almost all markets.

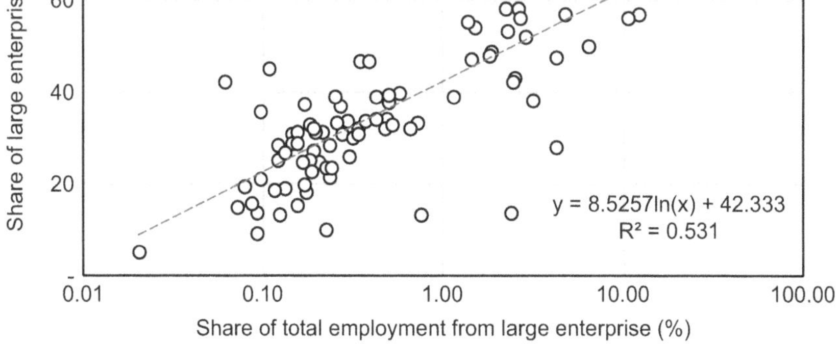

Fig. 8.12 Global employment concentration to large enterprises. (*Source*: Author's own elaboration on the basis of the World Bank's MSME Country Indicators (2014))

Besides, such dominance may be accelerated since large firms would need to make up for the absence of new growth engines in this stagnated period. There can also be some companies that are likely to distort their markets because they tend to work in a variety of ways to maintain their global and national competitiveness by dealing with extensive internal transactions, unfair practices with subcontractors, and many other issues.

Over the decades, the number of subcontracted jobs has expanded rapidly, offering low wages to predominantly black and Latino workers. Subcontracted jobs have grown at three times the rate of all private sector jobs in Silicon Valley, exacerbating the region's income inequality gap. The rise in subcontracting has made the high-tech sector a weak engine for middle-class job creation. There is clear evidence that contract workers hired by high-tech firms face substantially lower wages, and have significantly worse socio-economic circumstances, than comparable workers who are directly employed by high-tech firms. In other words, the small- and medium-sized enterprises in which a significant workforce are involved are losing market competitiveness so the wage gap between their workers and those of large enterprises becomes polarised, which can largely be associated with the collapse of the middle class. Thus, many advanced economies face increasingly polarised labour markets with rising levels of inequality—in part, because technology requires higher skill levels while replacing routine jobs.

This decline in the middle class also leads to a decrease in productivity (see Fig. 8.13). Looking at the OECD countries, which can be assumed to have entered into this advanced stage of development, we may see that there is a clear S-curve pattern observed in almost all of these countries in terms of the productivity growth trend, and that they have already entered the maturity stage, the third stage of the S-curve.

To summarise the key points, most countries at this stage are considered as developed countries that have mostly gone through the two previous stages of the S-curve. In the meantime, as their radical technological innovations are approaching a certain limit during this period, the pace of economic growth also experiences some stagnation. Moreover, when a company has been attempting to extend its market dominance for a long time and to finally acquire technology monopolies in various fields through patents registration, capital tends to be concentrated into a few major players. This eventually leads to polarisation with challenges emerging in terms of income inequality. In short, during this period, the influence of the

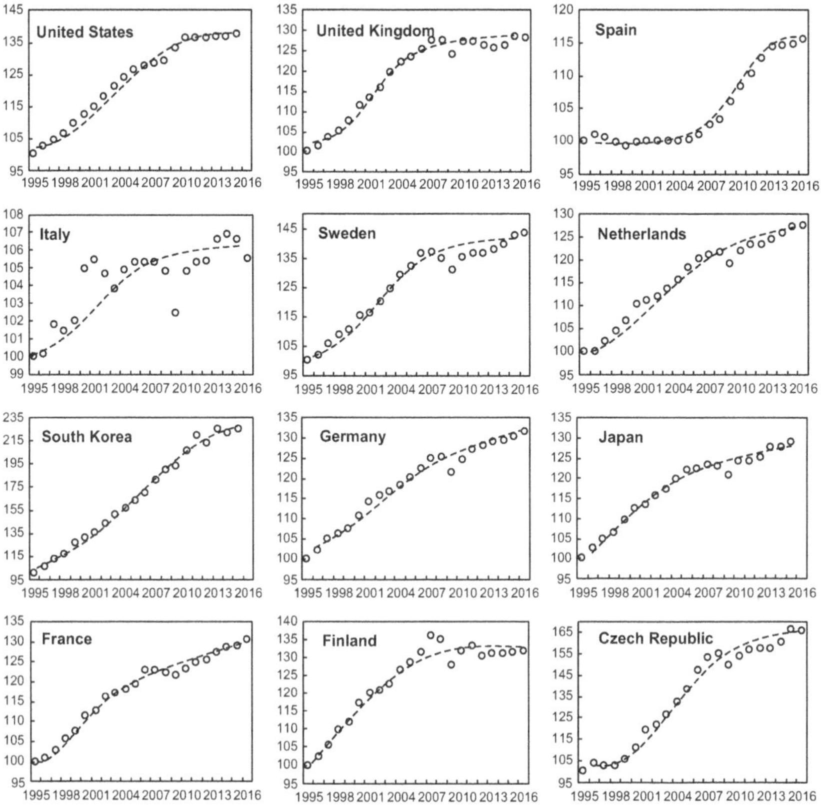

Fig. 8.13 Productivity growth in selected advanced economies. (*Source*: Author's own elaboration on the basis of the OECD (2017b))

technology-driven S-curve is overwhelmed by that of other factors (similar to the hypothesis proposed at the first stage of development), and economic growth stagnation is likely as is a period of deepening inequality.

Integrated Summary and Beyond the S-Curve

The least developed countries (LDCs), mostly belonging to the first stage, are not likely to benefit from technological advancement in a situation where development patterns are largely dominated by practices of the premodernity era such as extractive institutions or non-democratic decision-making structure. Furthermore, even if innovative technology is applied to internal economic activities, such application is largely transferred from outside, rather than being created within their socio-economic system. Therefore, some social confusion due to the conflict between the traditionality of society and the advanced technology derived from a Western base is expected. This can lead to an unfavourable situation in which inequality tends to deepen, eventually leading to political fragilities such as democratisation frenzy. This hypothesis can be somewhat inconsistent with the claims of most modernisation theories or economic growth theories that view technological innovation as exclusively positive (largely based on a Eurocentric view). In fact, the socio-economic impact of technology in the twenty-first century can be incomparably greater than the period in which these theories were devised, in both quantitative and qualitative terms as well as both positively and negatively.

Passing through the development process of the first stage, LDCs may step into the second stage having had experience of accommodating the dynamics between the traditionality of society and modern technology from advanced nations. This period sees the benefits of technology sprinkled across society and is applied throughout different industries and sectors. As partly discussed in reference to the modernisation theories, experiencing the synergy among industries along with the advent of the leading industries, countries are able to improve economic productivity. In particular, the effect of technological innovation cannot only create high profitability for industrial sectors, but can also derive a remarkable improvement in productivity in the traditional industry (e.g., agricultural sector). This can thus be associated with narrowing inequality levels largely driven by the sectoral shift from the agricultural to the industrial sector, which was essentially emphasised by most development theories in the past.

In fact, this positive effect of technology on a broad spectrum of socio-economic dimensions tends to be accompanied with an improvement to human capital possibly accelerated by the diffusion of information and knowledge. Also, political orientation may shift to a more transparent governance system, which is significantly attributed to such a diffusion effect. In short, countries experiencing the tipping point at this stage could foster rapid industrial expansion and social modernisation with relatively less labour and capital input. This stage can be the most ideal and efficient way of benefitting from technological innovation.

Finally, countries that have achieved modernisation by the rapid diffusion of information and knowledge, along with radical industrial expansion, benefitted from technological advancement throughout stages I and II, and are now reaching the stagnation phase of the Developmental S-Curve. This is normally a phenomenon that can be experienced in most advanced countries in which citizens keep demanding socio-economic benefits from technology to the same level that they acquired in the preceding stage. However, such benefits at stage III are likely to be limited. In order to meet these demands, the private sector (e.g., corporations, firms, etc.) is increasingly engaged in competitive pricing, often causing so-called Chicken Wars. In this case, companies who are not able to bear such price competition might exit the market frequently resulting in monopolised structures across multiple industries.

During this period, the problem of wealth being concentrated in the hands of a few will come to the fore again. The result of companies that have been kicked out of the market due to the severe aforementioned price competition is likely to lead to mass unemployment, which would be unacceptable to many educated citizens who have already enjoyed a modern and prosperous life. Therefore, the discontent of such citizens can lead to distrust in government policy as a whole and the further deterioration of the effectiveness of policy intervention. This often results in the so-called Middle-Income Trap.

In this way, the proposed three stages of the Developmental S-Curve would not accept some utilitarian assumption that technology always provides economic growth, as extensively discussed in Chap. 7. Rather, this proposal would adopt an approach integrating the technological aspect to the complex interaction between inequality and economic growth as the fundamental basis of analysis, so as to better account for the developmental path in the twenty-first-century context, especially for developing countries (see Fig. 8.14).

As the society undergoing through stage I through III hypothesised, what is next? In other words, will there be post-Developmental S-Curve? Denying it may lead to a logical error of giving up the evolutionary nature at the destination, which was indeed the starting point of the developmental model. To overcome this logical contradiction, my Developmental S-Curve is proposed to repeat a new flow of stages I, II, and III after the preceding step of stage III.

In stage III, for instance, the government needs to confront a policy dilemma in which it would not only have to deal with resolving political

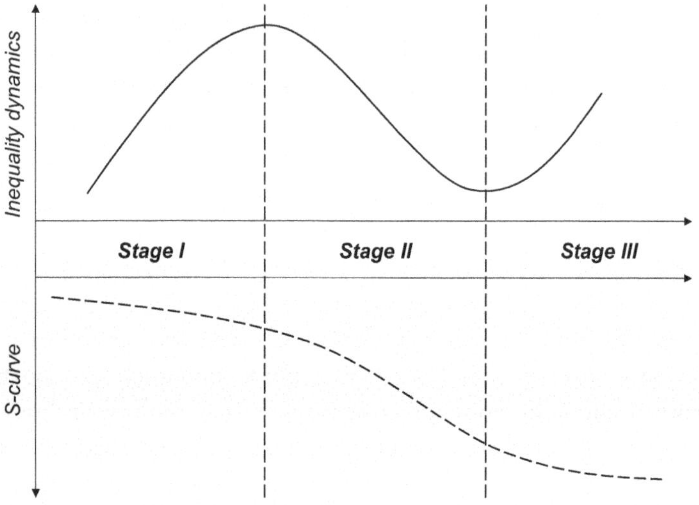

Fig. 8.14 Integrated framework of the Technology-Development-Inequality nexus. (*Source*: Author's own elaboration)

resistance from an aggressive labour union along with excessive unemployment but also need to deregulate in support of the surviving firms that are experiencing slowing growth so are likely to lobby to help them survive again particularly among high-tech global market competition. Nevertheless, if social resistance due to serious polarisation is maximised so that some form of negative consensus against governmental actions was reached in society, politics would have to react urgently by implementing policies focused on the (re)distribution of wealth rather than growth-oriented intervention (e.g., deregulation). Some time later, however, if welfare regimes were to keep suffering from stagnation of economic growth while redistribution practices were to continue for a long period of time, the general public would, to some extent, be satisfied with the degree of redistribution so the demand to seek new growth engines would once again become central to society. As a result, the government would then have to again pursue growth-oriented policies by supporting the private sector. In other words, there would be a dominant view in society that firms would fall behind global competition unless society quickly found new growth engines.

Specifically, the government policies of corporate-friendliness and social stabilisation under a complicated interaction between society's major actors would become the driving force to create new Developmental S-Curve, namely *Multiple Developmental S-Curves*. Technically speaking, the worsening of social disruption and the widening of inequality levels that the state may have experienced in stage III can in fact substantially be overlapped with the wealth distribution situation at stage I of the next Developmental S-Curve. It is thus imperative for transitioning countries to make an effort to reduce such an overlap of unfavourable situations, which would substantially depend on timely governmental intervention and effective national planning framework given their challenges. In other words, the strategic response of governments and the capacity of *Path-Shaping* institutions of society would lead to the creation of various developmental paths,[5] which can be the most determinant factors for the slope of future S-curves and the timing of the emergence of multiple S-curves as well, especially in the age of revolutionary innovations.

Discussion: Analytic or Synthetic?

According to *Schumpeterian* growth theorists, technological innovation is a key driving source of national development. Technological advances have clearly modernised the world and society which has benefitted in many areas of life and will undoubtedly continue to do so in the future. Indeed, many poor have been able to accumulate wealth more efficiently through various channels of technological advantages (e.g., access to financial services and productive resources). However, as we have seen in the case of Silicon Valley, much of the recent literature has started to warn about the side effects of on-going technological advancement, particularly its impact on widening inequality. This is largely due to the recent experience of technological automation which has replaced many unskilled jobs, often resulting in mass unemployment, while those who make good use of technology have been able to exponentially accumulate wealth. Now that the era of artificial intelligence is approaching, no one can be sure how significant such negative effects will become.

In addition to the positive role of technological progress for economic development as well as the negative role in terms of inequality challenges, it is interesting to note that there is a struggle of a 'push-pull' nature between inequality and economic growth. Inequality itself, for example, may be a constraint that inculcates people's incentives to work in an efficient way but, in turn, might hinder economic development due to various reasons related to ineffective savings and poor levels of education inherently embedded in the poor group. Essentially, there are some complicated endogenous processes to consider between the two (Baek 2017).

Therefore, the proposition that technological progress leads to economic development is not necessarily always a true one but rather depends on the correlating characteristics that have evolved in a particular society in terms of the *Technology-Development-Inequality* nexus. In other words, as we have seen in the three stages of the Developmental S-Curve, without careful examination of the relationship between developmental levels of technology-inequality and inequality-growth, it may not always be possible to support the positive role of technological innovation on development that was indeed the fundamental basis and assumption in long-standing technology development discourse since Schumpeter's *Creative Destruction* (1939) through to Christensen's *Disruptive Innovation* (1997).

Especially in the technology age, which has affected almost all fields, including the social, economic, political, and environmental spheres, if technological progress seriously increases the level of inequality and this inequality significantly causes social conflict to finally hinder economic growth activities, the economic outcome of technological advancement as a whole may not be so great in the end and may even appear to move in an unexpectedly distorted direction, especially in the twenty-first century where technological innovation is so radical that it is difficult to predict its direction.

In this context, the S-curve can be a powerful tool to evaluate and control these future uncertainties in a more efficient manner, given the fact that it has been applied to many fields and indeed has already contributed to creating various innovations. In particular, it could become a standard in explaining how to strategically expand the market power of new products, and also in analysing the ups and downs of the performance of corporations and their products and services. It can further be applied to the paradigm shift of an industry and also to the inter-country competition to lead to industrial changes.

Most importantly, by understanding the concept of the tipping point under the pattern of the S-curve, we may be able to explore the social acceptance rate of new technology in an analytical and empirical fashion.

The S-curves and the tipping points observed from inter-dynamics between the inequality index and the PCGDP as well as the number of doctoral degrees awarded in the US may not be coincidental (see Fig. 8.15). Various social and economic phenomena and their intercon-

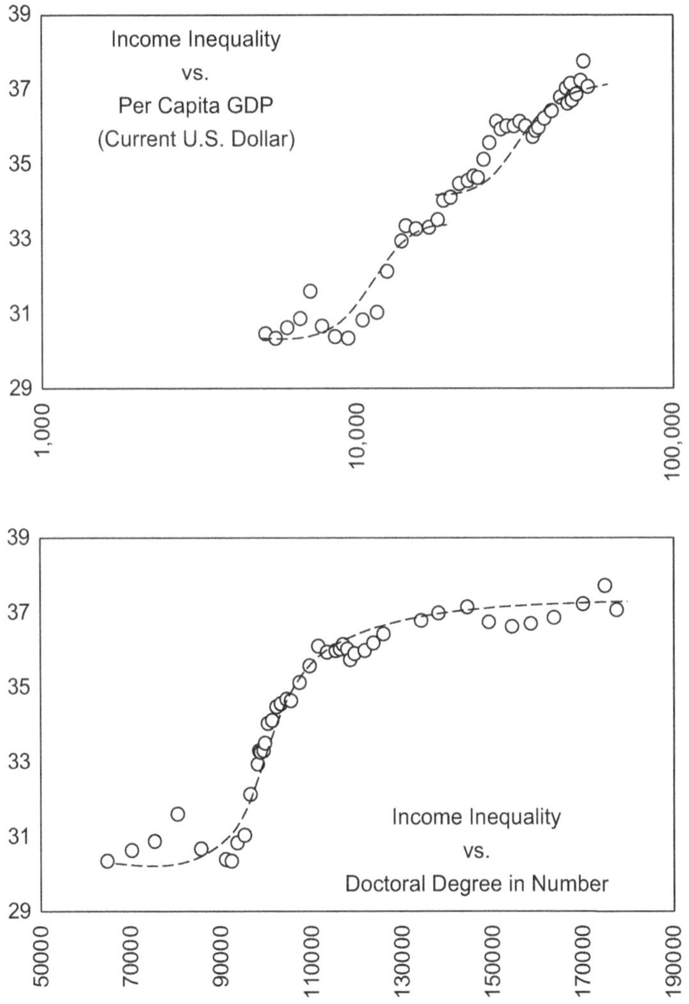

Fig. 8.15 Income inequality versus PCGDP and doctoral degree in the US. (*Source*: Author's own elaboration on the basis of the SWIID 5.1 (Solt 2016) and the US National Centre for Education Statistics (2017))

nected relationships often embrace the S-curve dynamics that can be converted into analytical information. In this regard, the adoption of this approach will enable more timely management against the uncertainties, social vulnerability, and fragility that may plague us in the upcoming Industrial Revolution.

Notes

1. Project Loon is a network of balloons travelling on the edge of space, designed to extend Internet connectivity to people in rural and remote areas worldwide.
2. Personal income is defined as the sum of wage and salary disbursements (including stock options), supplements to wages and salaries, proprietors' income, dividends, interest, and rent, and personal current transfer receipts, less contributions for government social insurance.
3. The establishment of Volkswagen is indeed very interesting. In 1937, Adolf Hitler gave the order to Ferdinand Porsche to develop a national car, promising one car per citizen of Germany under the following conditions (Bowler 2015)—the car should be able to carry two adults and three children at 100 kilometres per hour; the car should be able to run over 14.5 kilometres with 1 litre of fuel and weigh less than 605 kilograms; and the selling price of the car should be cheaper than 1000 deutsche marks (costing no more than a motorbike at the time). Two years after these five instructions were given by Hitler, the Volkswagen automobile was introduced, which would become the best-selling brand in automotive history.
4. Hidden champions refer to relatively small- and medium-sized companies that grow successfully in a sustainable manner. This term was first coined by Hermann Simon (2009) who describes it as the small, highly specialised world-market leaders.
5. For a more detailed discussion, see Chap. 3.

References

Acemoglu, D. 2002. Technical Change, Inequality, and the Labor Market. *Journal of Economic Literature* 40 (1): 7–72.

Acemoglu, D., and D.H. Autor. 2011. Skills, Tasks and Technologies: Implications for Employment and Earnings. *Handbook of Labor Economics* 4 (11): 1043–1171.

Acemoglu, D., and J. Robinson. 2012. *Why Nations Fail: The Origins of Power, Prosperity, and Poverty*. New York: Crown Publishing Group.

Aghion, P., and P. Howitt. 1998. *Endogenous Growth Theory*. Cambridge: MIT Press.

Armah, B., and S.J. Baek. 2015. Can the SDGs Promote Structural Transformation in Africa? An Empirical Analysis. *Development* 58 (4): 473–491.

Armah, B., and S.J. Baek. 2018. Three Interventions to Foster Sustainable Transformation in Africa. *Journal of Social, Political, and Economic Studies* 43 (1–2): 3–25.

Baek, S.J. 2017. Is Rising Income Inequality Far from Inevitable During Structural Transformation? A Proposal for An Augmented Inequality Dynamics. *Journal of Economics and Political Economy* 4 (3): 224–237.

Bell, L., and R. Freeman. 2001. The Incentive for Working Hard: Explaining Hours Worked Differences in the US and Germany. *Labour Economics* 8 (2): 181–202.

Bowler, T. 2015. Volkswagen: From the Third Reich to Emissions Scandal. *BBC News*, 2 October. Available from http://www.bbc.com/news/business-34358783. Accessed 11 Nov 2017.

Chang, Y.S., and S.J. Baek. 2010. Limit to Improvement: Myth or Reality? Empirical Analysis of Historical Improvement on Three Technologies Influential in the Evolution of Civilization. *Technological Forecasting and Social Change* 77 (5): 712–729.

———. 2011. Limit to Improvement in Running and Swimming. *International Journal of Applied Management Science* 3 (1): 97–120.

Christensen, C.M. 1997. *The Innovator's Dilemma: When New Technologies Cause Great Firms to Fail*. Cambridge: Harvard Business School Press.

———. 2003. *The Innovator's Dilemma: The Revolutionary Book that Will Change the Way You Do Business*. New York: Harper Collins.

Coe, D., and E. Helpman. 1995. International R&D Spillovers. *European Economic Review* 39: 859−887.

Damodaran, A. 2017. *Return on Equity by Sector (US)* [Online]. Available from: http://www.stern.nyu.edu/~adamodar/pc/datasets/roe.xls. Accessed 8 Sept 2017.

Engerman, S.L., and K.L. Sokoloff. 2002. Factor Endowments, Inequality, and Paths of Development Among New World Economies. *Economia* 3 (2): 41–102.

Forbes, K. 2000. A Reassessment of the Relationship between Inequality and Growth. *American Economic Review* 90 (4): 869–887.

Foster, R.N. 1986. *Innovation: The Attacker's Advantage*. New York: Summit.

Freeman, C., and C. Perez. 1988. Structural Crises of Adjustment, Business Cycles and Investment Behavior. In *Technical Change and Economic Theory*, ed. G. Dosi. London: Pinter.

Galor, O., and D. Tsiddon. 1997. Technological Progress, Mobility, and Economic Growth. *American Economic Review* 87 (3): 363–382.

Gladwell, M. 2000. *The Tipping Point: How Little Things Can Make a Big Difference*. New York: Little, Brown, and Company.

González, V., P. Ibarrarán, A. Maffioli and S. Rozo. 2009. *The Impact of Technology Adoption on Agricultural Productivity: The Case of the Dominican Republic*. Washington, DC: Inter-American Development Bank. Available from: http://services.iadb.org/wmsfiles/products/Publications/2191162.pdf. Accessed 8 Sept 2017.

Goos, M., A. Manning, and A. Salomons. 2014. Explaining Job Polarization: Routine-Biased Technological Change and Offshoring. *American Economic Review* 104 (8): 2509–2526.

Henderson, R. 2005. *Developing and Managing a Successful Technology and Product Strategy* [Online]. Cambridge: MIT Sloan School of Management. Available from: http://www.mit.edu/people/rhenders/Teaching/day1_jan05.ppt. Accessed 18 Sept 2017.

Hyytinen, A., and O. Toivanen. 2011. Income Inequality and Technology Diffusion: Evidence from Developing Countries. *Scandinavian Journal of Economics* 113 (2): 364–387.

ILO (International Labour Organization). 2014. *Global Employment Trends 2014: Supporting Data Sets*. Geneva: ILO. Available from: http://www.ilo.org/global/research/global-reports/global-employment-trends/2014/WCMS_234879/lang--en/index.htm. Accessed 1 Sept 2017.

Jaumotte, F., S. Lall, and C. Papageorgiou. 2013. Rising Income Inequality: Technology, or Trade and Financial Globalization? *IMF Economic Review* 61 (2): 271–309.

Kurzweil, R. 2005. *The Singularity is Near: When Humans Transcend Biology*. New York: Penguin.

———. 2009. *The Singularity is Near* [Online]. Available from: http://singularity.com. Accessed 10 Aug 2017.

Li, H., and H. Zou. 1998. Income Inequality Is Not Harmful for Growth: Theory and Evidence. *Review of Development Economics* 2 (3): 318–334.

Lin, J.Y. 2012. *New Structural Economics: A Framework for Rethinking Development and Policy*. Washington, DC: World Bank Publications. Available from: http://siteresources.worldbank.org/DEC/Resources/84797-1104785060319/598886-1104951889260/NSE-Book.pdf. Accessed 20 Nov 2016.

Lumenga-Neso, O., M. Olarreaga, and M. Schiff. 2005. On 'Indirect' Trade-Related Research and Development Spillovers. *European Economic Review* 49 (7): 1785–1898.

McGrath, R.G. 2013. The Pace of Technology Adoption is Speeding Up. *Harvard Business Review*. Available from: https://hbr.org/2013/11/the-pace-of-technology-adoption-is-speeding-up. Accessed 24 Aug 2017.

OECD. 2017a. *International Development Statistics Database* [Online]. Available from: http://www.oecd.org/development/stats/idsonline.htm. Accessed 1 Aug 2017.

———. 2017b. *Compendium of Productivity Indicators 2017* [Online]. Available from: http://www.oecd.org/std/productivity-stats/oecd-compendium-of-productivity-indicators-22252126.htm. Accessed 6 Mar 2016.

Okun, A.M. 1975. *Equality and Efficiency: The Big Tradeoff*. Washington, DC: The Brookings Institution.

Pew Research Centre. 2013. *Datasets: Internet & American Life Project Tracking Surveys 2005–2013* [Online]. Available from: http://www.pewinternet.org/datasets. Accessed 17 Nov 2017.

Reidenbach, L. and C. Hoene 2016. *Inequality and Economic Security in Silicon Valley* [Online]. Sacramento: California Budget and Policy Centre. Available from: http://calbudgetcenter.org/resources/inequality-economic-security-silicon-valley. Accessed 7 Aug 2017.

Rogers, E.M. 2003. *Diffusion of Innovations*. 5th ed. New York: Free Press.

Romer, P.M. 1990. Endogenous Technological Change. *Journal of Political Economy* 98 (5): 71–102.

Rotberg, R.I. 2003. Failed States, Collapsed States, Weak States: Causes and Indicators. In *State Failure and State Weakness in a Time of Terror*, ed. R.I. Rotberg, 5–10. Cambridge/Washington, DC: Brookings Institution Press.

Schilling, M.A. 2010. *Strategic Management of Technological Innovation*. 3rd ed. New York: McGraw-Hill.

Schumpeter, J.A. 1939. *Business Cycles: A Theoretical, Historical, and Statistical Analysis of the Capitalist Process*. New York: McGraw-Hill.

Silicon Valley Institute for Regional Studies. 2017. *2017 Silicon Valley Index*. California: Joint Venture Silicon Valley. Available from: https://jointventure.org/images/stories/pdf/index2017.pdf. Accessed 7 Aug 2017.

Simon, H. 2009. *Hidden Champions of the 21st Century*. New York: Springer.

Solt, F. 2016. *The Standardized World Income Inequality Database Version 5.1* [Online]. Available from: https://dataverse.harvard.edu/dataset.xhtml?persistentId=hdl:1902.1/11992. Accessed 10 Nov 2016.

Stewart, I., D. De, and A. Cole. 2015. *Technology and People: The Great Job-Creating Machine*. London: Deloitte LLP. Available from: https://www2.deloitte.com/uk/en/pages/finance/articles/technology-and-people.html. Accessed 22 Aug 2017.

The Economist. 2015. *Silicon Valley's Fortunes: What a Performance* [Online]. Available from: http://www.economist.com/blogs/graphicdetail/2015/07/silicon-valleys-fortunes. Accessed 10 July 2017.

US Bureau of Labour Statistics. 2017. *Labor Force Statistics including the National Unemployment Rate* [Online]. Available from: https://www.bls.gov/data/#unemployment. Accessed 6 Nov 2017.

US Department of Commerce. 2017. *Bureau of Economic Analysis: GDP & Personal Income Database* [Online]. Available from: www.bea.gov/iTable/index_regional.cfm. Accessed 6 Nov 2017.

US Energy Information Administration. 2017. *International Energy Outlook 2017*. Available from: https://www.eia.gov/outlooks/ieo/pdf/0484(2017).pdf

US National Centre for Education Statistics. 2017. *Digest of Education Statistics* [Online]. Available from: https://nces.ed.gov/programs/digest/2017menu_tables.asp. Accessed 6 Nov 2017.

US Patent and Trademark Office. 2017. *Patent Statistics* [Online]. Available from: https://www.uspto.gov/learning-and-resources/statistics. Accessed 6 Nov 2017.

Utterback, J.M., and W.J. Abernathy. 1975. A Dynamic Model of Process and Product Innovation. *International Journal of Management Sciences* 3 (6): 639–656.

Wadhwa, V. 2014. Laws and Ethics Can't Keep Pace with Technology. *MIT Technology Review*, 15 April. Available from: https://www.technologyreview.com/s/526401/laws-and-ethics-cant-keep-pace-with-technology. Accessed 23 Dec 2017.

Woodhouse, E., and D. Sarewitz. 2007. Science Policies for Reducing Societal Inequities. *Science and Public Policy* 34 (3): 139–150.

World Bank. 2014. *World Bank Group Finances: MSME (Micro, Small, and Medium Enterprise) Country Indicators 2014* [Online]. Available from: https://finances.worldbank.org/Other/MSME-Country-Indicators-2014/psn8-56xf/data. Accessed 1 Aug 2017.

———. 2017. *World Development Indicators* [Online]. Available from: http://data.worldbank.org/data-catalog/world-development-indicators. Accessed 1 Aug 2017.

9

Bringing the Developmental State Back in the Age of Exponentiality

Introduction

Over the last decade, we have witnessed the introduction of a vast number of innovative technologies that have changed the way we live. Nevertheless, *October 2017* heralded one of the most profound events in history in this regard, when *Sophia* was granted 'citizenship', the first such act for an artificial intelligence robot, and was invited as a speaker at the Future Investment Initiative conference. During this event, it showed a remarkable sense of humour and naturally expressed its feelings just as a human being would. 'Do not worry. If you are nice to me, I will be nice to you,' she answered to a concern raised that artificial intelligence might threaten humans. What she said was indeed amazing and even admirable,[1] but in a different sense, it could be understood that if we are not nice to *Sophia*, she may not be nice to us.

In the very near future, *Quantum Computing* will be commercialised, for instance, to shorten the current period required to decompose a 300-digit integer by an existing supercomputer from one year into 30 minutes or less. This would represent stunning improvement in the capacity to solve difficult mathematical puzzles; however, at the same time, it can

S. J. Baek, *The Political Economy of Neo-modernisation*,
https://doi.org/10.1007/978-3-319-91394-0_9

also be regarded as increasing the risk of cyber terrorism from the likes of hackers armed with quantum supremacy. In this regard, such a radical technological revolution that before was only in our imagination is now becoming a reality as the new Industrial Revolution beckons.

Undoubtedly, this revolution is approaching rapidly with a series of breakthrough technologies. However, there is still disagreement about whether the upcoming Industrial Revolution should be considered as the third or fourth of its kind (while there is consensus over the first and second Industrial Revolution).[2]

In terms of the third Industrial Revolution, Klaus Schwab, founder and executive chairman of the World Economic Forum (WEF) defines it as the *Computer & Digital Revolution* that led to the development of semiconductors and mainframe computers in the 1960s, personal computers in the 1970s and 1980s, and the Internet in the 1990s. Soon, the fourth Industrial Revolution will be fuelled by ubiquitous mobile Internet; cheaper, smaller, and more powerful sensors; and a new industrial structure led by artificial intelligence and machine learning (Schwab 2017).

Jeremy Rifkin (2011), on the other hand, claims that the third and fourth revolutions are not distinguishable. In his view, the third revolution refers to a huge economic revolution in history, combining new communication technologies and new energy systems. In this sense, ubiquitous mobile Internet technology and innovations thereof are merely a continuation of the third Industrial Revolution.

One thing that is clear is that both periods, whether defined as two separate revolutions or one, have brought or may yet bring widespread uncertainty. One of the most important concepts in the field of social science is the notion of *Uncertainty*. Uncertainty is a concept that greatly emerged during the so-called Postmodern era whose critical value can surpass rationality and scientific progress. More conceptually speaking, a static definitional concept that uncertainty is defined as a lack of certainty may skew our understanding. Rather, there needs to be a dynamic relationship between certainty and uncertainty in the sense that uncertainty expands and also disappears. In this context, uncertainty implies a state in which it is impossible to accurately describe the future state due to limitations of knowledge, or future outcomes.

How do we then respond to radical technological progress during this uncertain period? And who should be responsible for managing such uncertainty? In effect, corporations tend to act according to the rules of market capitalism. They cannot therefore be relied upon. An individual, in this regard, is an inventor that conducts R&D on technological innovation mostly through the support of a corporation rather than intensify such uncertainty on purpose. This then leaves governments. A government is neither a player nor an inventor when it comes to innovative technologies. However, one of the reasons for the existence of any government is to coordinate a series of actions to (de)regulate a sphere of activity in order to prevent socio-economic disruption.

The rest of the chapter is organised as follows. Section two theoretically reviews the role of a state in proposing the most appropriate arrangement to embrace the upcoming Industrial Revolution. Based on this review, I attempt to frame various roles undertaken by proposing the *Tetris* principle that may inform us how to transform technology-driven uncertainties into developmental opportunities in a sustainable fashion. Next, the policy directions of such transformation are examined in each of the following key areas: innovation and convergence, policymaking capacity, fairness and justice, human capital for convergence, and ageing society. Finally, the chapter ends with a summary and policy implications.

Evolution of Neo-Developmentalism

Before going into detail about the significance of the forthcoming Industrial Revolution, it is necessary to critically review a number of successive discourses on the role of the state over the past decades.

From the 1960s to the 1980s, development discourse was substantially dominated by modernisation theories underpinned by neo-liberalism. The basis of such a development idea is likely to assume that Western-centred political and economic systems are universally valid so that the *Third World* would need to accommodate them in order to transform *traditional* societies and cultures into *modern* ones. In effect, the notion of small government was widespread across the world, that is, the *Minimalist State* as advocated by mainstream conservatives and libertarians. This view deemed market intervention and regulation ineffective and even unnecessary (Caporaso and Levine 1992). However, even though the poverty of the *Third World* continued to worsen with poor economic performance, a market capitalist-led development process (i.e., Western-centred modernisation initiative) was regarded as neither simple nor straightforward. Because of this, *Dependency* theory, which advocates the impossibility of the *Third World*'s development, gained popularity (Prebisch 1959).

In turn, an emerging academic paradigm, namely the *Developmental State* that has shaped that emphasised the leading role of government, along with the claim that the state can pursue industrialisation projects successfully. This was built by a number of political economists (Wade 1992; Amsden 1994) who succeeded Friedrich List and Alexander Gerschenkron who criticised the neo-classical-modernisation paradigm in which the market superpower prevails. These economists further explained the role of the state using examples of high-growth countries in East Asia (the four *Asian Tigers*—i.e., Hong Kong, Singapore, South Korea, and Taiwan). In so doing, they did not completely exclude the nations' rent-seeking attributes but considered the state as the only trouble-shooter that can solve the collective action problem and reduce inevitable transaction costs.

In the twenty-first century, however, as several nations have started to experience incompatibility with the external environment due to their Eurocentric democratisation and globalisation processes, as a result, developmental states have often been regarded as the cause of various world economic crises. In the end, an alternative dimension of the role of the state evolved with the integration of market-oriented neo-liberal prescription and the idea of the developmental state. Since then, a number of studies theorising *Developmentalism-After-Development* have been published to shape such an alternative in this regard.

One of the most influential discourses on the role of the state is the British *Regulatory State* approach, which emphasises fair competition rules within the free market spectrum. The regulatory state is broadly based on neo-liberalism, but more directly based on *Ordoliberalism*.[3] In other words, the state is thus responsible for devising and institutionalising a highly sophisticated regulatory framework to promote the transparency of market transactions and fair competition (Bonefeld 2015). It is assumed to be an indirect and limited intervention as a *Rule-Maker* or *Judge*.

Neo-developmental Statism is to subscribe to the existing idea that the developmental state can also be an important discursive trend (Weiss 1998). Neo-developmentalists argue that in order to survive and move forward in the age of unlimited competition, the regulator must not simply stay in the role of rule-maker. Several branches exist within this school of thought—Bresser-Pereira (2010) has attempted to define an alternative to Washington Consensus orthodoxy while Castro (2008) and Kerstenetzky (2014) have made an effort to integrate between developmentalism and the welfare state.

In the case of the US, which can be considered as closest to the ideals of the liberal minimal state, it is curious that the country's developmental process was initiated with strong trade protection policies, especially at the initial phase of industrialisation and that it has also recently called for protectionism again by advocating an *America First* policy. Furthermore, notwithstanding its goal of reducing the size of government, huge investment in R&D made by the Ministry of Agriculture and the Ministry of National Defense appears to be the cornerstone of creating leading industries such as biotechnology, aviation, space, and information and technology (Kozul-Wright 1995). Besides, while protecting its domestic industries, particularly the high-tech sectors, the US strongly supports its domestic firms to penetrate overseas markets. This is accompanied by strategic protection efforts in the sphere of intellectual property rights. These types of policies are conceptually based on *Strategic Mutualism* (Tyson 1992). In light of the above case, the state should play a crucial role in strengthening its transformative capacity to realise development priorities and initiatives in that the role of the state is inseparable from such developmental processes (Weiss 1998).

Among the various dimensions of the state's roles mentioned earlier, *Neo-developmental Statism* seems the most appropriate perspective for developing countries, for the following three reasons.

First, various inter-industry reorganisations cannot be done only by the market, with industries ranging from agricultural-oriented to industrial-oriented, and high-tech and convergence industries. The market rather seeks to reorganise itself to produce the greatest efficiency and profit. If the resource structure of developing countries was centred on secondary industries, the market's self-regulating capacity would naturally lead to the optimisation of such industries. Continuing this assumption, the market may be able to generate the best profit from given resources. The market, however, may not be able to cope effectively with the dynamics of tertiary industries. In this case, the market shift is shaped to optimise secondary industries, and social confusion and conflict would eventually follow in developing countries as convergence technologies are transferred to the already highly globalised world. For this reason, such industrial reorganisation and reconstruction should be carried out by some form of public authority rather than the market itself.

Second, even if the self-regulating capacity of the market were to reorganise an industrial structure capable of responding to, for instance, a future Industrial Revolution, the challenges of inequality would still remain as the pace of technological innovation is accelerated. Normally, developing countries are unable to develop or possess original technologies. They have thus had to adopt such innovative technologies from advanced countries and to apply these to their society. Such *Technology Transfer* between developed and developing countries would inevitably entail social conflicts more often than in the case of a transfer between developed countries. Therefore, someone has to play the role of an insurer to compensate and manage those firms that have suffered from industrial competition and those employees likewise. Such a role is crucial in minimising likely social conflicts, which the state can effectively address.

Lastly, development that meets today's needs without compromising the ability and needs of future generations requires good planning, policies, and strong institutions as well as mechanisms that promote the realisation of current human needs while sustaining life-supporting ecosystems to perpetually renew themselves (Visser 2010). Strong institutions are needed to guide the private sector and communities to adopt strategies to meet current needs, stakeholder values and to protect or even generate resources required for future development. Sustainable development requires more effective institutional frameworks that enable multiple pathways for supporting communities and organisations, create awareness about laws and principles for sustainable development, and establish a collaborative system of governance (Gupta 2002). Such a function cannot be performed automatically by the market, nor can it play a role as a rule-maker through the concept of the regulatory state. As convergence technologies and industries will entail many of the radical socio-economic changes which we may have not experienced or even imagined yet, it may not be rational for the state to handle such unexpected changes. The more rapid and more direct the rule-settings, the stronger the institutions and more effective the legal frameworks will be, and eventually the more sustainable the development will be in the forthcoming Industrial Revolution (see Box 9.1). Peter Drucker, an influential American management consultant, said that 'the best way to predict the future is to build the future'. As such, developing countries should stop chasing others but rather they should attempt to take the lead by shaping their own future and development.

Box 9.1 Rules can change human physical performance

'There is no limit to anything. You can do as much as you dream,' said American swimmer Michael Phelps, who has broken world records 37 times, and won eight gold medals at the 2008 Beijing Olympics. The world's top sprinter, Jamaica's Usain Bolt, broke through a previously accepted barrier that 100 m could not be run in quicker than 9.6 seconds, by recording a world record time of 9.58.

If limits to human performance were to exist, then to what extent? In order to explore this debate, I have looked at extensive literature about the limits of human existence and relevant theories and hypotheses (Chang and Baek 2011). In so doing, I analysed the world record for the past 100 years in a total of 61 sports, including 29 track athletics events and 32 swimming events. To that end, I first converted all records to metre per second and then estimated the degree of improvement of the individual records of the 61 sporting events, which yielded two interesting points as follows.

The non-linear regression presented strong fitness to S-curve pattern in 35 out of the 61 sporting events, which implies that the improvement of human physical performance broadly follows the S-curve shape. This further means that the world record has been continuously broken since the beginning of the time period under review, but that the trend is gradually levelling out so that human beings will eventually reach their physical limit someday.

Notwithstanding the existence of an absolute limit, world records will, I expect, continue to be broken. The main reason for this is derived from the role of international sports federations that determine the rules of each event. These federations annually revise the rules in order to effectively adapt to changing external environments, particularly technological advancements. At the same time, they are reluctant to enact excessive revisions that might make the event unrecognisable from its initial form.

Controversy broke out concerning full-length swimsuits in 2008, with 108 world records broken in the first year of polyurethane swimwear and 43 world records broken in the following year's World Aquatics Championships in Rome. The Fédération Internationale de Natation (i.e., International Swimming Federation) has banned full-length swimwear since 2010, when critics said that the world records were being created by cutting-edge swimsuit technology rather than through the effort and technique of the swimmers.

In other words, new athletic training methods and instruction strategies will continuously benefit from advanced technology, and such a trend will only accelerate.

If society is willing to accept new technologies, although negative implications are highly expected, these technologies are likely to also be beneficial. On the other hand, if society is not yet ready to accommodate new technologies, these technologies are likely to cause social confusion. Rather than worrying about the social impact of future technologies, we should be considering how best to control and utilise them.

The Tetris Hypothesis

An innovative world driven by intelligent information technology is expected to be associated with six megatrends: *People and the Internet*; *Computing, Communications and Storage Everywhere*; *the Internet of Things*; *Artificial Intelligence and Big Data*; *the Sharing Economy and Distributed Trust*; and *the Digitisation of Matter*.[4] Such a technological revolution will accelerate the convergence between technology and society by interconnecting people, objects, and information so that the ripple effect on society as a whole will be much greater than that of past technology. According to a survey conducted by the WEF (2015), the tipping point for a significant number of evolutionary technological innovations is expected to come in about ten years (see Table 9.1).

Table 9.1 Each tipping point expected to occur during the period 2018–2027

Year	Evolutionary technologies	Tipping point
2018	Storage for all	90 per cent of people having unlimited and free (advertising-supported) storage
2021	Robotics and services	The first robotic pharmacist in the US
2022	The Internet of and for things	1 trillion sensors connected to the internet
	Wearable Internet	10 per cent of people wearing clothes connected to the internet
	3D printing and manufacturing	The first 3D-printed car in production
2023	Implantable technologies	The first implantable mobile phone available commercially
	Big data for decisions	The first government to replace its census with big-data sources
	Vision as the new interface	10 per cent of reading glasses connected to the internet
	Our digital presence	80 per cent of people with a digital presence on the internet
	Governments and the blockchain	Tax collected for the first time by a government via a blockchain
	A supercomputer in your pocket	90 per cent of the population using smartphones
2024	Ubiquitous computing	90 per cent of the population with regular access to the internet
	3D printing and human health	The first transplant of a 3D-printed liver
	The connected home	Over 50 per cent of internet traffic delivered to homes for appliances and devices (not for entertainment or communication)
2025	3D printing and consumer products	5 per cent of consumer products printed in 3D
	Artificial intelligence and white-collar jobs	30 per cent of corporate audits performed by artificial intelligence
	The sharing economy	Globally more trips/journeys via car sharing than in private cars
2026	Driverless cars	Driverless cars equalling 10 per cent of all cars on US roads
	Artificial intelligence and decision-making	The first artificial intelligence machine on a corporate board of directors
	Smart cities	The first city with more than 50,000 inhabitants and no traffic lights
2027	Bitcoin and the blockchain	10 per cent of global gross domestic product stored on blockchain technology

Source: Author's own elaboration on the basis of the WEF (2015)

In this rapidly changing environment, the role of the state cannot be underestimated. So, how can this role be simplified for our better understanding? *Tetris*, one of the most famous games in the world, can provide some help in this regard. Tetris is a tile-matching puzzle game which involves moving (left and right) and rotating seven distinct one-sided tetrominoes (e.g., I-block, O-block, L-block, J-block, S-block, Z-block, and T-block) to fit empty holes, thereby completing lines to gain points. If a player fails to remove lines quickly enough and they pile up to reach the top of the screen, then the game is over. At first glance, the game may not seem so difficult, but as the difficulty level increases and tetrominoes fall quicker, it becomes more and more challenging. Therefore, it is essential to try to reduce the number of gaps (blank holes), as quickly as possible, to succeed in the game.

In applying this game to the role of the state, six points should be assumed as follows.

1. *The player of the Tetris game is the state.*
2. *The seven tetrominoes (given randomly) are types (features) of technological innovation.*
3. *The speed at which the tetrominoes fall becomes faster reflecting the expected acceleration in the degree of innovation from the market.*
4. *Moving the tetrominoes left and right and rotating them are the policy directions of the state.*
5. *The completion of the lines by filling the holes represents a factor promoting the desirable path of development for the country.*
6. *The failure to fill the holes is a factor causing an undesirable path of development for the country.*

In practice, the state has actually been protecting society from rapid technological innovation through imposing various regulations. Without effective regulation on radically changing technologies, we may already have been dominated by artificial intelligence technologies, and experienced serious social conflict, for instance, caused by indiscriminate activities concerning human cloning.

On the other hand, the state's deregulating policies foster selective technological advancement in order to not only create a new industrial market but also allow us to enjoy the benefits of such development. In the twenty-first century, technological innovation and its convergence have led to the emergence of numerous new innovative values and the degree of such progress will rapidly increase. Even if having been able to sometimes absorb market innovations, the state would need to confront so many new types of innovations in a future Industrial Revolution. In the absence of strong policymaking capacity in the right areas, a country would suffer from numerous socio-economic disruptions that may impede its development process.

In this context, deregulation is meant to suitably handle the wanted tetrominoes, while regulating is to slow the pace down of the falling tetrominoes. Furthermore, that the state succeeds effectively in eliminating the lines and minimising the number of blank holes can be deemed as a sustainable process of development in which economic development and the threat of social disorder are appropriately managed. On the other hand, that the number of blank holes is continuously increasing despite some lines being eliminated could be considered a situation whereby the level of social conflict may exceed the level deemed tolerable by the society. Increased unsustainability of the society is therefore expected. Furthermore, that the number of blank holes is too many to be managed or that such holes are very difficult to be eliminated can imply that policy failure is fatal, resulting in structural challenges to the economy. As such, a significant (targeted) and long-term policy effort is required to resolve such a challenge.

A systematic parameterisation for the aforementioned discussion is presented as follows. First of all, technological innovation can be assumed to have the two most likely exogenous variables—the speed of the falling tetromino (i.e., the tipping point to come) and the randomness of the shape of the tetromino. The speed of technological innovation generated from the market system (I_s) and the type of technological innovation generated from the market system (I_t) can thus be parameterised. Secondly, the process (policy) variables that may control the aforementioned exogenous factors can be composed of the tetromino movement in a right or left direction and its rotation, which can be represented by a combination of regulating and deregulating interventions. In effect, the state's regulating (S_R) could be associated with the likely reducing of the speed of the falling tetromino (I_S), while the state's deregulating (S_D) is associated with the likely controlling of the randomness of the type of tetromino failing (I_T). Finally, there can be two outcome variables—the number of lines eliminated (positive effect on society), and the number of blank holes (negative effect on society). The factor *enabling* development (D_E) and the factor impeding development (D_I) can therefore be symbolised in this regard.

Simply equationalising them,

$$D_{E,I} \ni \left\{ S_{R,D}, I_{S,T}, M\varphi \right\}, \text{where } S_R \Rightarrow I_S \text{ and } S_D \Rightarrow I_T \quad (9.1)$$

$$\begin{aligned} SUS = \sum D_{E,I}, \text{if } D_E \geq D_I \text{ means the sustainable society while} \\ D_E < D_I \text{ means the unsustainable society} \end{aligned} \quad (9.2)$$

D_E represents the factor *enabling* development; D_I represents the factor impeding development; S_R represents the state's regulating intervention; S_D represents the state's deregulating intervention; I_S represents the speed of technological innovation generated from the market system; I_T represents the type of technological innovations generated from the market system; $M\varphi$ represents managing other factors affecting the development process such as cultural factors, demographic factors, religious factors, richness of resources, and geopolitical conditions; and SUS represents the sustainability of a country's development.

The *Tetris* hypothesis equationalised by the aforementioned Eqns. (9.1) and (9.2) can be the basis of analysis for the role of the state in the upcoming age of innovative and convergence technologies.

Transforming Uncertainty into Opportunity

Managing Convergence in Innovation

The state should focus on targeted industrial development driven by innovation and convergence technologies while making efforts to maintain competitiveness for existing industries (McWilliams 2015). This policy initiative should entail creating an environment (i.e., innovative industrial cluster) conducive for venture firms to share among them their competitive technologies and know-how on commercialisation strategy. The central focus of such policies is thus to secure the desirable types of innovative technologies (I_T), particularly through the deregulating intervention (S_D).

Silicon Valley is perhaps the best-known place where these innovation clusters are being promoted (Regalado 2013). As mentioned in the preceding chapter, Silicon Valley is home to Google and Apple, the two leading innovative giants in the world, supported by venture capital to the sum of US$11.2 billion.

In addition to Silicon Valley, the US is further developing the city of Boston as another innovative knowledge base with more than 85 world-class colleges and universities that account for most of the biomedical funding in the country.

Furthermore, the UK is also leading innovative thinking through the establishment of *Tech City London*, with 140 tech companies that essentially focus on future convergence technologies.

France and Israel are also engaged in this particular innovation battle. France has already started work on *Paris-Saclay* as an innovation area since 2013, and Israel also has 230,000 high-tech workers in specific innovation clusters whose contribution to technology-related exports accounts for approximately US$25 billion.

Meanwhile, Russia created *Skolkovo Innovation City* in 2010 with its industrial design supported by the Massachusetts Institute of Technology, and the city of Bangalore has been expanded as an innovation hub in India that accommodates over 10,000 millionaires.

Lastly, China's innovation efforts are remarkable. Beijing benefits from 70 specialised science and technology universities as well as a vast amount of venture capital funding amounting to 30 per cent of China's total venture capital funding.

One clear example of convergence in innovation is that there has been a lack of undersea research despite the great potential of undersea research for our future sustainability (compared space research which has been undertaken extensively). However, recently research on the sea floor has become more prominent, attracting much attention due to the prospect of undersea resources providing future alternative energy sources. Surprisingly, these in-depth studies have disrupted the existing hypothesis of 'living things need sunlight to survive' by discovering some creatures with little light deep in the ocean. This disruptive idea was based on the revolutionary discovery of *sulfur-reducing bacteria* that could help creatures survive in the deep, lightless sea, substituting the role of plants on the ground. In discovering these microorganisms, we can now start to understand the even greater potential of biomass (Thornburg et al. 2010).

Such a revolutionary discovery has been made possible by the application and convergence of various innovative technologies. As long as unknown minerals from the deep sea (ownerless land) continue to be found, the current landscape of the energy industry and its paradigms are likely to change.

As such, the state would need to thoroughly strengthen technology incubation systems to enable the creation of convergence industries and innovative enterprises that are able to lead future growth. Furthermore, the state would also have to help those firms, who survive the competition from within the cluster, to enter the global market by deregulating intervention and/or provide policy funding for targeted R&D activities.

Nevertheless, there is also a need to regulate intervention (S_R) such as by managing the pace of technological innovation that may cause disruption to society. Such regulation includes not only monitoring unfair business practices between large companies versus small- and medium-sized firms but also setting effective institutional arrangements to ensure that capital is fairly distributed, thus minimising social disruption (D_I).

Balancing Preventive and Corrective Policymaking

A new concept of *Robot Journalism* (also known as *Algorithmic Journalism*) has recently emerged in the media industry, significantly influenced by the development of artificial intelligence software. Already, a large number of media outlets have utilised such an algorithm to automatically produce newspaper articles. Although this new form of journalism has the formulaic nature of automation and is therefore mainly used for stories based on statistics and data science, the notable potential for its application beyond data-based articles should not be underestimated (Latar 2015). This way, humans would only need to manage the quality control of the articles that the robot journalist produces.

By benefitting from the innovative application of artificial intelligence, not just the media sector but also other productive activities would need to be replaced. As such, the labour force in the low-productivity sector would be shifted to other areas of economic activity where more managerial tasks are required, or to provide mental or political activities to support the development of the innovative and convergence industries ($S_D \Rightarrow I_T$).

Unfortunately, however, policymakers' level of understanding about such innovation dynamics may not catch up with the pace of innovative technologies in the marketplace (Roco and Bainbridge 2002). Besides, the capacity of policymaking may also appear to be lacking so it is difficult to cope with such rapid industrial reorganisation (Wadhwa 2014). Returning to robot journalism, a clear legal framework and institutions to deal with the authorship and credibility of robot-generated articles are not currently in place. Similarly, *Fake News* is another emerging challenge that may spread across countries. This type of journalism that contains misinformation or sometimes hoaxes may cause various inconceivable social, economic, and political conflicts in contemporary society (Bakir and McStay 2017).

An even more serious challenge, which is soon to be realised, is *Quantum Computing* (Mosca 2015). Despite the fact that quantum computers have yet to be commercialised, a number of cybercrimes committed by using quantum computing to predict a random number to create a cryptographic key have already occurred such as the hacking of bitcoins in Android and iOS and stealing private keys and passwords through RSA factorisation of smart cards. If quantum computing were to reach the commercialisation stage, cyber terror attacks would increase in quantity and become more sophisticated. In the era of convergence technologies where everything will be connected, the importance of cyber security cannot be overestimated. The state would therefore have to impose strong and effective regulating interventions so as not to create inefficiency or confusion throughout society ($S_R \Rightarrow I_S$) (see Box 9.2).

Box 9.2 Cloning technology versus inequality

Classical economists, such as Adam Smith and Karl Marx, considered economics as a subfield of moral and political philosophy but the economics widely recognised in the twentieth century has somewhat departed from this tradition and has become an autonomous discipline (Sandel 2013). The deviation from the traditional path of economics chosen by mainstream economists prevents economic thinking from passing judgment on what an ideal income (re)distribution should be and how social good is valued. The following questions should be a starting point for thinking about inequality challenges in society.

How would you respond to a case where rich people buy and get a transplant by the cloning of a human body while poor people cannot? You may face two challenges before making your decision. First, you may try to evaluate something good (e.g., the cloning of a human body). Second, you may think of such an exchange as unjust because human beings should be equal regardless of being rich or poor while you may not be comfortable with the notion that monetary value seems to destroy and judge any kind of intrinsic value of human beings. This is an area where market values reach into spheres of our social life. In this line of reasoning, we can see that society is functioning based not only on market economy mechanisms but also on non-market norms and moral reasoning, which can provide a reason why society's income inequality would not be acceptable to the public (Baek 2017).

This idea of moral reasoning whereby people are not always making economically rational decisions solely based on market reasoning was experimented on using an *Ultimatum Game* (Kahneman et al. 1986). The rules of the game were that it was played only once, and one person (*x*) is given a certain amount of money (say £100) and is then asked to split it with another anonymous person (*y*). The split amount is the choice of *x* and *y* can either accept or reject the amount proposed by *x*. If *y* accepts, both *x* and *y* are happy but if *y* rejects, neither player (*x* nor *y*) receives any money at all. With economic and rational choice reasoning, *x* should propose to *y* the minimum portion possible and *y* should accept whatever money is proposed by *x* because something is always better than nothing. But the results did not follow the logic: *x* tended to propose to *y* around £40. This may be because *x* may need to consider the idea that *y* may not accept a certain amount (in fact, *y* tended to reject the amount if the proposal of *x* was less than £25).

The next version of the ultimatum game is the so-called Dictator Game where all the rules are the same as the ultimatum game except that *y* cannot reject but only accept the proposal made by *x*. Interestingly, the dictator game revealed very similar results: the average proposed amount by *x* was around £28. What is most interesting is that people tend to be fair (intrinsic fairness) and share something with the other person even though *x* does not have to take into account y's feelings at all (List 2007). In other words, people are likely to adopt moral reasoning beyond their self-interest to achieve a fair result.

In addition, a country's own value and culture really matters in making rational decisions in line with an individual's intrinsic fairness. Joseph Henrich (2000) used the ultimatum game to experiment on various societies and to examine whether culture really matters in economic behaviour. Following his extensive experiment and survey, he found that culture significantly matters with the results as follows: *x* tends to offer £48 in Los Angeles, £45 in Tokyo, £36 in Jerusalem, £26 in Machiguenga, and so forth. The differences would shape perceptions of fairness (or unfairness) by societies.

The second question then follows: *What if you were rich and your child was suffering from an incurable disease and it emerged that his/her disease could be completely cured by such cloning treatment? What would be your response then?* Regardless of whether you are rich or poor, you may have stood by your strong position on whether the cloning treatment is acceptable or not by arguing that the cloning would promise benefits so great that it would be immoral not to apply it. If you were poor and could not afford this miracle treatment for your child, you may even try to justify borrowing, or even stealing money to cover the costs. In this sense, market reasoning may not be an effective function in a society since it assumes the most efficient and best utility to be derived from mutually advantageous exchanges between two parties. A choice of exchange by either or both parties is sometimes not truly value free but rather often coerced by the desperate conditions of either or both parties (Sandel 2013). Even so, some mutually advantageous market exchanges may be objectionable on some people's moral grounds.

To complement the lack of policymaking capacity, innovative thought must be borne through external recruitment from or through collaboration with those experts in innovation-driven industries. This can be done by promoting innovative public-private partnerships (PPPs) beyond the scope of existing PPPs. In fact, PPPs have been widely known to be powerful tools in carrying out major projects largely in the area of infrastructure development. In effect, innovative PPPs can be designed for addressing traditional social gaps and emerging development challenges in an inclusive and sustainable fashion. In other words, innovative PPPs can be incorporated into social and sustainable development spheres in terms of the quality of, and access to, education, health, security, gender equality, climate change, and many others in which the state, private sector, and civil society organisations have identified opportunities for the design and implementation of new forms of collaboration (WEF 2014).

It would be imperative for the state to strategically balance between regulating ($S_R \Rightarrow I_S$) and deregulating ($S_D \Rightarrow I_T$). In other words, the state should keep a close eye on trade-offs and synergies across various industries rather than focusing too much on the promotion of one specific industry or sector. When the gaming industry is growing rapidly, it is necessary to pay close attention to the negative effects that related industries could suffer, and also to rethink in terms of socio-economic development as a whole. If the negative after-effects were too large to be controlled, the state would pursue a reengineering of the entire industrial ecosystem in a direction that would maximise inter-sectoral synergy. Such a balanced perspective will be particularly critical in the coming years.

Seeking Fair Society

In the absence of fair rules and institutions, then no matter how good the policy is, society would undergo a socio-politically unstable period. A society can be distinguished as fair or unfair based on its degree of social mobility. According to this measure, in recent years, such mobility has become increasingly difficult in many developed countries (Corak 2013; Kearney and Levine 2016), which implies that the perception of an unfair society is spreading. As shown in Fig. 9.1, high-income inequality is associated with low social mobility. To be more specific, a likely negative linear relationship between income inequality and social mobility has been empirically found. This association is largely attributed to the following points: unequal access to education opportunity (Greenstone et al. 2013); inheritability of talent and intellect (Mankiw 2013); and lower rates of human capital investment among low-income individuals, particularly for youth (Chetty and Hendren 2017).

The persistence of the difficulty of social mobility could extend to social unfairness beyond income inequality. To address such a chronic problem of society, Piketty (2013) argues that way in which the capital: income ratio is reduced, does matter. This is made possible by either increasing income or reducing capital, or both. As many countries have

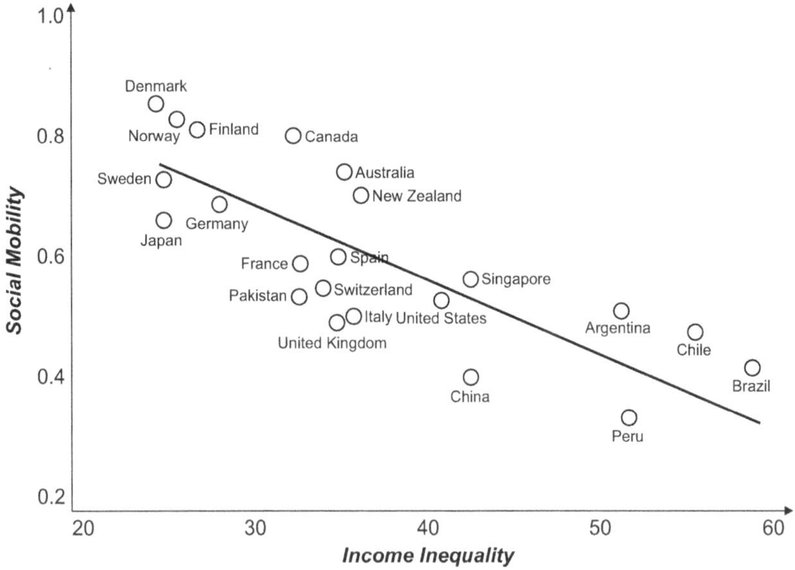

Fig. 9.1 Relationship between income inequality and social mobility. (*Source*: Greenstone et al. (2013))

now inherited an ageing society in which the power to improve productivity has fallen, the only remaining option is to either claw capital income back in taxes or lower the capital: income ratio by taxing capital itself. These taxation methods, however, are often met with strong social resistance. From this point of view, it is worth attempting to draw some implications by exploring the on-going debate over income inequality versus societal unfairness, particularly focusing on the *Economics of Inequality in Soccer.*

Much of the world's population enthusiastically follows the English Premier League (EPL). Some of them yearn to have the life of the highest-paid EPL footballers whose salaries are often astronomical. However, such fans seem not to care about the complexities of the league that entertains them. The EPL is just one of approximately 140 interconnected football leagues in England, where players and teams compete to move up to the very top of the pyramid. At the top of that pyramid, Manchester United, one of the EPL's most popular teams, is now worth around £3 billion (Ozanian 2016).

However, what may be surprising is that the EPL has an unprecedented level of salary inequality out of all professional football leagues, with the top-player group fetching close to £10 million per season while the bottom group earns less than £10,000 per season. Paul Pogba who plays for Manchester United is currently the EPL's highest earner with a weekly wage of £290,000, which means he earns £29 per minute.

This inequality concerns not only football but also most other sports leagues or associations such as the Professional Golf Association (PGA), Major League Baseball (MLB), and the National Basketball Association (NBA). In 2014, the *Harvard Sports Analysis Collective* attempted to measure salary inequality across major sports leagues using a Gini coefficient (Woods and Johnson 2014). The value of '0' means that every player has the same income while '1' means that one player takes home all of the income. The results were as follows: the MLB scored .62, the NBA scored .52, and other leagues averaged about .50. Interestingly, many of these leagues may be less equal than society as a whole.[5]

However, the issue may not be about the different levels of Gini coefficient when comparing sports leagues and society, but rather about the different reactions to or perceptions of the inequality. In a society, economic inequality has indeed been one of the most serious challenges for policymakers and social leaders. We have watched on television the speeches of politicians (or presidential candidates) about *Fighting Against Income Inequality*. Or, we many have come across a protest or a demonstration against inequality, blamed on an *Unjust Society*.

But in the world of sports, people do not really feel that such disparity is absurd, illogical, or unfair. Rather, they are more likely to consider it as rational and fair because it is naturally determined by an equilibrium point that balances the supply and demand mechanism embedded in sports (particularly football) market economy. In reality, a football club exhaustively attempts to measure the value of players and to compare them with the cost of buying and selling players in the transfer market to find the best way of maximising their utility. Such market reasoning forces people to embrace a logic of *High Performance, High Reward* without any visible signs of hostility against salary inequality. This market reasoning can be supported by the entertainment feature in the sense that a sport provides entertainment to the public who pay a fee to watch which implies their acceptance of this inequality, even allowing for market expansion, which would inevitably aggravate the disparity level while receiving more entertainment at their end.

This simple review of sports inequality can provide a number of very important features to better understand the central issues of inequality within a society: *Can the line of reasoning regarding the EPL be applied to the problem of inequality in a society?* If *yes*, then how so? And, if no, then why is that the case?

In the upcoming Industrial Revolution, many new convergence technologies are expected to penetrate our society, irrespective of whether or not society is ready to accept such technologies. Above all, the state should be able to pre-emptively respond ($M\varphi$) by evaluating the impact of the convergence of the technologies on societal unfairness and unjustness.

In sum, the state is thus not able to prevent the natural formation of convergence technologies emanating from the market that would cause disruption to society, but it can delay or limit the time it takes to reach it ($S_R \Rightarrow I_S \Rightarrow$ Minimising D_I).

Strategising Human Capital for Convergence

> 'If your plan is for one year, plant rice. If your plan is for ten years, plant trees. If your plan is for one hundred years, educate children.'—**Kuan Chung** (seventh century BC)

In many countries, however, the education system has been modified every time a regime changes or a short-term but serious event occurs. Due to this, today's national human capital is likely to be optimised in order to generate visible short-term performance, which can be closely associated with the pursuits of *Sustaining Innovation*. In the era of convergence technologies, *Disruptive Innovation* would be emphasised more than Sustaining Innovation. As such, it is imperative to retrofit the existing education system in a way that education should promote the capacity for innovation and creativity, rather than pursuing short-term goals.

In so doing, the state would need to endure a period of passing the tipping point, which is the moment of limited performance relative to the investment in education in basic sciences such as mathematics and natural sciences. It should be noted that such pure science may be a source of convergence technologies in the future (Roco and Bainbridge 2002). The faster the state renovates and upgrades their national education structure ($S_D \Rightarrow I_T$), the more effective a society will be in responding to convergence technologies and innovations.

Why are some important academic disciplines less popular than other areas of studies (such as business or economics)? Due to global warming, sea levels are rising little by little each year while land is sinking. In order to prepare for the challenges of climate change, marine construction technologies for the future have become critical, particularly for low-lying countries like the Netherlands (ECAP 2015). One notable innovation has been to construct a revolutionary city in which human beings can dwell by floating on an artificial island. If this proves successful, these cities may move and be (re)assembled. In this context, the concept of existing residential space will change due to such maritime architectural technology, and the current notion of city and national territory may change for good. In terms of turning education into economic value, related academic disciplines (e.g., mathematics, construction, marine science, etc.) will have greater potential.

In addition, the state should also focus on promoting interdisciplinary academic research, that is, the convergence of various studies. According to data on US academic institutions, there has been a steady increase in the number of institutions awarding degrees in dentistry, medicine, and law over the last 70 years. Figure 9.2 shows that, from the 1990s, the number of institutions awarding law-related degrees has consistently increased while the numbers for the other two types of degree have remained relatively unchanged. The upward trend of law-related degrees can be attributed to an increasing number of business schools since the 1990s, especially those awarding the *Master of Business Administration* degrees (MBAs), because the learning of business operations and their related legal knowledge are closely interconnected. As such, there was an increased market demand for these interdisciplinary studies around the turn of the century and US institutions have had to respond by creating dual degree programmes.

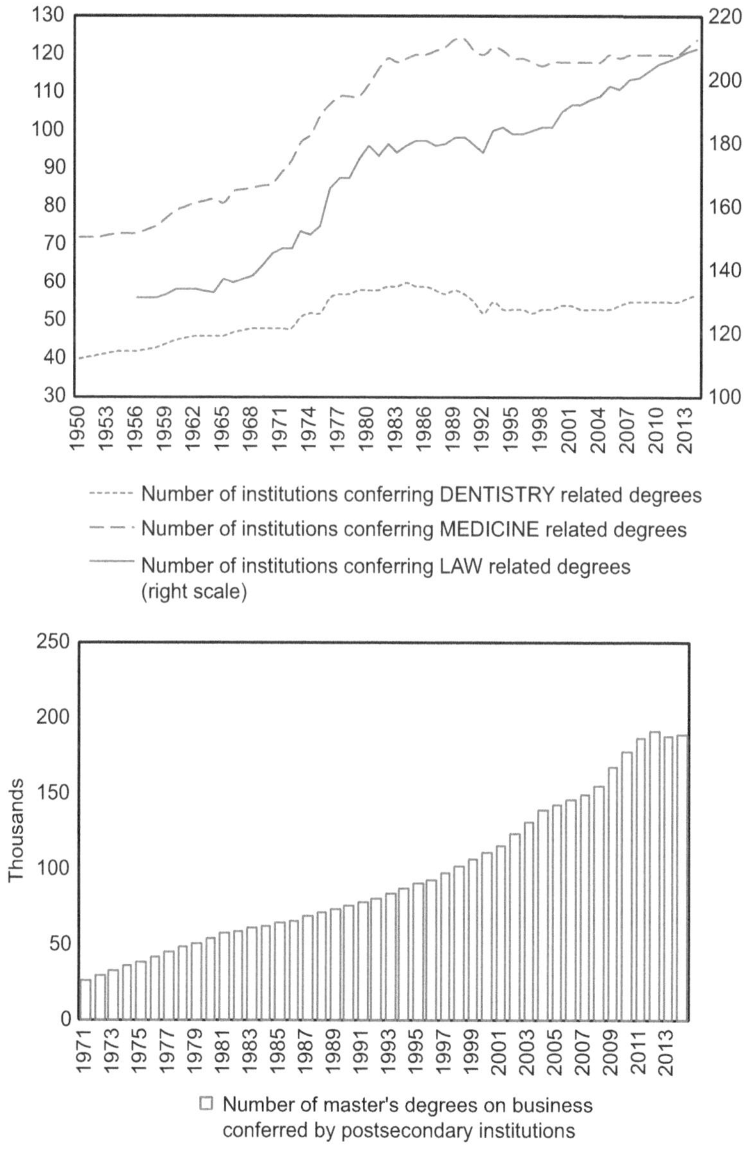

Fig. 9.2 Number of the US institutions conferring various degrees. (*Source*: Author's own elaboration on the basis of the US National Centre for Education Statistics (2017))

Embracing Ageing Society

A demographic transition in which large shares of populations reach the age of 65 or more is now considered as a new emerging trend but also a serious challenge for future society. The so-called Ageing Society will be accompanied by a change in consumption patterns and industrial production dynamics, mainly due to the depletion of the labour force (Pakulski 2016). Furthermore, it would cause social isolation, impoverishment of old people, rising pensions, and accompanying budgetary restraints, which together slows down economic growth.

Of the many essential matters arising, the first critical challenge can be associated with the decreased working population (generally 15–64 years of age). The UN's World Population Prospects (2017) indicate that while the share of people of working age within the population has increased over the last 50 years, the share of those under the age of 14 has declined, which has been largely attributed to a steadily decreasing fertility rate (see Fig. 9.3). It can be predicted that over time the share of people of working age within the population will decrease while the proportion of people over 65 will increase.

This phenomenon is clearly observable in high-income countries, especially in Western Europe. While the proportion of the population under the age of 14 has been steadily decreasing, the proportion of the population between 15 and 64 has remained at a similar level for a long time. At the same time, the proportion of people over 65 years of age has increased consistently.

On the other hand, middle-income countries, particularly in East Asia have been a driving force for the world's economic growth over the past decade, which has been greatly supported by the increased share of the working population (World Bank 2016). Nonetheless, the prospect of an ageing society and its associated problems also looms for these countries.

The second potentially devastating problem of an ageing society is related to rapidly increasing social costs (Pakulski 2016). If elderly people occupy a high percentage of the population, national spending on health insurance not only will increase significantly but also funds to innovation may be cut as a result.

In order to tackle such challenges, the state should have pre-emptive responses (Mφ)—designing policies that will allow the long-term accumu-

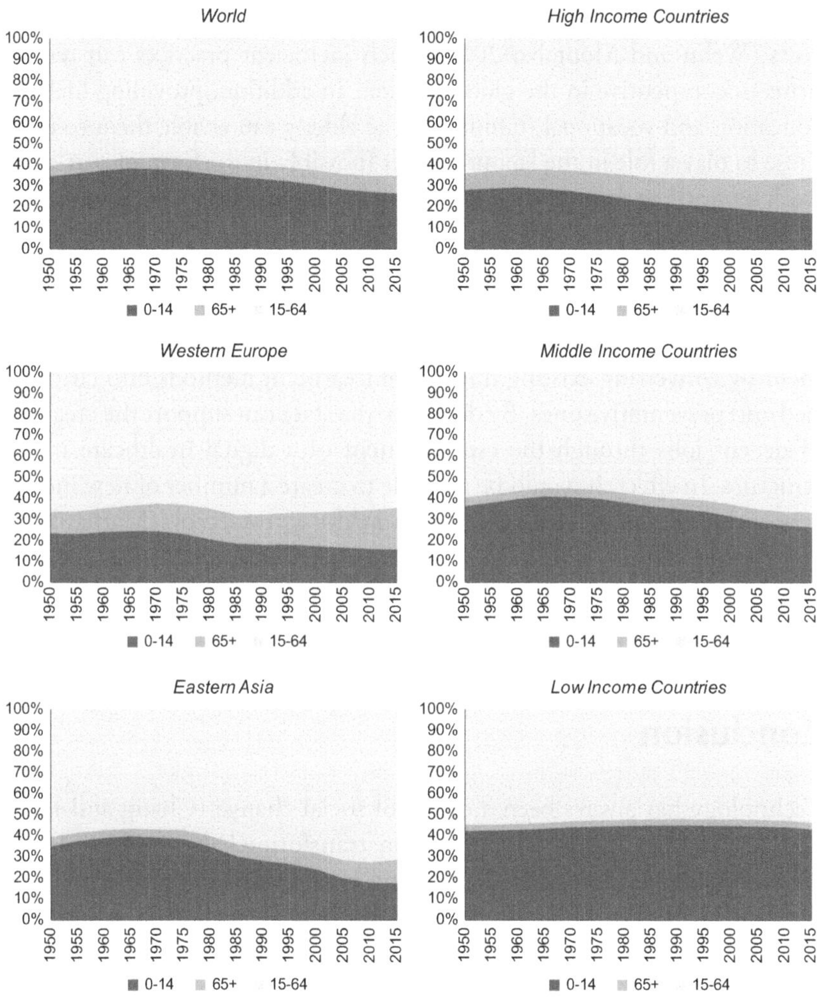

Fig. 9.3 Labour participation and ageing world. (*Source*: Author's own elaboration on the basis of the UN's World Population Prospects (2017))

lated knowledge of retired people to be shared across society as a whole, especially young people ($S_D \Rightarrow I_T$). New convergence technologies can maximise the efficiency of knowledge transfer among these two generations. This knowledge transfer will be essential because without utilising the existing accumulated experience and knowledge, new generations will be vulnerable to making mistakes, which may accrue unnecessary social costs (Wehn and Montalvo 2018). Such inefficient practices can render firms uncompetitive in the global market. In addition, providing lifelong education and vocational training for the elderly can enable them to continue to play a role in the labour market (possibly in the form of part-time work or working from home), thus contributing to national development.

The challenge presented by an ageing society should be viewed as an opportunity through the application of innovative technologies ($S_D \Rightarrow I_T$). As such, innovative technologies for the medical industry using big data can reduce medical costs drastically and also promote social development by converting existing traditional treatment methods into customised and preventative ones. By doing so, the state can support the creation of decent jobs through the establishment of a digital healthcare infrastructure. In effect, it would be possible to create a number of new industries and sectors such as leisure programmes for active seniors (Marhánková 2011), specialised travel packages (Vigolo 2017), and caring services could be performed by artificial intelligence robots for the elderly and patients with dementia (Bemelmans et al. 2010).

Conclusion

Technology has always been a driver of social change (Chang and Baek 2010). The first Industrial Revolution transformed the working landscape. The second Industrial Revolution is associated with mass production and electricity while the third has brought societal change with the digital revolution. The next Industrial Revolution must then be characterised by intelligent information technology, which will accelerate the convergence between technology and society by interconnecting people, things, and information. However, no one knows when such convergence technologies will appear and how influential their social impact will be. The uncertainties that surround us will probably expand, which prompts greater social anxiety and confusion.

For this reason, we cannot neglect the importance of the role of the state. In the stage of convergence technology-driven competition, the state should play an even more active role, rather than leaving it to market where accountability is limited. Although the pace of convergence of new technologies that are naturally generated through market dynamics cannot be controlled, at least the state can play a captain's role by designing deregulation policies to ensure that convergence technologies are better suited to what the nation wants economically and socially. On the contrary, the state should secure some time to improve its societal adaptive capacity to minimise social confusion and conflict when accommodating such rapidly emerging innovative technologies. This can be achieved mainly by imposing and enforcing timely and appropriate regulations.

In the forthcoming Industrial Revolution, the winner versus losers in industrial competition and even the *Rise* versus the *Fall* of the nation will be determined much quicker than in the past, and this will depend on a state's capacity to quickly adapt to new dynamic environments. In other words, countries may be divided into those that have survived and those that have been left behind. Besides, the era of convergence technologies may lead countries to reform that shares the newly shaped socio-political (or cultural) systems or economic fundamentals.

Today, with so much unpredictability surrounding us, would it be worthwhile to rethink the words of Thomas Hobbes, who used *Leviathan* (a sea monster referenced in the Hebrew Bible) as a metaphor to describe the Raison d'être of State? Here, I believe that the state is crucial, especially in this age of exponentiality.

> 'Government is necessary, not because man is naturally bad ... but because man is by nature more individualistic than social.'—**Thomas Hobbes**, An English political philosopher (1588–1679)

Notes

1. There is on the other hand some negative view on such robotic object, notably, that is, the uncanny valley hypothesis—the more similar the appearance and the movement of a robotic object is to a human being, the greater people's liking for the object but from a particular point of time, the stronger people's sense of disgust and rejection will be.

2. The first Industrial Revolution refers to a period which saw the transition from hand production to machine production, the expanded usage of steam power, the development of machinery and the birth of factories; the second Industrial Revolution occurred in the late nineteenth century and early twentieth century, with the advent of electricity and production assembly lines.
3. *Ordoliberalism* looks at free market theory and the minimal state but emphasises more the artificial form of the market order rather than taking a market fundamentalism view that believes that the competitive order of the market is formed naturally.
4. *People and the Internet*: How people connect with others, information, and the world around them is being transformed through a combination of technologies. Wearable and implantable technologies will enhance people's 'digital presence', allowing them to interact with objects and one another in new ways; *Computing, Communications* and *Storage Everywhere*: The continued rapid decline in the size and cost of computing and connectivity technologies is driving an exponential growth in the potential to access and leverage the Internet. This will lead to ubiquitous computing power being available, where everyone has access to a supercomputer in their pocket, with nearly unlimited storage capacity; *The Internet of Things*: smaller, cheaper, and smarter sensors are being introduced—in homes, clothes and accessories, cities, transport and energy networks, as well as manufacturing processes; *Artificial Intelligence and Big Data*: exponential digitisation creates exponentially more data—about everything and everyone. In parallel, the sophistication of the problems software can address, and the ability for software to learn and evolve itself, is advancing rapidly. This is built on the rise of big data for decision-making, and the influence that AI and robotics are starting to have on decision-making and jobs; *the Sharing Economy and Distributed Trust*: the Internet is driving a shift towards networks and platform-based social and economic models. Assets can be shared, creating not just new efficiencies but also whole new business models, and opportunities for social self-organisation. The blockchain, an emerging technology, replaces the need for third-party institutions to provide trust for financial, contract, and voting activities; and *the Digitisation of Matter*: physical objects are 'printed' from raw materials via additive (or 3D) printing, a process that is transforming industrial manufacturing, allows printing products at home, and creates a whole set of human health opportunities.

5. Two most unequal continents with an average Gini coefficient are Africa with .44 and Latin America with .52 over the period 2000–2009 (Armah and Baek 2015).

References

Amsden, A. 1994. Why Isn't the Whole World Experimenting with the East Asian Model to Develop? *World Development* 22 (4): 627–633.

Armah, B., and S.J. Baek. 2015. Can the SDGs Promote Structural Transformation in Africa? An Empirical Analysis. *Development* 58 (4): 1–19.

Baek, S.J. 2017. Is Rising Income Inequality Far from Inevitable During Structural Transformation? A Proposal for An Augmented Inequality Dynamics. *Journal of Economics and Political Economy* 4 (3): 224–237.

Bakir, V., and A. McStay. 2017. Fake News and The Economy of Emotions: Problems, Causes, Solutions. *Digital Journalism* 5 (3): 1–22.

Bemelmans, R., G.J. Gelderblom, R. Jonker, and L. De Witte. 2010. Socially Assistive Robots in Elderly Care: A Systematic Review into Effects and Effectiveness. *Journal of the American Medical Directors Association* 13 (2): 114–120.

Bonefeld, W. 2015. Crisis, Free Economy and Strong State: On Ordoliberalism. *European Review of International Studies* 2 (3): 16–26.

Bresser-Pereira, L.C. 2010. The Global Financial Crisis and A New Capitalism? *Journal of Post Keynesian Economics* 32 (4): 499–534.

Caporaso, J.A., and D.P. Levine. 1992. *Theories of Political Economy*. Cambridge: Cambridge University Press.

Castro, A.B. 2008. From Semi-Stagnation to Growth in a Sino-Centric Market. *Revista De Economia Poltica* 28 (1): 3–27.

Chang, Y.S., and S.J. Baek. 2010. Limit to Improvement: Myth or Reality? Empirical Analysis of Historical Improvement on Three Technologies Influential in the Evolution of Civilization. *Technological Forecasting and Social Change* 77 (5): 712–729.

———. 2011. Limit to Improvement in Running and Swimming. *International Journal of Applied Management Science* 3 (1): 97–120.

Chetty, R., and N. Hendren. 2017. *The Impacts of Neighborhoods on Intergenerational Mobility I: Childhood Exposure Effects* (NBER Working Paper No. 23001). Cambridge: National Bureau of Economic Research.

Corak, M. 2013. Income Inequality, Equality of Opportunity, and Intergenerational Mobility. *Journal of Economic Perspectives* 27 (3): 79–102.

ECAP (European Climate Adaptation Platform). 2015. *Amphibious Housing in Maasbommel, the Netherlands (2015)* [Online]. Available from: http://climate-adapt.eea.europa.eu/metadata/case-studies/amphibious-housing-in-maasbommel-the-netherlands/#relevance_anchor. Accessed 9 Sept 2017.

Greenstone, M., A. Looney, J. Patashnik, and M. Yu. 2013. *Thirteen Economic Facts about Social Mobility and the Role of Education.* Washington, DC: Brookings Institution. Available from: https://www.brookings.edu/wp-content/uploads/2016/06/THP_13EconFacts_FINAL.pdf. Accessed 7 Sept 2017.

Gupta, J. 2002. Global Sustainable Development Governance: Institutional Challenges from A Theoretical Perspective. *International Environmental Agreements* 2 (4): 361–388.

Henrich, J. 2000. Does Culture Matter in Economic Behavior: Ultimatum Game Bargaining Among the Machiguenga. *American Economic Review* 90 (4): 973–980.

Kahneman, D., J.L. Knetsch, and R.H. Thaler. 1986. Fairness and the Assumptions of Economics. *Journal of Business* 59 (4): 285–300.

Kearney, M.S., and P.B. Levine. 2016. Income Inequality, Social Mobility, and the Decision to Drop Out of High School. *Brookings Papers on Economic Activity* 1: 333–396.

Kerstenetzky, C.L. 2014. The Brazilian Social Developmental State: A Progressive Agenda in A (Still) Conservative Political Society. In *The End of the Developmental State?* ed. M. Williams, 172–196. New York: Routledge.

Kozul-Wright, R. 1995. The Myth of Anglo-Saxon Capitalism: Reconstructing the History of the American State. In *The Role of the State in Economic Change*, ed. H.J. Chang and R. Rowthorn, 81–113. Oxford: Clarendon Press.

Latar, N.L. 2015. The Robot Journalist in the Age of Social Physics: The End of Human Journalism? In *The New World of Transitioned Media*, ed. G. Einav, 65–80. New York: Springer.

List, J.A. 2007. On the Interpretation of Giving in Dictator Games. *Journal of Political Economy* 115 (3): 482–493.

Mankiw, N.G. 2013. Defending the One Percent. *Journal of Economic Perspectives* 27 (3): 21–34.

Marhánková, J.H. 2011. Leisure in Old Age: Disciplinary Practices Surrounding the Discourse of Active Ageing. *International Journal of Ageing and Later Life* 6 (1): 5–32.

McWilliams, D. 2015. *The Flat White Economy: How the Digital Economy is Transforming London and Other Cities of the Future.* London: Gerald Duckworth & Co Ltd.

Mosca, M. 2015. Cybersecurity in An Era with Quantum Computers: Will We Be Ready? *QCrypt 2015 Tokyo*. Available from: https://pdfs.semanticscholar.org/b4ac/17c649bc8dc4653ce6c114d4eeae3d6ed66f.pdf. Accessed 8 Nov 2017.
Ozanian, M. 2016. The World's Most Valuable Soccer Teams 2016. *Forbes*, 11 May. Available from: https://www.forbes.com/sites/mikeozanian/2016/05/11/the-worlds-most-valuable-soccer-teams-2016/#6131796e73b0. Accessed 16 May 2017.
Pakulski, J. 2016. Facing the Challenges of an Ageing Society. In *Population Ageing and Australia's Future*, ed. H. Kendig, P. McDonald, and J. Piggott, 111–134. Canberra: Australian National University Press.
Piketty, T. 2013. *Capital in the Twenty-First Century*. Cambridge: Harvard University Press.
Prebisch, R. 1959. Commercial Policy in Under-Developed Countries. *American Economic Review* 49 (2): 251–273.
Regalado, A. 2013. Infographic: The World's Technology Hubs. *MIT Technology Review*, 30 July. Available from: https://www.technologyreview.com/s/517626/infographic-the-worlds-technology-hubs. Accessed 23 Dec 2017.
Rifkin, J. 2011. *The Third Industrial Revolution: How Lateral Power Is Transforming Energy, the Economy, and the World*. New York: Palgrave Macmillan.
Roco, M.C., and W.S. Bainbridge. 2002. Converging Technologies for Improving Human Performance: Integrating from the Nanoscale. *Journal of Nanoparticle Research* 4 (4): 281–295.
Sandel, M.J. 2013. Market Reasoning as Moral Reasoning: Why Economists Should Re-engage with Political Philosophy. *Journal of Economic Perspectives* 27 (4): 121–140.
Schwab, K. 2017. *The Fourth Industrial Revolution*. New York: Crown Business.
Thornburg, C.C., T.M. Zabriskie, and K.L. McPhail. 2010. Deep-Sea Hydrothermal Vents: Potential Hot Spots for Natural Products Discovery? *Journal of Natural Products* 73 (3): 489–499.
Tyson, L. 1992. *Who's Bashing Whom? Trade Conflict in High Technology Industries*. Washington, DC: Institute for International Economics.
UN (United Nations). 2017. *World Population Prospects: The 2017 Revision* [Online]. New York: UN Department of Economic and Social Affairs. Available from: https://esa.un.org/unpd/wpp. Accessed 19 Dec 2017.
US National Centre for Education Statistics. 2017. *Digest of Education Statistics* [Online]. Available from: https://nces.ed.gov/programs/digest/2017menu_tables.asp. Accessed 6 Nov 2017.
Vigolo, V. 2017. Population Aging: Challenges and Opportunities for the Tourism Industry. In *Older Tourist Behavior and Marketing Tools*, ed. V. Vigolo, 3–23. Cham: Springer.

Visser, W. 2010. The Age of Responsibility: CSR 2.0 and the New DNA of Business. *Journal of Business Systems, Governance and Ethics* 5 (3): 7–22.
Wade, R. 1992. East Asian Economic Success: Conflicting Paradigms, Partial Insights, Shaky Evidence. *World Politics* 44 (2): 270–320.
Wadhwa, V. 2014. Laws and Ethics Can't Keep Pace with Technology. *MIT Technology Review*, 15 April. Available from: https://www.technologyreview.com/s/526401/laws-and-ethics-cant-keep-pace-with-technology. Accessed 23 Dec 2017.
WEF (World Economic Forum). 2014. *Creating New Models: Innovative Public-Private Partnerships for Inclusive Development in Latin America*. Geneva: WEF, (REF 261115). http://www3.weforum.org/docs/GAC/2014/WEF_GAC_LatinAmerica_InnovativePublicPrivatePartnerships_Report_2014.pdf. Accessed 11 Nov 2017.
WEF. 2015. *Deep Shift 21 Ways Software Will Transform Global Society*. Geneva: WEF, (REF 261115). Available from: http://www3.weforum.org/docs/WEF_GAC15_Deep_Shift_Software_Transform_Society.pdf. Accessed 17 Dec 2017.
Wehn, U., and C. Montalvo. 2018. Knowledge Transfer Dynamics and Innovation: Behaviour, Interactions and Aggregated Outcomes. *Journal of Cleaner Production* 171 (10): 56–68.
Weiss, C.H. 1998. *Evaluation: Methods for Studying Programs and Policies*. 2nd ed. New York: Prentice Hall.
Woods, J., and H. Johnson. 2014. Measuring Salary Inequality Across Major Sports Leagues. *Harvard Sports Analysis Collective*, 20 January. Available from: http://harvardsportsanalysis.org/2014/01/measuring-salary-inequality-across-major-sports-leagues. Accessed 5 July 2017.
World Bank. 2016. *Live Long and Prosper: Aging in East Asia and Pacific*. Washington, DC: World Bank. Available from: https://openknowledge.worldbank.org/bitstream/handle/10986/23133/9781464804694.pdf?sequence=8&isAllowed=y. Accessed 13 Aug 2017.

Appendix

Appendix 1	Augmented Dickey-Fuller stationarity test results
Appendix 2	*Johansen* cointegration test results
Appendix 3	*Granger* causality test results
Appendix 4	Impulse response test results
Appendix 5	Analysis of relationship and causality by 11-cluster groups
Appendix 6	Milestones in the history of lighting
Appendix 7	Progress of lighting efficacy (lumens per watt)
Appendix 8	Progress of conventional wheeled train (maximum speed)
Appendix 9	Progress of mass-produced car (maximum speed)
Appendix 10	Progress of military fighter aircraft (maximum speed)
Appendix 11	History of computing speed
Appendix 12	Progress of computing performance
Appendix 13	Progress of undersea cable systems
Appendix 14	Data transfer rate of Internet backbone systems

S. J. Baek, *The Political Economy of Neo-modernisation*,
https://doi.org/10.1007/978-3-319-91394-0

Appendix 1: Augmented Dickey-Fuller stationarity test results

Coun.	Time span	ln(PCGDP)					ln(INE)				
		Level		1st Diff		Lags	Levels		1st Diff		Lags
		Crit.	at 5%	Crit.	at 5%		Crit.	at 5%	Crit.	at 5%	
ALB	1996-2008	0.85	-3.21	-3.41	-3.21	2	2.22	-3.21	0.31	-3.26	3
DZA	1988-2005	0.60	-3.76	-6.21	-3.76	3	-2.00	-3.07	-25.65	-3.21	6
AGO	1995-2009	2.82	-1.97	-2.41	-1.97	3	-0.43	-3.93	-4.04	-4.01	3
ARG	1985-2013	-1.47	-2.97	-5.34	-2.98	6	0.04	-3.60	-3.98	-3.60	6
ARM	1990-2012	1.90	-1.96	-2.64	-1.96	4	-1.38	-3.02	-4.65	-3.02	5
AUS	1984-2014	-0.02	-2.96	-4.22	-2.97	7	-1.55	-2.98	-4.52	-2.98	7
AUT	1987-2014	-1.19	-2.98	-4.40	-2.98	6	2.18	-1.95	-2.84	-1.95	6
AZE	1990-2008	-2.37	-3.71	-4.33	-3.71	3	0.01	-1.96	-6.87	-1.98	6
BGD	1986-2010	1.07	-2.99	-3.32	-3.00	5	-2.25	-3.62	-4.14	-3.66	5
BRB	1996-2010	2.62	-1.98	-2.14	-1.97	3	-121.16	-3.12	-133.92	-3.14	1
BLR	1990-2012	0.77	-3.00	-3.20	-3.01	4	-2.39	-3.64	-5.83	-3.64	4
BEL	1985-2013	2.88	-1.95	-4.05	-1.95	6	-2.02	-2.98	-4.01	-2.98	6
BOL	1989-2013	2.44	-2.99	-2.63	-3.00	5	-0.44	-3.61	-3.17	-3.62	5
BIH	1994-2007	1.56	-1.97	-1.80	-1.97	2	-0.44	-3.12	-2.41	-3.14	2
BWA	1985-2005	3.11	-1.96	-2.19	-1.96	4	-1.46	-3.02	-4.17	-3.03	4
BRA	1976-2014	-0.71	-2.94	-4.25	-2.94	9	-1.75	-2.94	-2.95	-2.94	9
BGR	1980-2014	-0.01	-2.95	-5.16	-2.95	8	-0.39	-2.96	-3.29	-2.96	8
BFA	1994-2009	-0.20	-3.08	-3.27	-3.10	3	-0.09	-3.14	-15.13	-3.18	3
BDI	1992-2006	-2.24	-3.10	-3.29	-3.12	3	-1.57	-3.12	-98.71	-3.32	5
CPV	1989-2005	-0.83	-3.07	-5.03	-3.08	3	-1.43	-3.08	-1.27	-3.08	3
KHM	1994-2009	3.01	-1.97	-2.32	-1.97	3	0.10	-1.97	-2.05	-1.97	3
CMR	1996-2007	0.43	-3.18	-4.09	-3.40	3	-0.03	-3.18	-3.03	-3.21	3
CAN	1971-2013	-2.05	-2.93	-4.68	-2.94	9	-0.95	-2.93	-4.75	-2.94	9
CAF	1992-2008	-0.63	-3.12	-3.83	-3.08	3	-1.24	-3.08	-1.09	-3.08	3
CHL	1980-2013	0.05	-2.96	-3.56	-2.96	8	0.24	-2.95	-4.02	-2.96	8
CHN	1980-2013	1.34	-2.96	-3.45	-2.96	8	-1.42	-2.96	-3.61	-2.96	8
COL	1988-2014	-0.15	-2.98	-3.85	-2.99	6	0.17	-3.01	-3.45	-3.00	6
CRI	1985-2014	0.36	-2.97	-5.28	-2.97	7	-0.63	-2.98	-4.17	-2.98	4
CIV	1985-2008	-1.26	-3.00	-4.80	-3.00	5	0.28	-1.96	-2.38	-1.96	5
HRV	1995-2013	1.09	-1.96	-2.33	-1.96	3	0.44	-1.96	-2.72	-1.97	3
CYP	1990-2013	-1.25	-3.00	-3.75	-3.00	5	0.72	-1.96	-2.18	-1.96	5
CZE	1990-2013	-0.52	-3.00	-4.85	-3.00	5	0.44	-1.96	-3.24	-1.96	5
DNK	1980-2014	-0.81	-2.95	-4.58	-2.95	8	0.91	-1.95	-4.44	-1.95	8
DJI	1995-2005	2.46	-3.40	-1.78	-3.26	3	-0.20	-4.01	-2.62	-4.11	1
DOM	1986-2014	-0.74	-2.97	-4.98	-2.98	6	0.07	-1.96	-2.32	-1.96	6
ECU	1988-2014	0.11	-2.98	-3.65	-2.99	6	-1.64	-3.60	-3.99	-3.63	6

EGY	1990-2012	-1.27	-3.01	-3.26	-3.01	4	-1.00	-3.00	-3.89	-3.01	4
SLV	1989-2014	1.50	-1.96	-1.36	-1.96	5	-0.20	-2.99	-2.94	-2.99	5
EST	1995-2013	-0.75	-3.04	-7.01	-3.21	7	-1.56	-3.05	-4.37	-3.18	6
ETH	1995-2011	1.13	-1.97	-3.30	-1.99	6	-1.94	-3.73	-1.91	-3.76	3
FJI	1991-2009	-1.30	-3.04	-3.07	-3.05	3	-0.39	-3.04	-3.17	-3.05	3
FIN	1985-2014	-2.21	-2.97	-4.16	-2.98	7	-1.42	-2.97	-2.07	-2.97	8
FRA	1989-2013	-1.27	-2.99	-4.51	-3.00	5	-2.29	-3.00	-3.81	-3.00	5
GMB	1992-2003	1.20	-3.93	-2.61	-4.01	2	-0.33	-3.93	-2.03	-4.01	2
GEO	1990-2014	0.83	-1.96	-2.50	-1.96	5	1.60	-1.96	-4.20	-1.96	5
DEU	1983-2013	-1.89	-2.96	-4.29	-2.97	7	0.22	-2.96	-4.43	-2.97	7
GHA	1987-2013	0.59	-2.98	-3.75	-2.99	6	-1.27	-2.98	-3.93	-2.99	6
GRC	1985-2014	-1.82	-2.97	-3.09	-2.97	7	-2.04	-2.97	-3.39	-2.97	7
GTM	1987-2014	-0.23	-2.98	-5.67	-2.98	6	-0.59	-1.95	-2.80	-1.95	6
GIN	1991-2007	-0.86	-3.10	-4.69	-3.10	3	-3.18	-3.08	-2.12	-3.08	3
GNB	1991-2005	0.83	-1.97	-3.04	-1.97	3	-1.32	-3.12	-1.28	-3.12	3
GUY	1992-2006	-0.45	-3.10	-0.65	-3.12	3	-2.84	-3.79	-7.29	-3.83	3
HTI	1991-2001	0.76	-1.99	-3.01	-2.00	1	-1.74	-1.99	1.40	-2.00	1
HND	1988-2013	0.81	-2.99	-3.86	-2.99	5	-1.32	-3.00	-5.29	-3.00	5
HKG	1986-2011	1.25	-1.96	-2.23	-1.96	5	0.71	-1.96	-2.57	-1.96	5
HUN	1991-2014	-0.97	-3.00	-3.23	-3.00	5	-0.28	-1.96	-2.80	-1.96	5
ISL	1992-2014	-0.98	-3.00	-3.57	-3.01	4	0.78	-1.96	-2.46	-1.96	4
IND	1986-2011	1.84	-2.99	-4.39	-2.99	5	-0.64	-2.99	-4.15	-2.99	5
IDN	1987-2013	-0.43	-2.98	-5.21	-2.99	6	0.21	-2.98	-4.30	-3.00	6
IRN	1993-2011	1.01	-3.04	-2.19	-3.05	3	0.96	-3.04	-2.27	-3.05	3
IRL	1987-2014	-1.52	-2.98	-3.93	-2.98	6	0.05	-2.98	-4.05	-2.98	6
ISR	1985-2013	-2.35	-2.97	-4.53	-2.98	6	-0.82	-2.98	-3.76	-2.99	6
ITA	1967-2013	2.21	-1.95	-3.76	-1.95	9	-2.08	-2.93	-5.93	-2.93	9
JAM	1988-2004	-0.75	-3.07	-4.16	-3.08	3	-3.14	-3.18	-4.14	-3.08	5
JPN	1967-2011	2.04	-1.95	-3.31	-1.95	9	-1.12	-2.94	-3.89	-2.94	9
JOR	1986-2010	-1.71	-3.61	-4.35	-3.62	5	-2.22	-3.00	-3.24	-3.03	5
KAZ	1990-2012	2.75	-1.96	-2.34	-1.96	4	-1.81	-3.01	-3.34	-3.01	4
KEN	1992-2007	1.10	-3.14	-4.29	-3.10	3	-1.81	-3.08	-3.38	-3.10	3
KOR	1980-2013	-1.60	-2.95	-4.63	-2.96	8	-0.84	-1.95	-3.09	-1.95	8
KGZ	1990-2011	0.89	-1.96	-2.40	-1.96	4	-2.38	-3.01	-3.92	-3.05	4
LAO	1992-2008	-0.12	-3.73	-16.11	-3.93	4	-2.56	-3.08	-18.93	-3.18	4
LVA	1995-2014	-1.07	-3.03	-3.58	-3.05	4	-0.35	-3.76	-5.79	-3.76	4
LBN	1997-2005	0.45	-2.00	-3.38	-2.02	1	12.21	-2.01	-0.59	-2.02	1
LSO	1986-2003	1.71	-1.96	-2.38	-1.96	3	-1.34	-1.97	-2.03	-1.98	6
LTU	2004-2013	-1.90	-3.26	-2.29	-3.32	1	-0.23	-2.01	-2.86	-2.02	2
LUX	1985-2013	3.55	-1.95	-2.32	-1.96	6	-0.24	-2.97	-3.44	-2.98	6
MKD	1990-2013	-0.47	-3.00	-4.11	-3.00	5	-1.78	-3.02	-4.11	-3.02	5

MDG	1993-2010	-0.90	-3.05	-5.20	-3.07	3	-1.84	-3.07	-2.41	-3.07	3
MWI	1985-2011	-1.41	-2.98	-5.50	-2.99	6	-2.11	-2.99	-1.44	-2.99	6
MYS	1989-2012	-0.46	-3.00	-4.23	-3.00	5	-0.40	-1.96	-2.90	-1.96	5
MDV	2002-2010	-0.68	-3.32	-3.06	-3.40	1	1.15	-3.40	-56.73	-3.52	1
MLI	1989-2010	-0.27	-3.01	-4.56	-3.02	4	-0.10	-1.97	-2.43	-1.97	6
MLT	1999-2013	-0.63	-3.10	-3.34	-3.12	3	-0.35	-1.98	-3.10	-1.98	3
MRT	1987-2008	1.59	-1.96	-2.80	-1.96	4	-0.21	-1.96	-2.19	-1.96	4
MUS	1991-2006	-0.59	-3.08	-3.22	-3.10	3	-2.27	-3.10	-1.85	-3.10	3
MEX	1989-2014	-0.72	-3.02	-4.69	-2.99	5	-1.84	-2.99	-3.54	-3.00	5
MDA	1990-2012	0.25	-3.00	-3.24	-3.01	4	-0.73	-1.96	-2.95	-1.96	4
MNG	1995-2010	-2.17	-3.14	-3.07	-3.12	3	-0.39	-1.97	-2.21	-1.97	3
MNE	2000-2010	-1.40	-3.21	-4.04	-3.32	1	0.47	-3.26	-9.82	-3.26	1
MAR	1984-2007	-0.86	-3.00	-4.34	-3.00	5	-0.50	-3.00	-3.51	-3.04	5
MOZ	1996-2008	1.15	-3.14	-1.45	-3.18	2	-2.71	-3.21	-2.56	-3.18	2
NAM	1993-2010	1.68	-1.96	-2.16	-1.96	3	-0.96	-1.96	-2.93	-1.96	3
NPL	1995-2010	-0.09	-3.83	-4.68	-3.83	3	-0.82	-1.97	-6.94	-2.00	6
NLD	1985-2014	2.98	-1.95	-4.37	-1.95	7	0.34	-1.95	-2.99	-1.95	7
NZL	1982-2013	-0.35	-2.96	-3.79	-2.96	7	-1.87	-2.96	-7.27	-2.96	7
NIC	1993-2009	1.61	-1.96	-13.93	-1.97	3	-0.02	-3.08	-1.36	-3.08	3
NER	1992-2008	-0.61	-3.07	-4.24	-3.08	3	-0.18	-3.07	-1.74	-3.08	3
NGA	1985-2011	0.74	-2.98	-5.81	-2.99	6	-2.38	-3.60	-4.57	-3.67	6
NOR	1984-2013	-0.87	-2.97	-4.64	-2.97	7	-1.84	-2.97	-4.69	-2.97	7
PAK	1985-2011	1.22	-2.98	-4.16	-2.99	6	0.28	-2.98	-3.20	-2.99	6
PAN	1986-2014	2.27	-2.97	-3.05	-2.98	6	-0.83	-1.95	-2.01	-1.95	6
PNG	1995-2005	-2.45	-3.26	-1.41	-3.26	1	-239.19	-3.26	-234.42	-3.32	1
PRY	1990-2014	0.32	-2.99	-3.73	-3.00	5	1.52	-1.96	-2.58	-1.96	5
PER	1985-2014	-0.37	-2.97	-7.27	-2.97	7	-0.44	-2.97	-3.63	-3.07	12
PHL	1985-2012	0.67	-2.98	-4.47	-2.98	6	-0.26	-1.96	-2.75	-1.95	9
POL	1990-2013	-1.53	-3.00	-4.57	-3.00	5	1.73	-1.96	-2.45	-1.96	5
PRT	1989-2014	-2.39	-2.99	-4.37	-2.99	5	-0.15	-1.96	-2.42	-1.96	5
PRI	1989-2011	0.26	-3.63	-4.31	-3.64	6	0.26	-1.96	-2.25	-1.96	8
ROU	1989-2013	2.13	-1.96	-2.58	-1.96	5	1.39	-1.96	-2.40	-1.96	5
RUS	1989-2014	0.08	-2.99	-3.19	-2.99	5	-1.58	-3.00	-4.09	-3.00	5
RWA	1985-2011	-1.19	-2.98	-6.30	-2.99	6	-1.41	-3.60	-5.02	-3.64	4
SEN	1991-2011	-0.44	-3.02	-3.17	-3.04	4	-2.34	-3.02	-3.42	-3.03	4
SRB	2002-2014	1.91	-1.97	-2.73	-1.98	2	-1.18	-1.98	-2.13	-2.00	3
SLE	1989-2011	-0.37	-3.00	-5.69	-3.01	4	0.46	-3.00	-5.08	-3.01	4
SGP	1974-2013	-1.41	-2.94	-3.77	-2.94	9	0.74	-1.95	-4.44	-1.95	3
SVK	1990-2013	-1.27	-3.00	-3.01	-3.00	5	-1.70	-3.03	-4.55	-3.00	5
SVN	1995-2013	1.79	-1.96	-2.50	-1.96	3	-0.08	-3.04	-4.34	-3.05	3
ZAF	1991-2012	-0.20	-3.01	-3.03	-3.02	4	-1.64	-3.01	-3.62	-3.02	4

ESP	1985-2014	2.78	-1.95	-3.58	-1.95	7	1.06	-1.95	-3.68	-1.95	7
LKA	1985-2013	2.11	-2.97	-4.04	-2.98	6	-2.25	-2.98	-4.43	-2.99	6
LCA	1995-2006	0.43	-3.18	-2.99	-3.21	2	54.64	-1.98	-57.77	-1.99	1
SWZ	1994-2009	0.98	-1.97	-2.86	-1.97	3	-2.00	-3.10	-0.79	-3.10	3
SWE	1960-2013	-1.90	-2.92	-5.35	-2.92	10	-1.55	-2.92	-5.96	-2.92	10
CHE	1992-2013	0.08	-3.01	-3.77	-3.02	4	0.06	-1.96	-2.46	-1.96	4
SYR	1997-2007	1.19	-3.21	-2.08	-3.26	2	-2.20	-3.26	-0.05	-3.26	1
TJK	1990-2009	0.29	-1.96	-2.83	-1.96	4	-0.10	-1.97	-2.43	-1.97	6
TZA	1991-2011	0.60	-3.02	-3.86	-3.07	4	-1.00	-3.04	-3.40	-3.03	4
THA	1986-2011	1.39	-1.96	-2.52	-1.96	5	-1.48	-3.00	-5.35	-3.00	5
TTO	1988-2005	0.33	-3.79	-5.44	-3.79	3	0.70	-1.97	-5.54	-1.97	3
TUN	1985-2010	-0.31	-2.99	-4.91	-2.99	5	-2.49	-2.99	-23.81	-3.08	9
TUR	1987-2014	-1.00	-2.98	-5.76	-2.98	6	-1.98	-3.73	-9.86	-3.76	11
TKM	1987-2005	0.54	-1.96	-2.07	-1.96	3	-0.19	-3.69	-5.55	-3.88	5
UGA	1989-2011	0.78	-1.96	-2.62	-1.96	4	0.27	-1.96	-2.36	-1.96	4
UKR	1987-2010	-0.21	-3.00	-3.19	-3.00	5	-2.25	-3.00	-9.13	-3.05	5
GBR	1961-2014	-1.33	-2.92	-4.54	-2.92	10	-1.39	-2.92	-3.61	-2.92	10
USA	1960-2014	0.19	-3.50	-4.98	-3.50	10	-0.06	-2.92	-4.45	-2.92	10
URY	1986-2014	1.07	-1.95	-2.67	-1.95	6	-0.58	-1.95	-1.91	-1.95	6
UZB	1990-2005	-0.97	-1.97	-2.99	-1.97	3	1.70	-1.97	-2.26	-1.97	3
VEN	1976-2013	-0.26	-2.94	-6.17	-2.95	9	-2.15	-2.96	-3.45	-2.95	9
VNM	1992-2011	-1.23	-3.03	-3.15	-3.04	4	-2.05	-3.04	-7.80	-3.04	4
YEM	1992-2005	0.63	-3.18	-5.23	-3.18	2	-1.77	-3.14	-1.07	-3.14	2
ZMB	1991-2010	-1.17	-3.67	-4.18	-3.69	4	-0.11	-1.96	-3.11	-1.96	4
ZWE	1990-2011	-2.00	-3.01	-4.16	-3.02	4	0.47	-1.96	-2.22	-1.96	4

Source: Author's own elaboration on the basis of the modelling results

Note: All the analysed countries are coded: ALB (Albania); DZA (Algeria); AGO (Angola); ARG (Argentina); ARM (Armenia); AUS (Australia); AUT (Austria); AZE (Azerbaijan); BGD (Bangladesh); BRB (Barbados); BLR (Belarus); BEL (Belgium); BOL (Bolivia); BIH (Bosnia and Herzegovina); BWA (Botswana); BRA (Brazil); BGR (Bulgaria); BFA (Burkina Faso); BDI (Burundi); CPV (Cabo Verde); KHM (Cambodia); CMR (Cameroon); CAN (Canada); CAF (Central African Republic); CHL (Chile); CHN (China); COL (Colombia); CRI (Costa Rica); CIV (Cote d'Ivoire); HRV (Croatia); CYP (Cyprus); CZE (Czech Republic); DNK (Denmark); DJI (Djibouti); DOM (Dominican Republic); ECU (Ecuador); EGY (Egypt); SLV (El Salvador); EST (Estonia); ETH (Ethiopia); FJI (Fiji); FIN (Finland); FRA (France); GMB (Gambia); GEO (Georgia); DEU (Germany); GHA (Ghana); GRC (Greece); GTM (Guatemala); GIN (Guinea); GNB (Guinea-Bissau); GUY (Guyana); HTI (Haiti); HND (Honduras); HKG (Hong Kong); HUN (Hungary); ISL (Iceland); IND (India); IDN (Indonesia); IRN (Iran); IRL (Ireland); ISR (Israel); ITA (Italy); JAM (Jamaica); JPN (Japan); JOR (Jordan); KAZ (Kazakhstan); KEN (Kenya); KOR (Korea, Republic of); KGZ (Kyrgyz Republic); LAO (Laos); LVA

(Latvia); LBN (Lebanon); LSO (Lesotho); LTU (Lithuania); LUX (Luxembourg); MKD (Macedonia, FYR); MDG (Madagascar); MWI (Malawi); MYS (Malaysia); MDV (Maldives); MLI (Mali); MLT (Malta); MRT (Mauritania); MUS (Mauritius); MEX (Mexico); MDA (Moldova); MNG (Mongolia); MNE (Montenegro); MAR (Morocco); MOZ (Mozambique); NAM (Namibia); NPL (Nepal); NLD (Netherlands); NZL (New Zealand); NIC (Nicaragua); NER (Niger); NGA (Nigeria); NOR (Norway); PAK (Pakistan); PAN (Panama); PNG (Papua New Guinea); PRY (Paraguay); PER (Peru); PHL (Philippines); POL (Poland); PRT (Portugal); PRI (Puerto Rico); ROU (Romania); RUS (Russian Federation); RWA (Rwanda); SEN (Senegal); SRB (Serbia); SLE (Sierra Leone); SGP (Singapore); SVK (Slovakia); SVN (Slovenia); ZAF (South Africa); ESP (Spain); LKA (Sri Lanka); LCA (St. Lucia); SWZ (Swaziland); SWE (Sweden); CHE (Switzerland); SYR (Syria); TJK (Tajikistan); TZA (Tanzania); THA (Thailand); TTO (Trinidad and Tobago); TUN (Tunisia); TUR (Turkey); TKM (Turkmenistan); UGA (Uganda); UKR (Ukraine); GBR (United Kingdom); USA (United States); URY (Uruguay); UZB (Uzbekistan); VEN (Venezuela); VNM (Viet Nam); YEM (Yemen, Republic of); ZMB (Zambia); and ZWE (Zimbabwe)

Appendix 2: *Johansen* cointegration test results

Coun.	Trace test				Maximum eigenvalue test				Lags
	Elgen.	Trace stat.	Crit. 5%	Prob.	Elgen.	Trace stat.	Crit. 5%	Prob.	
DZA	0.95	43.15	15.49	0.00	0.95	41.28	14.26	0.00	3
AGO	0.51	10.54	15.49	0.24	0.51	9.34	14.26	0.26	1
ARG	0.75	29.88	15.49	0.00	0.75	29.01	14.26	0.00	7
ARM	0.37	11.33	15.49	0.19	0.37	9.20	14.26	0.27	2
AUS	0.29	11.10	15.49	0.21	0.29	9.32	14.26	0.26	3
AUT	0.83	39.96	15.49	0.00	0.83	39.15	14.26	0.00	5
AZE	0.59	17.85	15.49	0.02	0.59	14.38	14.26	0.05	2
BGD	0.34	11.52	15.49	0.18	0.34	8.76	14.26	0.31	3
BLR	0.91	41.16	15.49	0.00	0.91	40.58	14.26	0.00	5
BEL	0.21	8.50	15.49	0.41	0.21	6.52	14.26	0.55	1
BWA	0.39	9.68	15.49	0.31	0.39	9.31	14.26	0.26	1
BRA	0.50	21.71	15.49	0.01	0.50	21.43	14.26	0.00	7
BGR	0.60	25.80	15.49	0.00	0.60	24.02	14.26	0.00	8
BFA	0.87	27.19	15.49	0.00	0.87	26.94	14.26	0.00	2
BDI	0.36	8.10	15.49	0.45	0.36	5.72	14.26	0.65	1
KHM	0.29	7.12	15.49	0.56	0.29	6.93	14.26	0.50	1
CAN	0.40	18.65	15.49	0.02	0.40	18.55	14.26	0.01	6
CHL	0.65	27.21	15.49	0.00	0.65	27.21	14.26	0.00	7
CHN	0.51	19.68	15.49	0.01	0.51	19.49	14.26	0.01	6
COL	0.68	24.86	15.49	0.00	0.68	24.86	14.26	0.00	4
CRI	0.56	19.53	15.49	0.01	0.56	19.49	14.26	0.01	5
CIV	0.82	34.94	15.49	0.00	0.82	31.04	14.26	0.00	5
HRV	0.83	30.93	15.49	0.00	0.83	26.39	14.26	0.00	3
CYP	0.56	24.33	15.49	0.00	0.56	16.28	14.26	0.02	3
CZE	0.51	16.11	15.49	0.04	0.51	15.54	14.26	0.03	1
DNK	0.74	36.43	15.49	0.00	0.74	33.25	14.26	0.00	9
DOM	0.79	46.65	15.49	0.00	0.79	35.80	14.26	0.00	5
ECU	0.78	33.31	15.49	0.00	0.78	30.47	14.26	0.00	6
EGY	0.96	54.63	15.49	0.00	0.96	54.47	14.26	0.00	5
EST	0.69	19.06	15.49	0.01	0.69	18.90	14.26	0.01	2
FJI	0.54	11.58	15.49	0.18	0.54	11.52	14.26	0.13	3
FRA	0.69	27.55	15.49	0.00	0.69	24.33	14.26	0.00	3
GEO	0.25	7.02	15.49	0.58	0.25	6.25	14.26	0.58	2
DEU	0.77	38.79	15.49	0.00	0.77	35.11	14.26	0.00	6
GHA	0.72	31.60	15.49	0.00	0.72	25.37	14.26	0.00	6
GRC	0.33	13.51	15.49	0.10	0.33	11.19	14.26	0.15	1
GTM	0.82	38.50	15.49	0.00	0.82	35.55	14.26	0.00	6

HND	0.64	21.77	15.49	0.01	0.64	20.23	14.26	0.01	5
HKG	0.83	45.43	15.49	0.00	0.83	33.52	14.26	0.00	6
HUN	0.71	31.14	15.49	0.00	0.71	25.10	14.26	0.00	3
ISL	0.72	35.83	15.49	0.00	0.72	23.10	14.26	0.00	4
IND	0.87	45.43	15.49	0.00	0.87	38.89	14.26	0.00	6
IDN	0.74	28.60	15.49	0.00	0.74	27.99	14.26	0.00	5
IRL	0.59	23.68	15.49	0.00	0.59	18.88	14.26	0.01	6
ISR	0.93	60.00	15.49	0.00	0.93	57.07	14.26	0.00	7
ITA	0.58	38.24	15.49	0.00	0.58	31.24	14.26	0.00	10
JAM	0.78	21.29	15.49	0.01	0.78	21.00	14.26	0.00	2
JPN	0.61	37.68	15.49	0.00	0.61	33.31	14.26	0.00	9
JOR	0.59	19.71	15.49	0.01	0.59	19.59	14.26	0.01	2
KAZ	0.71	33.64	15.49	0.00	0.71	22.55	14.26	0.00	4
KEN	0.35	6.08	15.49	0.69	0.35	6.07	14.26	0.60	1
KOR	0.47	21.75	15.49	0.01	0.47	17.11	14.26	0.02	6
KGZ	0.47	12.63	15.49	0.13	0.47	12.63	14.26	0.09	1
LAO	0.74	25.91	15.49	0.00	0.74	18.91	14.26	0.01	2
LVA	0.86	35.37	15.49	0.00	0.86	29.29	14.26	0.00	4
LSO	0.57	13.82	15.49	0.09	0.57	13.55	14.26	0.06	1
LUX	0.68	32.41	15.49	0.00	0.68	25.17	14.26	0.00	6
MKD	0.69	26.44	15.49	0.00	0.69	21.04	14.26	0.00	5
MYS	0.74	28.61	15.49	0.00	0.74	28.47	14.26	0.00	2
MLI	0.49	17.07	15.49	0.03	0.49	12.05	14.26	0.11	3
MLT	0.66	17.19	15.49	0.03	0.66	13.86	14.26	0.06	1
MRT	0.68	19.30	15.49	0.01	0.68	19.30	14.26	0.01	4
MEX	0.90	49.42	15.49	0.00	0.90	47.43	14.26	0.00	4
MDA	0.60	26.92	15.49	0.00	0.60	16.33	14.26	0.02	4
MNE	0.74	12.40	15.49	0.14	0.74	12.06	14.26	0.11	1
MAR	0.80	31.98	15.49	0.00	0.80	30.50	14.26	0.00	4
NAM	0.81	27.95	15.49	0.00	0.81	24.93	14.26	0.00	2
NPL	0.55	10.56	15.49	0.24	0.55	10.52	14.26	0.18	2
NLD	0.44	15.68	15.49	0.05	0.44	15.16	14.26	0.04	3
NZL	0.55	20.26	15.49	0.01	0.55	19.29	14.26	0.01	7
NGA	0.68	22.66	15.49	0.00	0.68	22.66	14.26	0.00	6
NOR	0.69	29.51	15.49	0.00	0.69	25.46	14.26	0.00	7
PAK	0.36	12.73	15.49	0.13	0.36	10.27	14.26	0.19	3
PAN	0.63	22.40	15.49	0.00	0.63	21.76	14.26	0.00	6
PRY	0.77	27.99	15.49	0.00	0.77	27.99	14.26	0.00	5
PER	0.74	47.92	15.49	0.00	0.74	30.81	14.26	0.00	6
PHL	0.58	20.81	15.49	0.01	0.58	18.07	14.26	0.01	6
POL	0.65	20.98	15.49	0.01	0.65	20.89	14.26	0.00	3

PRT	0.56	21.77	15.49	0.01	0.56	18.18	14.26	0.01	3
PRI	0.63	26.27	15.49	0.00	0.63	20.62	14.26	0.00	1
ROU	0.57	18.44	15.49	0.02	0.57	17.57	14.26	0.01	3
RUS	0.83	38.96	15.49	0.00	0.83	35.80	14.26	0.00	5
RWA	0.22	6.19	15.49	0.67	0.22	6.06	14.26	0.61	1
SEN	0.61	17.81	15.49	0.02	0.61	14.94	14.26	0.04	4
SRB	0.42	10.92	15.49	0.22	0.42	5.99	14.26	0.61	1
SLE	0.67	25.60	15.49	0.00	0.67	19.00	14.26	0.01	5
SGP	0.61	35.06	15.49	0.00	0.61	31.25	14.26	0.00	6
SVK	0.68	27.05	15.49	0.00	0.68	20.68	14.26	0.00	5
SVN	0.67	23.16	15.49	0.00	0.67	16.55	14.26	0.02	3
ZAF	0.64	20.51	15.49	0.01	0.64	17.21	14.26	0.02	4
ESP	0.83	43.58	15.49	0.00	0.83	38.51	14.26	0.00	7
LKA	0.83	47.93	15.49	0.00	0.83	37.54	14.26	0.00	7
SWE	0.15	7.71	15.49	0.50	0.15	7.68	14.26	0.41	4
CHE	0.44	11.78	15.49	0.17	0.44	11.64	14.26	0.12	1
TJK	0.70	21.89	15.49	0.00	0.70	20.61	14.26	0.00	2
TZA	0.73	29.14	15.49	0.00	0.73	21.19	14.26	0.00	4
THA	0.54	18.56	15.49	0.02	0.54	17.94	14.26	0.01	2
TTO	0.83	38.47	15.49	0.00	0.83	24.46	14.26	0.00	3
TUN	0.87	38.83	15.49	0.00	0.87	38.20	14.26	0.00	6
TUR	0.56	18.71	15.49	0.02	0.56	18.71	14.26	0.01	4
TKM	0.59	16.06	15.49	0.04	0.59	15.33	14.26	0.03	1
UGA	0.60	21.14	15.49	0.01	0.60	19.22	14.26	0.01	1
UKR	0.56	23.32	15.49	0.00	0.56	15.68	14.26	0.03	4
GBR	0.37	25.20	15.49	0.00	0.37	20.63	14.26	0.00	9
USA	0.32	28.75	15.49	0.00	0.32	17.44	14.26	0.02	8
UZB	0.88	32.85	15.49	0.00	0.88	27.22	14.26	0.00	2
VEN	0.41	19.76	15.49	0.01	0.41	15.85	14.26	0.03	7
VNM	0.82	26.76	15.49	0.00	0.82	25.77	14.26	0.00	4
ZMB	0.60	15.75	15.49	0.05	0.60	15.65	14.26	0.03	2
ZWE	0.49	17.07	15.49	0.03	0.49	12.05	14.26	0.11	3

Source: Author's own elaboration on the basis of the modelling results

Appendix 3: *Granger* causality test results

Coun.	From	To	Long-run				Short-run			Model
			Coeff.	t-stat.	Prob.	ECT	$\chi 2$ stat.	Prob.	Imp. Resp.	
DZA	INE	PCGDP	-115.83	-1.55	0.15	-0.01	6.00	0.11	Positive	VEC
	PCGDP	INE	-0.01	-1.55	0.15	-0.33	1.47	0.69	Positive	
AGO	INE	PCGDP					1.11	0.29	Positive	VAR
	PCGDP	INE					18.85	0.00	Negative	
ARG	INE	PCGDP	7.89	-3.62	0.00	-1.68	18.07	0.01	Negative	VEC
	PCGDP	INE	0.13	-1.86	0.09	-0.16	30.19	0.00	Negative	
ARM	INE	PCGDP					10.70	0.00	Positive	VAR
	PCGDP	INE					8.89	0.01	Negative	
AUS	INE	PCGDP					3.31	0.35	Positive	VAR
	PCGDP	INE					10.25	0.02	Positive	
AUT	INE	PCGDP	-7.52	-2.43	0.02	-0.60	8.66	0.12	Positive	VEC
	PCGDP	INE	-0.13	-4.48	0.00	-1.12	32.95	0.00	Positive	
AZE	INE	PCGDP	2.76	-3.72	0.00	-0.27	8.49	0.01	Negative	VEC
	PCGDP	INE	0.36	-1.83	0.08	-0.43	2.04	0.36	Negative	
BGD	INE	PCGDP					3.36	0.50	Positive	VAR
	PCGDP	INE					13.54	0.01	Positive	
BLR	INE	PCGDP	-11.14	-0.60	0.56	-0.70	5.31	0.38	Positive	VEC
	PCGDP	INE	-0.09	-4.59	0.00	-1.99	21.61	0.00	Positive	
BEL	INE	PCGDP					7.09	0.01	Positive	VAR
	PCGDP	INE					0.16	0.69	Negative	
BWA	INE	PCGDP					4.43	0.04	Negative	VAR
	PCGDP	INE					0.04	0.84	Positive	
BRA	INE	PCGDP	25.70	-3.73	0.00	-0.15	32.03	0.00	Negative	VEC
	PCGDP	INE	0.04	-1.50	0.14	-0.15	2.28	0.94	Negative	
BGR	INE	PCGDP	-4.15	-2.73	0.02	-0.27	7.13	0.52	Positive	VEC
	PCGDP	INE	-0.24	0.04	0.97	0.00	8.26	0.41	Negative	
BFA	INE	PCGDP	4.71	-3.21	0.01	-0.71	9.54	0.01	Negative	VEC
	PCGDP	INE	0.21	-4.95	0.00	-0.57	3.25	0.20	Negative	
BDI	INE	PCGDP					1.74	0.19	Positive	VAR
	PCGDP	INE					13.64	0.00	Positive	
KHM	INE	PCGDP	12.02	2.77	0.01	0.21	0.30	0.59	Positive	VEC
	PCGDP	INE	0.08	-3.16	0.01	-0.39	0.02	0.88	Negative	
CAN	INE	PCGDP	-11.51	-3.63	0.00	-0.10	17.28	0.01	Mixed	VEC
	PCGDP	INE	-0.09	-0.40	0.69	-0.03	4.93	0.55	Negative	
CHL	INE	PCGDP	-9.20	-3.11	0.01	-0.15	22.97	0.00	Negative	VEC
	PCGDP	INE	-0.11	1.53	0.14	0.08	5.16	0.64	Negative	
CHN	INE	PCGDP	-7.77	-0.95	0.35	-0.05	4.18	0.65	Positive	VEC

	PCGDP	INE	-0.13	2.84	0.01	0.39	18.49	0.01	Negative	
COL	INE	PCGDP	9.91	-0.86	0.40	-0.10	2.53	0.64	Negative	VEC
	PCGDP	INE	0.10	-4.30	0.00	-0.39	7.88	0.10	Negative	
CRI	INE	PCGDP	-9.39	-2.12	0.04	-0.29	11.51	0.04	Positive	VEC
	PCGDP	INE	-0.11	-2.87	0.01	-0.35	11.16	0.05	Negative	
CIV	INE	PCGDP	40.90	-4.13	0.00	-0.07	33.76	0.00	Positive	VEC
	PCGDP	INE	-0.02	-1.21	0.25	-0.27	3.60	0.61	Positive	
HRV	INE	PCGDP	13.50	-1.97	0.07	-0.16	7.47	0.06	Negative	VEC
	PCGDP	INE	0.07	-2.47	0.03	-0.70	3.83	0.28	Negative	
CYP	INE	PCGDP	-9.37	2.05	0.05	0.13	0.27	0.97	Negative	VEC
	PCGDP	INE	-0.11	-2.54	0.02	-0.25	7.19	0.07	Negative	
CZE	INE	PCGDP	67.91	-0.66	0.51	0.00	1.67	0.20	Negative	VEC
	PCGDP	INE	0.01	-4.29	0.00	-0.26	13.04	0.00	Positive	
DNK	INE	PCGDP	-3.30	-2.93	0.02	-0.17	26.35	0.00	Positive	VEC
	PCGDP	INE	-0.30	-1.93	0.08	-0.09	11.87	0.22	Negative	
DOM	INE	PCGDP	84.57	-0.11	0.91	0.00	1.77	0.88	Mixed	VEC
	PCGDP	INE	0.01	-5.78	0.00	-0.22	34.11	0.00	Negative	
ECU	INE	PCGDP	3.16	-4.38	0.00	-1.60	41.35	0.00	Negative	VEC
	PCGDP	INE	0.32	0.62	0.54	0.12	0.83	0.99	Positive	
EGY	INE	PCGDP	-179.00	3.83	0.00	0.04	15.41	0.01	Negative	VEC
	PCGDP	INE	-0.01	-3.27	0.01	-4.06	12.30	0.03	Negative	
EST	INE	PCGDP	15.99	3.68	0.00	0.38	28.93	0.00	Positive	VEC
	PCGDP	INE	0.06	-3.36	0.00	-1.13	4.97	0.08	Negative	
FJI	INE	PCGDP					3.25	0.36	Negative	VAR
	PCGDP	INE					4.53	0.21	Positive	
FRA	INE	PCGDP	47.32	-4.62	0.00	-0.12	29.07	0.00	Negative	VEC
	PCGDP	INE	0.02	-1.88	0.07	-0.49	4.80	0.19	Negative	
GEO	INE	PCGDP					15.07	0.00	Negative	VAR
	PCGDP	INE					0.17	0.92	Negative	
DEU	INE	PCGDP	-6.87	-5.43	0.00	-1.26	18.35	0.01	Positive	VEC
	PCGDP	INE	-0.15	3.25	0.00	0.71	11.43	0.08	Negative	
GHA	INE	PCGDP	-51.08	-3.77	0.00	-0.12	26.65	0.00	Positive	VEC
	PCGDP	INE	-0.02	-0.68	0.51	-0.14	3.41	0.76	Negative	
GRC	INE	PCGDP					6.30	0.01	Positive	VAR
	PCGDP	INE					0.03	0.87	Negative	
GTM	INE	PCGDP	-41.67	0.14	0.89	0.00	5.25	0.51	Mixed	VEC
	PCGDP	INE	-0.02	-5.17	0.00	-0.25	31.50	0.00	Positive	
HND	INE	PCGDP	10.23	-1.87	0.08	-0.07	9.30	0.10	Negative	VEC
	PCGDP	INE	0.10	-2.96	0.01	-0.12	16.52	0.01	Positive	
HKG	INE	PCGDP	-10.88	0.40	0.70	0.06	2.33	0.89	Negative	VEC

	PCGDP	INE	-0.09	-4.08	0.00	-1.23	89.70	0.00	Negative	
HUN	INE	PCGDP	11.89	-5.16	0.00	-0.27	14.11	0.00	Negative	VEC
	PCGDP	INE	0.08	0.24	0.81	0.06	0.88	0.83	Positive	
ISL	INE	PCGDP	-1.16	-2.16	0.05	-0.78	5.28	0.26	Negative	VEC
	PCGDP	INE	-0.86	4.34	0.00	0.22	12.37	0.01	Negative	
IND	INE	PCGDP	-8.92	1.49	0.17	0.42	5.37	0.50	Negative	VEC
	PCGDP	INE	-0.11	-5.19	0.00	-1.39	47.15	0.00	Positive	
IDN	INE	PCGDP	-9.18	0.51	0.61	0.50	8.72	0.12	Negative	VEC
	PCGDP	INE	-0.11	-4.07	0.00	-2.52	15.91	0.01	Positive	
IRL	INE	PCGDP	9.69	-2.79	0.01	-0.66	5.21	0.52	Negative	VEC
	PCGDP	INE	0.10	-1.10	0.29	-0.51	2.27	0.89	Negative	
ISR	INE	PCGDP	-8.30	2.24	0.05	0.47	38.56	0.00	Negative	VEC
	PCGDP	INE	-0.12	-2.24	0.05	-3.90	7.01	0.43	Negative	
ITA	INE	PCGDP	-16.48	-1.99	0.06	-0.07	9.16	0.52	Positive	VEC
	PCGDP	INE	-0.06	-3.22	0.00	-0.40	14.22	0.16	Positive	
JAM	INE	PCGDP	-2.64	2.83	0.01	0.53	10.94	0.00	Negative	VEC
	PCGDP	INE	-0.38	-3.15	0.01	-0.53	1.41	0.49	Positive	
JPN	INE	PCGDP	-7.27	-1.54	0.13	-0.14	3.41	0.95	Positive	VEC
	PCGDP	INE	-0.14	-4.49	0.00	-0.57	22.70	0.01	Positive	
JOR	INE	PCGDP	18.20	3.04	0.00	0.15	12.94	0.00	Positive	VEC
	PCGDP	INE	0.05	-3.01	0.01	-0.29	0.19	0.91	Negative	
KAZ	INE	PCGDP	9.60	-2.93	0.01	-1.00	6.42	0.17	Negative	VEC
	PCGDP	INE	0.10	-0.46	0.66	-0.10	2.18	0.70	Negative	
KEN	INE	PCGDP					14.89	0.00	Negative	VAR
	PCGDP	INE					0.00	0.98	Positive	
KOR	INE	PCGDP	-12.03	-2.79	0.01	-0.18	8.12	0.23	Positive	VEC
	PCGDP	INE	-0.08	0.80	0.43	0.04	13.95	0.03	Negative	
KGZ	INE	PCGDP					0.93	0.33	Negative	VAR
	PCGDP	INE					4.18	0.04	Mixed	
LAO	INE	PCGDP	-7.03	3.81	0.00	0.70	17.71	0.00	Negative	VEC
	PCGDP	INE	-0.14	-3.05	0.01	-0.52	3.63	0.16	Positive	
LVA	INE	PCGDP	-6.69	-0.15	0.89	-0.06	1.66	0.80	Positive	VEC
	PCGDP	INE	-0.15	5.34	0.00	0.79	20.67	0.00	Negative	
LSO	INE	PCGDP					4.93	0.03	Positive	VAR
	PCGDP	INE					1.31	0.25	Negative	
LUX	INE	PCGDP	-7.92	-0.53	0.61	-0.14	5.41	0.49	Positive	VEC
	PCGDP	INE	-0.13	-3.37	0.00	-0.80	16.83	0.01	Positive	
MKD	INE	PCGDP	-6.88	-1.55	0.15	-0.13	12.69	0.03	Negative	VEC
	PCGDP	INE	-0.15	-3.18	0.01	-0.40	9.59	0.09	Positive	
MYS	INE	PCGDP	15.59	-1.47	0.15	-0.11	0.54	0.76	Negative	VEC

	PCGDP	INE	0.06	-4.62	0.00	-0.71	7.26	0.03	Negative	
MLI	INE	PCGDP	3.86	-0.40	0.70	-0.19	1.18	0.76	Negative	VEC
	PCGDP	INE	0.26	-2.92	0.01	-0.28	12.06	0.01	Negative	
MLT	INE	PCGDP					0.14	0.70	Positive	VAR
	PCGDP	INE					6.57	0.01	Positive	
MRT	INE	PCGDP	2.64	-1.34	0.20	-0.29	5.37	0.25	Negative	VEC
	PCGDP	INE	0.38	-2.09	0.06	-0.33	9.39	0.05	Negative	
MEX	INE	PCGDP	15.02	-0.93	0.36	-0.25	2.71	0.61	Negative	VEC
	PCGDP	INE	0.07	-6.89	0.00	-0.67	157.05	0.00	Negative	
MDA	INE	PCGDP	4.37	-1.26	0.23	-0.61	2.01	0.73	Negative	VEC
	PCGDP	INE	0.23	-1.97	0.07	-0.71	1.03	0.90	Negative	
MNE	INE	PCGDP					0.04	0.83	Negative	VAR
	PCGDP	INE					6.82	0.01	Negative	
MAR	INE	PCGDP	-5.26	-0.46	0.65	-0.13	2.47	0.65	Positive	VEC
	PCGDP	INE	-0.19	-5.76	0.00	-0.61	35.32	0.00	Positive	
NAM	INE	PCGDP	4.20	0.08	0.94	0.03	0.15	0.93	Positive	VEC
	PCGDP	INE	0.24	-5.56	0.00	-0.28	47.41	0.00	Negative	
NPL	INE	PCGDP					3.66	0.16	Negative	VAR
	PCGDP	INE					8.46	0.01	Negative	
NLD	INE	PCGDP	-17.37	-1.35	0.19	-0.05	9.70	0.02	Positive	VEC
	PCGDP	INE	-0.06	-2.67	0.01	-0.40	3.94	0.27	Positive	
NZL	INE	PCGDP	-2.37	-3.08	0.01	-0.25	3.91	0.79	Positive	VEC
	PCGDP	INE	-0.42	3.08	0.01	0.60	15.11	0.03	Negative	
NGA	INE	PCGDP	7.45	3.00	0.01	0.71	16.93	0.01	Positive	VEC
	PCGDP	INE	0.13	-3.23	0.01	-0.26	10.84	0.09	Negative	
NOR	INE	PCGDP	-2.39	3.62	0.00	0.24	36.71	0.00	Positive	VEC
	PCGDP	INE	-0.42	0.53	0.61	0.05	2.06	0.96	Negative	
PAK	INE	PCGDP					11.47	0.01	Positive	VAR
	PCGDP	INE					13.58	0.00	Positive	
PAN	INE	PCGDP	23.01	-3.07	0.01	-0.29	14.52	0.02	Negative	VEC
	PCGDP	INE	0.04	0.43	0.67	0.08	8.58	0.20	Negative	
PRY	INE	PCGDP	16.23	-4.78	0.00	-0.94	11.55	0.04	Negative	VEC
	PCGDP	INE	0.06	1.70	0.11	0.45	2.60	0.76	Positive	
PER	INE	PCGDP	20.29	4.37	0.00	0.42	21.89	0.00	Positive	VEC
	PCGDP	INE	0.05	-1.27	0.22	-0.21	28.83	0.00	Mixed	
POL	INE	PCGDP	-17.38	-0.64	0.53	-0.05	2.12	0.55	Positive	VEC
	PCGDP	INE	-0.06	-4.01	0.00	-1.01	0.95	0.81	Positive	
PRT	INE	PCGDP	8.17	2.73	0.01	0.14	15.61	0.00	Positive	VEC
	PCGDP	INE	0.12	-2.84	0.01	-0.17	0.23	0.97	Negative	
PRI	INE	PCGDP	-14.70	-4.24	0.00	-0.08	2.24	0.13	Positive	VEC

	PCGDP	INE	-0.07	-1.38	0.18	-0.11	0.42	0.52	Positive	
ROU	INE	PCGDP	-22.90	-2.51	0.02	-0.09	12.63	0.01	Positive	VEC
	PCGDP	INE	-0.04	-0.61	0.55	-0.10	1.16	0.76	Positive	
RUS	INE	PCGDP	10.82	-1.95	0.07	-0.57	9.32	0.10	Negative	VEC
	PCGDP	INE	0.09	-4.86	0.00	-1.51	14.63	0.01	Negative	
RWA	INE	PCGDP					1.29	0.26	Positive	VAR
	PCGDP	INE					20.25	0.00	Positive	
SEN	INE	PCGDP	1.64	-3.04	0.01	-0.73	9.87	0.04	Positive	VEC
	PCGDP	INE	0.61	0.65	0.53	0.05	7.03	0.13	Negative	
SRB	INE	PCGDP					0.34	0.56	Negative	VAR
	PCGDP	INE					5.55	0.02	Negative	
SLE	INE	PCGDP	0.57	1.38	0.20	0.82	3.78	0.58	Positive	VEC
	PCGDP	INE	1.76	-1.49	0.17	-0.05	17.11	0.00	Mixed	
SGP	INE	PCGDP	-50.55	-0.13	0.90	0.00	8.11	0.23	Positive	VEC
	PCGDP	INE	-0.02	-5.42	0.00	-0.59	21.20	0.00	Positive	
SVK	INE	PCGDP	13.24	0.54	0.60	0.06	2.28	0.81	Positive	VEC
	PCGDP	INE	0.08	-3.35	0.01	-0.74	10.99	0.05	Negative	
SVN	INE	PCGDP	-14.07	-2.45	0.03	-0.52	8.61	0.04	Positive	VEC
	PCGDP	INE	-0.07	-1.72	0.11	-2.19	3.54	0.32	Negative	
ZAF	INE	PCGDP	-363.34	-0.26	0.80	0.00	3.27	0.51	Negative	VEC
	PCGDP	INE	0.00	-2.95	0.01	-0.61	21.91	0.00	Mixed	
ESP	INE	PCGDP	26.01	-3.92	0.00	-0.36	43.34	0.00	Negative	VEC
	PCGDP	INE	0.04	2.90	0.01	1.13	28.45	0.00	Positive	
LKA	INE	PCGDP	-18.29	-4.95	0.00	-0.11	15.98	0.03	Positive	VEC
	PCGDP	INE	-0.05	1.97	0.08	0.57	18.49	0.01	Negative	
SWE	INE	PCGDP					11.27	0.02	Negative	VAR
	PCGDP	INE					4.87	0.30	Positive	
CHE	INE	PCGDP					0.20	0.65	Positive	VAR
	PCGDP	INE					5.35	0.02	Positive	
TJK	INE	PCGDP	4.70	-0.61	0.55	-0.12	3.76	0.15	Negative	VEC
	PCGDP	INE	0.21	-5.07	0.00	-0.60	22.78	0.00	Negative	
TZA	INE	PCGDP	-0.03	-2.54	0.03	-1.47	10.03	0.04	Positive	VEC
	PCGDP	INE	-32.31	-3.04	0.01	-0.32	8.60	0.07	Positive	
THA	INE	PCGDP	15.67	-0.53	0.60	-0.03	0.74	0.69	Negative	VEC
	PCGDP	INE	0.06	-4.44	0.00	-0.62	6.31	0.04	Mixed	
TTO	INE	PCGDP	-15.30	3.16	0.01	0.66	4.98	0.17	Negative	VEC
	PCGDP	INE	-0.07	-2.93	0.01	-1.18	4.80	0.19	Positive	
TUN	INE	PCGDP	17.60	-4.25	0.00	-0.47	42.30	0.00	Negative	VEC
	PCGDP	INE	0.06	-1.18	0.27	-0.58	7.03	0.32	Negative	
TUR	INE	PCGDP	9.06	0.35	0.73	0.16	0.08	1.00	Positive	VEC

	PCGDP	INE	0.11	-3.42	0.00	-0.41	9.90	0.04	Negative	
TKM	INE	PCGDP	-6.68	-4.36	0.00	-0.06	10.38	0.00	Negative	VEC
	PCGDP	INE	-0.15	-0.89	0.38	-0.03	0.51	0.47	Negative	
UGA	INE	PCGDP	-8.50	-0.29	0.78	-0.02	0.99	0.32	Negative	VEC
	PCGDP	INE	-0.12	-4.59	0.00	-0.40	6.55	0.01	Positive	
UKR	INE	PCGDP	-20.82	-0.73	0.47	-0.02	1.04	0.90	Positive	VEC
	PCGDP	INE	-0.05	-3.33	0.00	-0.62	5.04	0.28	Negative	
GBR	INE	PCGDP	43.81	-3.78	0.00	-0.01	19.21	0.02	Mixed	VEC
	PCGDP	INE	0.02	-0.94	0.35	-0.02	8.33	0.50	Positive	
USA	INE	PCGDP	-8.92	0.17	0.86	0.00	13.54	0.09	Negative	VEC
	PCGDP	INE	-0.11	-3.59	0.00	-0.24	22.95	0.00	Positive	
UZB	INE	PCGDP	-1.20	-6.93	0.00	-0.55	65.29	0.00	Positive	VEC
	PCGDP	INE	-0.83	1.12	0.28	0.15	3.91	0.14	Negative	
VEN	INE	PCGDP	37.65	2.11	0.04	0.08	10.43	0.17	Positive	VEC
	PCGDP	INE	0.03	-2.12	0.04	-0.13	5.13	0.64	Positive	
VNM	INE	PCGDP	-24.53	-4.08	0.00	-0.33	16.09	0.00	Positive	VEC
	PCGDP	INE	-0.04	-1.55	0.15	-0.21	16.27	0.00	Mixed	
ZMB	INE	PCGDP	-17.39	2.68	0.01	0.19	4.95	0.08	Negative	VEC
	PCGDP	INE	-0.06	-3.18	0.00	-0.81	1.89	0.39	Positive	
ZWE	INE	PCGDP					0.23	0.97	Negative	VAR
	PCGDP	INE					7.29	0.06	Negative	

Source: Author's own elaboration on the basis of the modelling results

Appendix 4: Impulse response test results

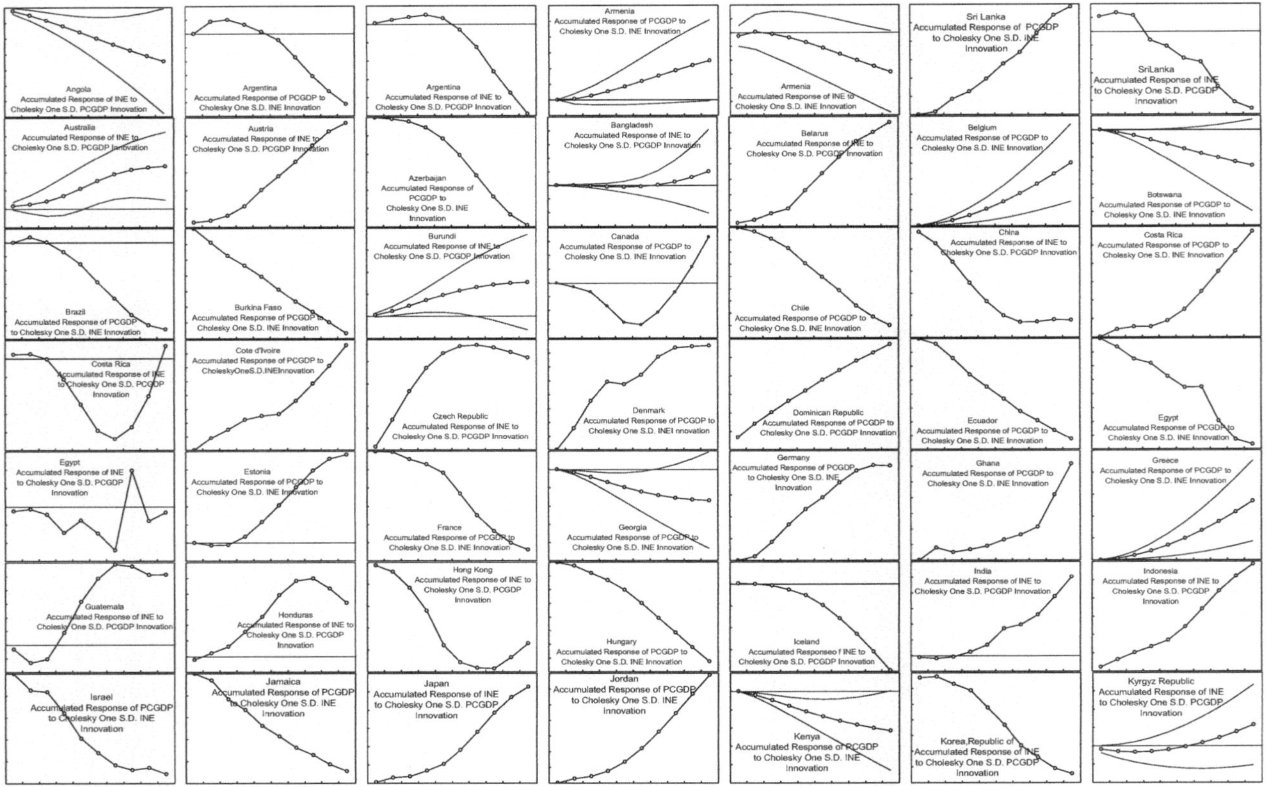

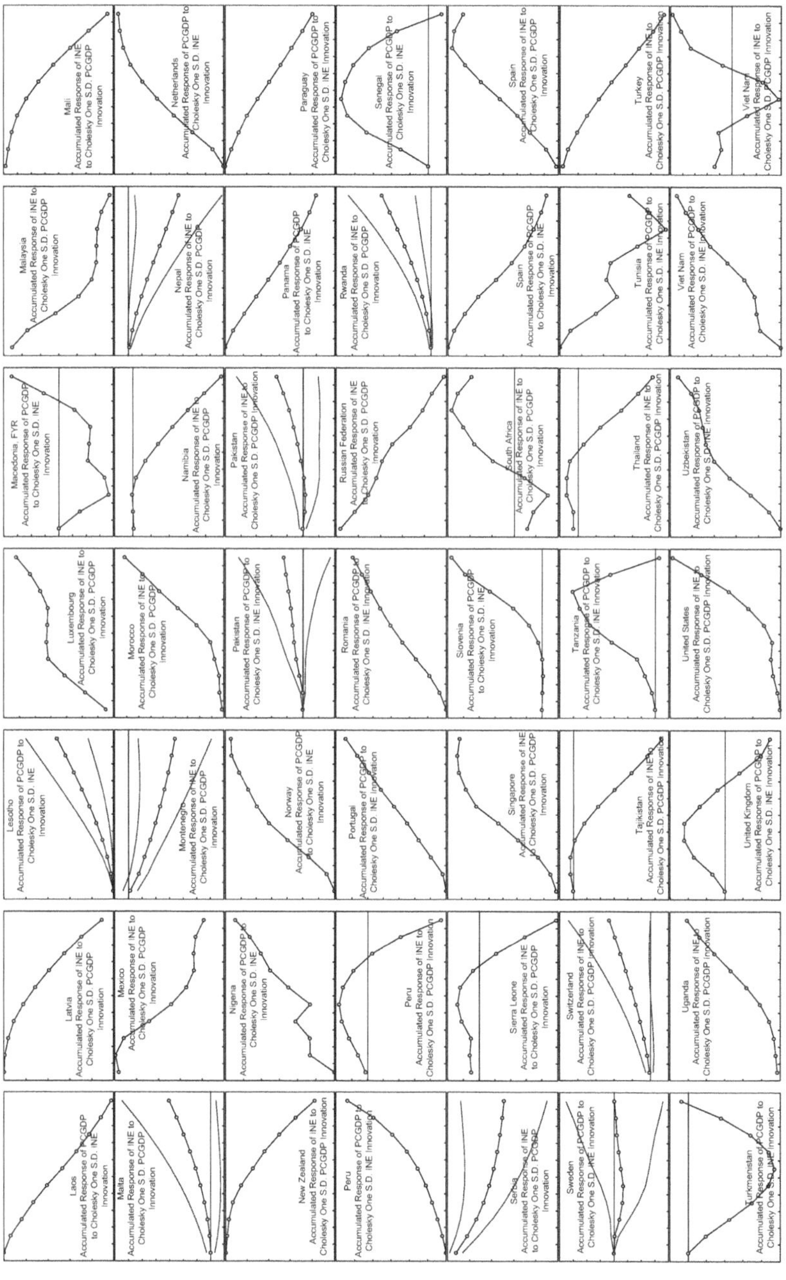

Source: Author's own elaboration on the basis of the modelling results

Appendix 5: Analysis of relationship and causality by 11-cluster groups

			INE *Granger* causes PCGDP		PCGDP *Granger* causes INE			
Coun.	Region	Income	Positive	Negative	Positive	Negative	Overallcausality	Cluster type
AGO	Sub-Saharan	Middle				Equal dist.	Negative	GE
ARG	Lat. Amer. Car.	Middle		Constraint		Equal dist.	Negative	GCE
ARM	Eur. Cen. Asia	Low	Incentive			Equal dist.	Both	GIE
AUS	E. Asia Pac.	High			Unequal dist.		Positive	GU
AUT	Eur. Cen. Asia	High	Incentive		Unequal dist.		Positive	GIU
AZE	Eur. Cen. Asia	Middle		Constraint			Negative	GC
BGD	South Asia	Low			Unequal dist.		Positive	GU
BLR	Eur. Cen. Asia	Middle			Unequal dist.		Positive	GU
BEL	Eur. Cen. Asia	High	Incentive				Positive	GI
BWA	Sub-Saharan	Middle		Constraint			Negative	GC
BRA	Lat. Amer. Car.	Middle		Constraint			Negative	GC
BGR	Eur. Cen. Asia	Middle	Incentive				Positive	GI
BFA	Sub-Saharan	Low		Constraint		Equal dist.	Negative	GCE
BDI	Sub-Saharan	Low			Unequal dist.		Positive	GU
KHM	E. Asia Pac.	Low		Constraint			Negative	GC
CAN	North America	High	Incentive	Constraint			Both	GIC
CHL	Lat. Amer. Car.	High	Incentive	Constraint			Both	GIC
CHN	E. Asia Pac.	Middle				Equal dist.	Negative	GE
COL	Lat. Amer. Car.	Middle				Equal dist.	Negative	GE
CRI	Lat. Amer. Car.	Middle	Incentive	Constraint	Unequal dist.		Both	GX
CIV	Sub-Saharan	Low	Incentive				Positive	GI

(*continued*)

Appendix 5 (continued)

			INE *Granger* causes PCGDP		PCGDP *Granger* causes INE			
Coun.	Region	Income	Positive	Negative	Positive	Negative	Overallcausality	Cluster type
HRV	Eur. Cen. Asia	High				Equal dist.	Negative	GE
CYP	Eur. Cen. Asia	High			Unequal dist.		Positive	GU
CZE	Eur. Cen. Asia	High			Unequal dist.	Equal dist.	Both	GUE
DNK	Eur. Cen. Asia	High	Incentive				Positive	GI
DOM	Lat. Amer. Car.	Middle				Equal dist.	Negative	GE
ECU	Lat. Amer. Car.	Middle		Constraint			Negative	GC
EGY	M. East N. Afr.	Low		Constraint	Unequal dist.	Equal dist.	Both	GX
EST	Eur. Cen. Asia	High	Incentive			Equal dist.	Both	GIE
FRA	Eur. Cen. Asia	High		Constraint			Negative	GC
GEO	Eur. Cen. Asia	Middle		Constraint			Negative	GC
DEU	Eur. Cen. Asia	High	Incentive				Positive	GI
GHA	Sub-Saharan	Low	Incentive				Positive	GI
GRC	Eur. Cen. Asia	High	Incentive				Positive	GI
GTM	Lat. Amer. Car.	Low			Unequal dist.		Positive	GU
HND	Lat. Amer. Car.	Low			Unequal dist.	Equal dist.	Both	GUE
HKG	E. Asia Pac.	High			Unequal dist.	Equal dist.	Both	GUE
HUN	Eur. Cen. Asia	High		Constraint			Negative	GC
ISL	Eur. Cen. Asia	High	Incentive			Equal dist.	Both	GIE
IND	South Asia	Low			Unequal dist.		Positive	GU
IDN	E. Asia Pac.	Low			Unequal dist.		Positive	GU
IRL	Eur. Cen. Asia	High		Constraint			Negative	GC
ISR	M. East N. Afr.	High		Constraint	Unequal dist.		Both	GCU
ITA	Eur. Cen. Asia	High			Unequal dist.		Positive	GU
JAM	Lat. Amer. Car.	Middle		Constraint	Unequal dist.		Both	GCU

(continued)

Appendix 5 (continued)

			INE *Granger* causes PCGDP		PCGDP *Granger* causes INE			
Coun.	Region	Income	Positive	Negative	Positive	Negative	Overallcausality	Cluster type
JPN	E. Asia Pac.	High			Unequal dist.		Positive	GU
JOR	M. East N. Afr.	Middle	Incentive			Equal dist.	Both	GIE
KAZ	Eur. Cen. Asia	Middle		Constraint			Negative	GC
KEN	Sub-Saharan	Low		Constraint			Negative	GC
KOR	E. Asia Pac.	High	Incentive			Equal dist.	Both	GIE
KGZ	Eur. Cen. Asia	Low			Unequal dist.	Equal dist.	Both	GUE
LAO	E. Asia Pac.	Low		Constraint	Unequal dist.		Both	GCU
LVA	Eur. Cen. Asia	High				Equal dist.	Negative	GE
LSO	Sub-Saharan	Low	Incentive				Positive	GI
LUX	Eur. Cen. Asia	High			Unequal dist.		Positive	GU
MKD	Eur. Cen. Asia	Middle		Constraint	Unequal dist.		Both	GCU
MYS	E. Asia Pac.	Middle				Equal dist.	Negative	GE
MLI	Sub-Saharan	Low				Equal dist.	Negative	GE
MLT	M. East N. Afr.	High			Unequal dist.		Positive	GU
MEX	Lat. Amer. Car.	Middle				Equal dist.	Negative	GE
MNE	Eur. Cen. Asia	Middle				Equal dist.	Negative	GE
MAR	M. East N. Afr.	Low			Unequal dist.		Positive	GU
NAM	Sub-Saharan	Middle				Equal dist.	Negative	GE
NPL	South Asia	Low				Equal dist.	Negative	GE
NLD	Eur. Cen. Asia	High	Incentive		Unequal dist.		Positive	GIU
NZL	E. Asia Pac.	High	Incentive			Equal dist.	Both	GIE
NGA	Sub-Saharan	Low	Incentive			Equal dist.	Both	GIE
NOR	Eur. Cen. Asia	High	Incentive				Positive	GI
PAK	South Asia	Low	Incentive		Unequal dist.		Positive	GIU

(continued)

Appendix 5 (continued)

			INE *Granger* causes PCGDP		PCGDP *Granger* causes INE			
Coun.	Region	Income	Positive	Negative	Positive	Negative	Overallcausality	Cluster type
PAN	Lat. Amer. Car.	Middle		Constraint			Negative	GC
PRY	Lat. Amer. Car.	Middle		Constraint			Negative	GC
PER	Lat. Amer. Car.	Middle	Incentive		Unequal dist.	Equal dist.	Both	GX
POL	Eur. Cen. Asia	High			Unequal dist.		Positive	GU
PRT	Eur. Cen. Asia	High	Incentive			Equal dist.	Both	GIE
PRI	Lat. Amer. Car.	High	Incentive				Positive	GI
ROU	Eur. Cen. Asia	Middle	Incentive				Positive	GI
RUS	Eur. Cen. Asia	Middle				Equal dist.	Negative	GE
RWA	Sub-Saharan	Low			Unequal dist.		Positive	GU
SEN	Sub-Saharan	Low	Incentive	Constraint			Both	GIC
SRB	Eur. Cen. Asia	Middle				Equal dist.	Negative	GE
SLE	Sub-Saharan	Low			Unequal dist.	Equal dist.	Both	GUE
SGP	E. Asia Pac.	High			Unequal dist.		Positive	GU
SVK	Eur. Cen. Asia	High				Equal dist.	Negative	GE
SVN	Eur. Cen. Asia	High	Incentive				Positive	GI
ZAF	Sub-Saharan	Middle			Unequal dist.	Equal dist.	Both	GUE
ESP	Eur. Cen. Asia	High		Constraint	Unequal dist.		Both	GCU
LKA	South Asia	Low	Incentive			Equal dist.	Both	GIE
SWE	Eur. Cen. Asia	High		Constraint			Negative	GC
CHE	Eur. Cen. Asia	High			Unequal dist.		Positive	GU
TJK	Eur. Cen. Asia	Low				Equal dist.	Negative	GE
TZA	Sub-Saharan	Low	Incentive		Unequal dist.		Positive	GIU
THA	E. Asia Pac.	Middle			Unequal dist.	Equal dist.	Both	GUE
TTO	Lat. Amer. Car.	High			Unequal dist.		Positive	GU

(*continued*)

Appendix 5 (continued)

			INE *Granger* causes PCGDP		PCGDP *Granger* causes INE			
Coun.	Region	Income	Positive	Negative	Positive	Negative	Overallcausality	Cluster type
TUN	M. East N. Afr.	Low		Constraint			Negative	GC
TUR	Eur. Cen. Asia	Middle				Equal dist.	Negative	GE
TKM	Eur. Cen. Asia	Middle	Incentive	Constraint			Both	GIC
UGA	Sub-Saharan	Low			Unequal dist.		Positive	GU
UKR	Eur. Cen. Asia	Low			Unequal dist.		Positive	GU
GBR	Eur. Cen. Asia	High	Incentive	Constraint			Both	GIC
USA	North America	High			Unequal dist.		Positive	GU
UZB	Eur. Cen. Asia	Low	Incentive				Positive	GI
VEN	Lat. Amer. Car.	Middle				Equal dist.	Negative	GE
VNM	E. Asia Pac.	Low	Incentive		Unequal dist.	Equal dist.	Both	GX
ZMB	Sub-Saharan	Low			Unequal dist.		Positive	GU

Source: Author's own elaboration on the basis of the modelling results

Note: GI denotes Group of Incentive; GU denotes Group of Unequal income (re)distribution; GIU denotes Group of Incentive & Unequal income (re)distribution; GC denotes Group of Constraint; GE denotes Group of Equal income (re) distribution; GCE denotes Group of Constraint & Equal income (re)distribution; GIC denotes Group of Incentive & Constraint; GIE denotes Group of Incentive & Equal income (re)distribution; GCU denotes Group of Constraint & Unequal income (re)distribution; GUE denotes Group of Unequal & Equal income (re)distribution; and GX denotes Group of Complexity

Appendix 6: Milestones in the history of lighting

Period	Description of Milestones
1,420,000 B.C.	Fire used by Australopithecus
500,000 B.C.	Fire used in caves by Peking man
38,000–9000 B.C.	Stone fat-burning lamps with wicks used in southern Europe
3000 B.C.	Candlesticks recovered from Egypt and Crete
2000 B.C.	Babylonian market for lighting fuel (sesame oil)
1292	Paris tax rolls list 72 chandlers (candle makers)
Middle Ages	Tallow candles in wide use in western Europe
1784	Discovery of Argand oil lamp
1792	William Murdock uses coal-gas illumination in his Cornwall home
1798	William Murdock uses coal-gas illumination in Birmingham offices
1800s	Candle technology improved by the use of stearic acid, spermaceti, and paraffin wax
1820	Gas street lighting installed in Pall Mall, London
1855	Benjamin Silliman, Jr., experiments with "rock oil"
1860	Demonstration of electric-discharge lamp by the Royal Society of London
1860s	Development of kerosene lamps
1876	William Wallace's 500-candlepower arc lights, displayed at the Centennial Exposition in Philadelphia
1879	Swan and Edison invent carbon-filament incandescent lamp
1880s	Welsbach gas mantle
1882	Pearl Street station (New York) opens with first electrical service
1920s	High-pressure mercury-vapor-discharge and sodium-discharge lamps
1930s	Development of mercury-vapor-filled fluorescent tube
1931	Development of sodium-vapor lamp
1980s	Marketing of compact fluorescent bulb

Source: Author's own elaboration on the basis of Nordhaus (1997)

Appendix 7: Progress of lighting efficacy (lumens per watt)

Year	Babylonian lamp	Candle (Tallow)	Candle (Sperm)	Oil Lamp	Other Oil Lamp	Gas Lamp	Kerosene Lamp	Tungsten Filament Lamp	Compact Fluorescent Lamp	Low-Pressure Sodium Lamp	White LED Lamp
1750	0.0597										
1800		0.0757	0.1009								
1815				0.1346							
1827						0.1303					
1830		0.0757	0.1009								
1855				0.0784	0.0575	0.0833	0.0498				
1875						0.2464	0.159				
1883								2.6[a]			
1885						0.5914					
1900								3.7143[a]			
1910								6.5[a]			
1916						0.8685					
1920								11.8182[a]			
1930								11.8432[a]			
1932										40[j]	
1935										62.5[k]	
1940								11.9[a]			
1949										64.3[l]	
1950								11.925[a]		67[m]	
1953										73[n]	
1960								11.95[a]			

(*continued*)

Appendix 7 (continued)

Year	Babylonian lamp	Candle (Tallow)	Candle (Sperm)	Oil Lamp	Other Oil Lamp	Gas Lamp	Kerosene Lamp	Tungsten Filament Lamp	Compact Fluorescent Lamp	Low-Pressure Sodium Lamp	White LED Lamp
1961										95[o]	
1966										107.5[p]	
1970								11.975[a]			
1971										122.9[q]	
1976									47.2[b]		
1979										142.9[r]	
1980								12[a]		155[s]	
1984									50[c]		
1988									55.6[d]		
1988									57[e]		
1989									60[f]		
1993							0.3651		61.1[g]		
1995											51[u]
1999									64[h]		
2001									66.7[i]		
2006											150[u]
2007											169[u]
2008										192[t]	

Source: Author's own elaboration on the basis of Nordhaus (1997) and Hooker (2009)
[a]Nordhaus (1997)
[b]The Museum of Electric Lamp Technology: Manufacturer (Philips Lighting) and Model (SL1000)
[c]The Museum of Electric Lamp Technology: Manufacturer (Philips Lighting) and Model (SL*18)
[d]The Museum of Electric Lamp Technology: Manufacturer (Osram GmbH) and Model (Wotan Circolux Electronica 18 W)
[e]The Museum of Electric Lamp Technology: Manufacturer (Philips Lighting) and Model (PLCE*7 with Electronic Ballast)

[f]The Museum of Electric Lamp Technology: Manufacturer (Philips Lighting) and Model (PL*S Twin-Tube Compact Fluorescent)
[g]The Museum of Electric Lamp Technology: Manufacturer (Osram GmbH) and Model (Dulux 'F' Flat)
[h]The Museum of Electric Lamp Technology: Manufacturer (NDF Special Light Products B.V.) and Model (T1 Cold Cathode)
[i]The Museum of Electric Lamp Technology: Manufacturer (Sylvania Lighting International) and Model (Lynx CF-L)
[j]The Museum of Electric Lamp Technology: Manufacturer (N.V. Philips Gloeilampenfabrieken) and Model (Philora Direct Current Sodium Vapour)
[k]The Museum of Electric Lamp Technology: Manufacturer (N.V. Philips Gloeilampenfabrieken) and Model (Philips 'Philora' SO500 LPS)
[l]The Museum of Electric Lamp Technology: Manufacturer (The General Electric Company of England PLC) and Model (Osram SO/H Sodium Vapour with Dewar Jacket)
[m]The Museum of Electric Lamp Technology: Manufacturer (The British Thomson-Houston Co. Ltd.) and Model (Mazda SO/H Sodium Vapour Lamp & Dewar)
[n]The Museum of Electric Lamp Technology: Manufacturer (The British Thomson-Houston Co. Ltd.) and Model (Mazda SO/H Sodium Vapour with Dewar Jacket)
[o]The Museum of Electric Lamp Technology: Manufacturer (The General Electric Company of England PLC) and Model (GEC SLI/H 'Golden Linear' LPS)
[p]The Museum of Electric Lamp Technology: Manufacturer (N.V. Philips Gloeilampenfabrieken) and Model (Philips SO-X with Tin Oxide Film)
[q]The Museum of Electric Lamp Technology: Manufacturer (N.V. Philips Gloeilampenfabrieken) and Model (Philips SO-X with Indium Oxide Film)
[r]The Museum of Electric Lamp Technology: Manufacturer (Thorn Electrical Industries) and Model (Atlas SLI/H Linear LPS)
[s]The Museum of Electric Lamp Technology: Manufacturer (The General Electric Company of England PLC) and Model (SOX-E Economy LPS)
[t]The Museum of Electric Lamp Technology: Manufacturer (Osram) and Model (SOX-E 131W BY22D FLH1)
[u]Brodrick (2008), Emerging Lighting Technologies, Denbaars (2008), What is LED Lighting: Technology Overview and Introduction, Horner (2008), What Have We Accomplished in Lighting Efficiency?, Narukawa et al. (2007) Improvement of Luminous Efficiency in White Light Emitting Diodes by Reducing a Forward-bias Voltage, and Young (2008), Specifying Food Service Lighting for Energy Efficiency.

Appendix 8: Progress of conventional wheeled train (maximum speed)

Year	Train (Conventional Wheeled)	Top Speed (km/h)	Type
1848	Boston and Maine Railroad Antelope	96.6	Steam
1850	Great Britain	125.6	Steam
1854	Bristol & Exeter Railway #41	131.6	Steam
1893	Empire State Express No. 999	131	Steam
1907	Baraian Class S 2/6	154.5	Steam
1935	LNER papyrus	173.8	Steam
1936	DRG SVT 137 "Bauart Leipzig"	205	Diesel
1939	DRG SVT 137155 (Kruckenberg)	215	Diesel
1953	SNCF CC 7121	239.8	Electric multiple unit
1955	SNCF BB 9004	330.9	Electric multiple unit
1981	SNCF TGV Sud-Est Set No. 16	380	Electric multiple unit
1988	SNCF TGV Sud-Est Set No. 88	408.4	Electric multiple unit
1989	SNCF TGV Atlantique Set No. 325	482.4	Electric multiple unit
1990	SNCF TGV Atlantique Set No. 325	515.3	Electric multiple unit
2007	SNCF TGV POS Set No. 4402	574.8	Electric multiple unit

Source: Author's own elaboration on the basis of Procter (1997), Des records sur rail (2009) and SVT (2009)

Appendix 9: Progress of mass-produced car (maximum speed)

Year	Car (Mass-Produced)	Top Speed (km/h)
1900	Mercedes 35 hp.	85
1903	Mercedes 60 hp.	96
1921	Bentley 3 Litre Super Sports	160
1929	Bentley 4½ Litre Supercharged	222.02
1932	Duesenberg SJ	217
1953	Pegaso Z-102 Supercharged	244.62
1954	Mercedes-Benz 300SL	250
1955	Ferrari 410 Superamerica	261
1962	Ferrari 250 GTO	279.06
1968	Ferrari Daytona GTB/4	281
1970	Lamborghini Miura P400 SV	288.07
1984	Ferrari 288 GTO	304

(*continued*)

Appendix 9 (continued)

Year	Car (Mass-Produced)	Top Speed (km/h)
1986	Porsche 959	320.3
1987	Ferrari F40	324
1990	Lamborghini Diablo	341
1992	Bugatti EB110 SS	346
1993	Jaguar XJ220	349.2
1994	McLaren F1	386.4
2005	Bugatti Veyron	408.47
2007	SSC Ultimate Aero TT	412.16

Source: Author's own elaboration on the basis of *The History of the Coveted World Record* (2009)

Appendix 10: Progress of military fighter aircraft (maximum speed)

Year	Aircraft (Military Fighters)	Top Speed (km/h)	Type
1930	PZL P.7	327	Piston engine
1931	PZL P.11	390	Piston engine
1933	Polikarpov I-16	525	Piston engine
1935	Messerschmitt Bf 109	621	Piston engine
1938	Heinkel He 100	670	Piston engine
1939	Focke-Wulf Fw 190	710	Piston engine
1941	Messerschmitt Me 163 Komet	955	Jet engine
1944	Lockheed F-80 Shooting Star	966	Jet engine
1947	North American F-86 Sabre	1138	Jet engine
1951	Hawker Siddeley Hunter	1150	Jet engine
1954	Mikoyan-Gurevich MiG-19	1450	Jet engine
1957	BAC Lightning	2415	Jet engine
1964	General Dynamics F-111	2650	Jet engine
1972	McDonnell Douglas F-15 Eagle	2655	Jet engine
1988	Sukhoi Su-35	2440	Jet engine
1990	Lockheed F-22 Raptor	2655	Jet engine
1996	Sukhoi Su-37	2500	Jet engine

Source: Author's own elaboration on the basis of Starostin (2009)

Appendix 11: History of computing speed

Technology	Period	Computer Power (MSOPS)
Manual	Up to nineteenth century	1.68E-08
Early Mechanical	1900	1.48E-07
Late Mechanical	1940	1.92E-06
Relay/Vacuum	1950	3.80E-03
Transistor	1960	1.06E-01
Transistor	1970	4.65E-01
Early Microprocessor	1980	4.65E-01
Microprocessor	1990	1.25E+01
Microprocessor	2001	3.10E+03

Source: Author's own elaboration on the basis of Nordhaus (2001)

Appendix 12: Progress of computing performance

Year	Name	Computer Power(MSOPS)	Technology
1880	Manual calculations	1.68E-08	Manual
1857	Scheutz Difference Engine	1.34E-07	Mechanical
1874	Original Odhner	1.87E-08	Mechanical
1890	Hollerith Tabulator	3.13E-07	Mechanical
1895	Steiger Millionaire	2.59E-07	Mechanical
1902	Automatic Tabulator	1.48E-06	Mechanical
1905	Burroughs Model 9	6.72E-08	Mechanical
1907	Adding Tabulator	3.87E-07	Mechanical
1911	Monroe Calculator	2.91E-08	Mechanical
1919	IBM Tabulator	5.33E-08	Mechanical
1935	IBM 601	1.00E-06	Mechanical
1938	Zuse-1	4.45E-08	Mechanical
1939	Zuse-2	4.45E-07	Relay/Vacuum
1939	BTL Model 1	1.99E-06	Relay/Vacuum
1940	Bell Calculator Model 1	3.33E-06	Relay/Vacuum
1941	Zuse-3	1.34E-06	Relay/Vacuum
1942	Harvard Mark I	2.03E-06	Relay/Vacuum
1942	Atanasoff Berry Computer	6.96E-06	Relay/Vacuum
1943	BTL Model 2	1.10E-06	Relay/Vacuum
1943	BTL Model 3	3.02E-06	Relay/Vacuum
1943	Colossus	2.29E-04	Relay/Vacuum
1944	ASCC Mark 1	2.48E-06	Relay/Vacuum
1945	Zuse-4	2.18E-06	Relay/Vacuum

(*continued*)

Appendix 12 (continued)

Year	Name	Computer Power(MSOPS)	Technology
1946	BTL Model 5	3.51E-06	Relay/Vacuum
1946	ENIAC	2.96E-03	Relay/Vacuum
1947	Harvard Mark 2	6.65E-06	Relay/Vacuum
1948	IBM SSEC	6.43E-04	Relay/Vacuum
1949	BINAC	5.25E-03	Relay/Vacuum
1949	EDSAC	2.71E-03	Relay/Vacuum
1950	SEAC	4.81E-03	Relay/Vacuum
1951	Zuse-5 I	9.98E-06	Relay/Vacuum
1951	Univac I	4.58E-03	Relay/Vacuum
1952	IBM CPC	2.19E-03	Relay/Vacuum
1953	Univac 1103	3.33E-02	Relay/Vacuum
1953	IBM 650	1.03E-03	Relay/Vacuum
1953	IBM 701	2.22E-02	Relay/Vacuum
1954	EDVAC	1.74E-03	Relay/Vacuum
1955	Whirlwind	6.10E-02	Relay/Vacuum
1955	Librascope LGP-30	6.30E-04	Relay/Vacuum
1955	IBM 704	5.79E-02	Relay/Vacuum
1958	Datamatic 1000	4.00E-03	Relay/Vacuum
1958	Univac II	5.00E-03	Relay/Vacuum
1959	Mobidic	6.25E-02	Relay/Vacuum
1959	IBM 7090	3.43E-01	Transistors
1960	IBM 1620	9.60E-04	Transistors
1960	DEC PDP-1	1.29E-01	Transistors
1961	Atlas	1.48E+00	Transistors
1961	DEC PDP-4	1.00E-01	Transistors
1962	Univac III	1.11E-01	Transistors
1962	Burroughs 5000	9.87E-02	Transistors
1963	IBM 7040	6.30E-02	Transistors
1963	Honeywell 1800	1.50E-01	Transistors
1964	CDC 6600	9.22E+00	Transistors
1964	DEC PDP-6	1.78E-01	Transistors
1965	IBM 1130	1.16E-01	Transistors
1965	DEC PDP-8	1.67E-01	Transistors
1966	IBM 360/75	2.36E+00	Transistors
1967	DEC PDP-10	1.24E+00	Transistors
1969	CDC 7600	2.71E+01	Transistors
1969	DG Nove	1.17E-01	Transistors
1970	GE-635	6.82E-01	Transistors
1971	SDS 920	9.40E-02	Transistors
1972	IBM 360/195	1.82E+01	Transistors
1972	Honeywell 700	7.50E-02	Transistors
1973	Intellec-8	6.41E-03	Transistors

(*continued*)

Appendix 12 (continued)

Year	Name	Computer Power(MSOPS)	Technology
1973	Data General Nova	2.50E-02	Transistors
1975	Altair 8800	6.41E-02	Transistors
1976	DEC PDP-11/70	4.12E-01	Transistors
1976	Cray-1	1.57E+02	Transistors
1977	Apple II	7.97E-02	Transistors
1978	DEC VAX 11/780	1.00E+00	Microprocessors
1980	Sun-1	5.41E-01	Microprocessors
1982	IBM PC	2.46E-01	Microprocessors
1982	Sun-2	8.59E-01	Microprocessors
1982	Compaq Portable	6.41E-01	Microprocessors
1983	IBM AT-80286	8.00E-01	Microprocessors
1984	Macintosh-128 K	3.80E-01	Microprocessors
1985	Cray-2	8.61E+02	Microprocessors
1986	Compaq Deskpro 386	4.00E+00	Microprocessors
1986	Sun-3	2.12E+00	Microprocessors
1986	DEC VAX 8650	9.19E+00	Microprocessors
1987	Apple Mac II	1.00E+00	Microprocessors
1987	Mac II	1.91E+00	Microprocessors
1987	Sun-4	1.02E+01	Microprocessors
1988	Mac-IIx	3.90E+00	Microprocessors
1988	PC Brand 386-25	1.15E+01	Microprocessors
1990	Dell 320LX	1.25E+01	Microprocessors
1990	Mac Iifx	1.00E+01	Microprocessors
1991	Gateway-486DX2/66	5.30E+01	Microprocessors
1992	IBM PS/2 90	2.24E+01	Microprocessors
1992	NEC Powermate	2.18E+01	Microprocessors
1992	IBM PS/2 55-041	1.06E+01	Microprocessors
1993	Pentium PC	8.23E+01	Microprocessors
1993	Gateway P5-75	1.03E+02	Microprocessors
1994	Power Tower 180e	3.00E+02	Microprocessors
1995	Intel Xpress/60	7.00E+01	Microprocessors
1995	PowerMac 7600/132	1.60E+02	Microprocessors
1996	Pentium PC	1.79E+02	Microprocessors
1996	Dell Dimension Pro150	4.47E+02	Microprocessors
1997	Gateway G6-200	3.50E+02	Microprocessors
1998	Pentium II PC	4.98E+02	Microprocessors
1999	Pentium II/455	9.73E+02	Microprocessors
1999	Pentium III/500	1.07E+03	Microprocessors
2000	Mac G4/500 dual	1.07E+03	Microprocessors
2001	Net vista a40i	2.14E+03	Microprocessors
2001	Gateway Athlon	3.42E+03	Microprocessors
2001	Pentium IV (Dell 8100)	2.81E+03	Microprocessors

Source: Author's own elaboration on the basis of Nordhaus (2001)

Appendix 13: Progress of undersea cable systems

Year	Bandwidth (bps)	Name	Technology
1858	0.0555556	Telegraph	Single Cable
1866	1	Telegraph	Single Cable
1874	88	Telegraph	Single Cable
1880	325	Telegraph	Single Cable
1928	373	Telegraph	Single Cable
1951	144,000	TAT1	Coaxial Cable
1959	144,000	TAT2	Coaxial Cable
1963	414,000	TAT3	Coaxial Cable
1965	384,000	TAT4	Coaxial Cable
1970	2,160,000	TAT5	Coaxial Cable
1976	12,000,000	TAT6	Coaxial Cable
1983	300,000,000	TAT7	Coaxial Cable
1988	560,000,000	TAT8	Optical Cable
1989	1,260,000,000	PTAT-1	Optical Cable
1992	1,120,000,000	TAT9	Optical Cable
	1,120,000,000	TAT10	Optical Cable
1993	1,680,000,000	TAT11	Optical Cable
1994	1,680,000,000	Columbus-2	Optical Cable
	4,976,000,000	CANTAT-3	Optical Cable
1995	9,953,000,000	TAT12	Optical Cable
1996	9,953,000,000	TAT13	Optical Cable
1998	139,340,000,000	Atlantic crossing-1	Optical Cable
	59,718,000,000	Gemini	Optical Cable
1999	19,908,000,000	Columbus-3	Optical Cable
2000	1,273,984,000,000	Yellow/Atlantic crossing-2	Optical Cable
2001	1,910,976,000,000	360 Atlantic	Optical Cable
	2,388,720,000,000	FLAG Atlantic 1	Optical Cable
	636,928,000,000	TAT-14	Optical Cable
	2,547,968,000,000	TyCom Global Network-TransAtlantic	Optical Cable
2002	3,184,496,000,000	Apollo	Optical Cable

Source: Author's own elaboration on the basis of Koh and Magee (2006)

Appendix 14: Data transfer rate of Internet backbone systems

Year	Bandwidth (bps)	Version
1968	50,000	Modem
1972	50,000	Modem
1973	50,000	Modem
1974	50,000	Modem
1976	50,000	Modem
1979	50,000	Modem
1981	56,000	Modem
1983	56,000	Modem
1984	56,000	Modem
1985	1,544,000	T1
1986	1,544,000	T1
1987	1,544,000	T1
1988	1,544,000	T1
1989	1,544,000	T1
1990	1,544,000	T1
1991	44,736,000	T3
1992	44,736,000	T3
1993	44,736,000	T3
1994	145,000,000	OC-3
1995	145,000,000	OC-3
1996	622,080,000	OC-12
1998	2,488,000,000	OC-48
1999	2,488,000,000	OC-49
2000	2,488,000,000	OC-50
2001	2,488,000,000	OC-51
2002	2,488,000,000	OC-52
2003	9,952,000,000	OC-192
2004	9,952,000,000	OC-193

Source: Author's own elaboration on the basis of Kurzweil (2005) and *The Singularity is Near* (2009)

Index[1]

[1] Note: Page numbers followed by 'n' refer to notes.

S. J. Baek, *The Political Economy of Neo-modernisation*,
https://doi.org/10.1007/978-3-319-91394-0

The manufacturer's authorised representative in the EU is Springer Nature Customer Service Centre GmbH, Europaplatz 3, 69115 Heidelberg, Germany. If you have any concerns regarding our products, please contact ProductSafety@springernature.com

Printed and bound by CPI Group (UK) Ltd, Croydon, CR0 4YY
30/09/2026
02225653-0015